CLARE ISLAND

John Feehan

RIA

Clare Island

First published 2019

Royal Irish Academy, 19 Dawson Street, Dublin 2
www.ria.ie

ISBN 978-1-911479-13-0 (PB)
ISBN 978-1-911479-15-4 (pdf)
ISBN 978-1-911479-16-1 (epub)
ISBN 978-1-911479-17-8 (mobi)

British Library Cataloguing in Publication Data. A CIP catalogue record for this book is available from the British Library.

Design: Bernard Kaye
Cover design: Resonate Design
Index: Martin Steer

Cover photograph: Clare Island at sunset by P.J. Rankin. Reproduced by permission.

Printed in the UK by CPI

5 4 3

MIX
Paper | Supporting responsible forestry
FSC www.fsc.org FSC® C013604

Royal Irish Academy is a member of Publishing Ireland, the Irish book publishers' association.

This publication has received support from Heritage Office, Mayo Co. Council

MAYO.IE

Contents

NEW SURVEY OF CLARE ISLAND

New Survey of Clare Island Executive Committee 2019

Contributors to previous volumes in the New Survey of Clare Island series

Volume 1: History and Cultural Landscape: Timothy Collins, Críostóir Mac Cárthaigh (ed.), Nollaig Ó Muraíle, Kevin Whelan (ed.)

Volume 2: Geology: Peter Coxon, David Evans, John R. Graham (ed.), Kenneth T. Higgs, W.E.A. Phillips, C.J. Stillman, B.G.J. Upton, Michael Williams

Volume 3: Marine Intertidal Ecology: Michelle Cronin, Thomas Cross, Robert Cussen, Jane Delany, Michael Guiry, Louise Harrington, Christine Maggs, David McGrath, Alan Myers (ed.), Julia Nunn, Sandy O'Driscoll, Ruth O'Riordan, Anne Marie Power, Fabio Rindi

Volume 4: The Abbey: Ann Buckley, Ian Cantwell, Fergus Gillespie, Paul Gosling (ed.), Conleth Manning (ed.), Karena Morton, Micheál Ó Comáin, Christoph Oldenbourg, Roger Stalley, John Waddell (ed.)

Volume 5: Archaeology: Markus Casey, Michelle Comber, Paul Gosling (ed.), Maureen McCorry, Conleth Manning (ed.), Karen Molloy, Sharon Nestor, Adrian Phillips, Mike Williams, Paula King, John Waddell (ed.)

Volume 6: The Freshwater and Terrestrial Algae: Jenny Bryant, Michael Guiry (ed.), David John (ed.), T.K. McCarthy (ed.), Fabio Rindi (ed.), Patricia Sims, Brian Whitton, David Williamson

Volume 7: Plants and Fungi: Donal Synnott (ed.), Pete Coxon, Ryan Corcoran, Paul Gibson, Stephen McCarron, Timothy Ryle, Matthew Jebb, M.R.D. Seaward, D.H.S. Richardson and David Mitchel

Volume 8: Soils and soil association: by Wies Vullings, James F. Collins, George Smillie

Forthcoming volumes in the New Survey of Clare Island series

Volume 9: Birds: T. Kelly (ed.)

Volume 10: Zoology: R. Anderson, T. Bolger, K.G.M. Bond, J. Breen (ed.), M. Cawley, K. Creed, E. de Eyto, E. DeMilio, M. Gormally, C. Lawton, M. Cawley, M. Gormally, T.C. Kelly, E.S. Lenihan (ed.), N.J. Marley, K. McAney, G. McCall, T.K. McCarthy (ed.), S. McCormack, R. Mc Donnell, R. Møbjerg Kristensen, D.A. Murray, M. Nolan, A. O'Grady, R. Paxton, J. Petter Nilssen, A. Trowanska-Olichwer

Preface and Acknowledgments

> The only true voyage of discovery, the only fountain of youth, would be not to travel to new landscapes, but to possess other eyes, to behold the universe through the eyes of another, of a hundred others.
>
> Marcel Proust, *La prisonnière*[1]

Were it not for the great Irish naturalist Robert Lloyd Praeger most of us might never have heard of Clare Island, but as a result of his first visit in 1903 this detached fragment of the Irish mainland came to be studied in more detail than any island of its size at that time. A hundred years later the Royal Irish Academy, under whose auspices the Original Survey had been carried out, set about repeating the exercise with the intention of assessing and evaluating change on the island over the intervening century. As a result of specialist attention brought to bear by these two surveys, the natural and cultural history of Clare Island is known in exceptional detail, and reported in detail in the 67 fascicles of the Original Survey and 12 volumes of the New Survey.

Many of the volumes of the New Survey are targeted at a specialised readership and are not readily accessible to the general reader, on which account it was felt that a less technical 'composite' volume would be useful, this book being the result. I have tried to synthesise the results of the two great surveys of Clare Island and assess the significance of the differences between them in a way that is accessible to a wider readership than the specialised survey volumes. I have also tried to give a sense of the *achievement* they represent. Beyond this I have tried to bind the whole together against the background of a broader context, looking not only to the past and present of the island's natural and cultural heritage, but to its future, and in particular to the future of farming on the island.

However, for all their learning, the volumes of the New Survey of Clare Island do not of themselves offer us what enchanted Praeger, as it enchants any receptive visitor: the magic of an extraordinary place whose unique character is the result of the interaction of the people for whom Clare Island is home with a place of exceptional natural beauty and diversity. The island's landscape preserves in a uniquely rich way the imprint of that interaction over more than 60 centuries.

The wider audience for which this book is intended might include those in a position to use and develop the findings of the surveys for educational purposes at levels from young children to adults. It is also anticipated that the book can in some significant way be used to promote the island's tourism potential. At the same time I would hope that these introductory remarks will entice others to dig deeper into the detailed volumes, and perhaps explore one or other of the specialisms that can be studied and experienced so marvellously on Clare Island. I hope also that it will prove useful to the visitor who wants a more in-depth understanding of this extraordinary place than the current literature provides.

The book may help provide a realistic and properly informed direction as to how the findings of the surveys might be used to plan for the future of the community of the island. At the time of writing, Clare Island is in the early stages of an exercise that attempts to formulate a new model for farming on the island—one that is specifically adapted to its people's needs and explores the unique natural and cultural heritage so meticulously documented in the two surveys—while keeping a close eye on the new economic, social and environmental direction of the EU's farming policy.

The book is for me the fruit of a twin enchantment. I fell under the spell of Clare Island many decades ago, long before I ever set foot on the is-

[1] Marcel Proust, *La prisonnière*, Paris: Gallimard (1925), 69 (translation in Alister McGrath, *Inventing the Universe*, 202). 'Le seul véritable voyage, le seul bain de Jouvence, ce ne serait pas d'aller vers de nouveaux paysages, mais d'avoir d'autres yeux, de voir l'univers avec les yeux d'un autre, de cent autres.'

land, largely thanks to Adrian Phillips of Trinity College's Geology Department, and to the great survey organised by R.L. Praeger under the auspices of the Royal Irish Academy. The fabric of which the portrait of Clare Island sketched out in these pages is woven has its warp and its woof. The threads of the warp consist of my summary distillation of the contributions to the Original Survey as well as the New Survey, which brings to bear modern methodologies of greater precision and a more nuanced taxonomy in order to evaluate change in this microcosm over the century between the two surveys. The woof with which these individual threads are woven into the fabric of this portrait is the depth of my own feeling for the island.

*****

The book is of course indebted on every page to the authors of the various fascicles and volumes of the two surveys, but it is equally indebted to the people for whom Clare Island is home, and whose hospitality and help made the book such a privilege to write—and for whom indeed it is written in the first place. I would particularly like to thank the many New Survey participants and others who read and critically commented on my attempts to integrate some of their insights and findings into a general narrative that might have an appeal beyond the narrower readership the individual volumes may attract: Roy Anderson, Tom Bolger, Ken Bond, John Breen, Jim Collins, Pete Coxon, Hubert Fuller, Paul Gosling, John Graham, Michael Guiry, D.M. John, Tom Kelly, Kieran McCarthy, Críostóir MacCarthaigh, Con Manning, Declan Murray, Myles Nolan, Mark Seaward, John Sweeney, Marcus Sweeney, Donal Synnott and Kevin Whelan. The stream survey upon which the concluding section of Chapter 7 draws heavily was carried out by Bryan Kennedy in April 2019. A very special thank you also to the Academy's appointed readers of the volume, and to Michael Viney for contributing the Foreword.

I would like in a very special way to acknowledge the hospitality of the many people on the island who shared their knowledge or helped in other ways with my research for this volume. In this regard, I echo the heartfelt appreciation of the authors of the other volumes in the New Survey; sincere thanks to Joe and Mary O'Malley (Kill), Michael McNamara and Mary Finn (Fawnglass), Oliver and Mary O'Malley (Toormore), Martie O'-Malley, (Bunnamohaun), Carl and Chris O'Grady (Sailors), Ciara Cullen (Macalla Farm), Michael O'Grady, Helen and Alan O'Grady, Bridget and Humphrey O'Leary, Joe Toole, Micheál O'Malley, Martin Moran, Michael Oliver Moran, John Francis O'Malley and Sean O'Grady. I am very particularly indebted to Marcus and Irene Sweeney, who put their roof at Little Austie's in Glen over my head during my many stays on Clare Island with no more asked (in the spirit of the *bruidhean* of an earlier Ireland) than the completion of this volume. Marcus also provided valuable comment to the section on fishes and fishing, and critically reviewed the completed text.

The book owes its attractive visual appearance to the many people who have provided photographs and other illustrations, most notably Roy Anderson, William Clarke, Pete Coxon, Mike Guiry, Bernard Kaye, Steve McCarron, Críostóir MacCarthaigh and Michael Guiry. Where photographs are not individually acknowledged they are my own. A special debt is due to Philip Armstrong for his wonderful paintings of the cluster settlement at Glen and the Bronze Age farmstead in Toormore. The high standard of the end product owes much to the layout and design of Bernard Kaye.

The list of the friends, colleagues and acquaintances at University College Dublin and further afield who have helped with the project in all sorts of ways strains the edge of memory. I hope this will serve as some sort of excuse for names omitted: Paddy Ashe, Enda Bannon, Noel Campbell, Fr Ned Crosby, Donal Daly, Stephen Daly, Clodagh Doyle, Tom Doyle, Damian Egan, Les Fox, Peter Leonard, Richard Manton, Elizabeth McArdle, Eoin MacCraith, Brian McConnell, Sean McDonagh, Rebecca McGovern, Roland McHugh, Nigel Monaghan, Gráinne Ni Shé, Michael O'Connell, Jim O'Connor, Adrian Patterson, Andrea Patterson, Amanda Pedlow, Andy Ryder, Ray Scanlon, Olaf Schmidt, Ray Spain, Alan Stephens, Koen Verbruggen, Des Walshe and Robert Wilkes.

This concluding volume in the New Survey of Clare Island has been improved at every turn by the many suggestions of Martin Steer, who has given the book his meticulous care and attention over the several years of its gestation. At the very end, he undertook the task of preparing a detailed and comprehensive index that makes it possible to navigate a bewildering range of apparently unrelated subjects with ease and confidence. I am indebted also to the meticulous editorial supervision of Jonathan Dykes, and the direction and oversight of Ruth Hegarty and Róisín Jones at the Royal Irish Academy.

As the book goes to press, work is commencing on the research programme necessary to make the proposed new model of farming on Clare Island a reality. I am deeply grateful to the team that has made this possible and who will be involved in its future direction and implementation: Helen Sheridan, Jim Kinsella in the School of Agriculture and Food Science at University College Dublin, Jim McAdam at the Department of Agriculture and Rural Development at Queen's University, Daire O hUallacháin and Owen Fenton in Teagasc, and Joe O'Malley and Michael McNamara on Clare Island.

I am as always indebted beyond words to Róisín, Jane and Christiaan for their constant encouragement and support.

John Feehan
Little Austie's, Glen
Clare Island
Spring 2019

Usage of online imagery

Our thanks to the photographers who have made their images available via Wikimedia Commons and other online sites. License summary codes are given with the individual images. Full details of any particular license can be accessed via the web pages listed below.

CC0 1.0—The person who associated a work with this deed has dedicated the work to the public domain by waiving all of his or her rights to the work worldwide under copyright law, including all related and neighboring rights, to the extent allowed by law.

CC BY-SA 1.0— This work is licensed under the Creative Commons Attribution-ShareAlike 1.0 Generic License. To view a copy of this license, visit http://creativecommons.org/licenses/by-sa/1.0/

CC BY 2.0—This work is licensed under the Creative Commons Attribution 2.0 Generic License. To view a copy of this license, visit http://creativecommons.org/licenses/by/2.0/

CC BY-SA 2.0— This work is licensed under the Creative Commons Attribution-ShareAlike 2.0 Generic License. To view a copy of this license, visit http://creativecommons.org/licenses/by-sa/2.0/

CC BY-NC-ND 2.0— This work is licensed under the Creative Commons Attribution-NonCommercial-NoDerivs 2.0 Generic License. To view a copy of this license, visit http://creativecommons.org/licenses/by-nc-nd/2.0/

CC BY-SA 2.5— This work is licensed under the Creative Commons Attribution-ShareAlike 2.5 Generic License. To view a copy of this license, visit http://creativecommons.org/licenses/by-sa/2.5/

CC BY 3.0— This work is licensed under the Creative Commons Attribution 3.0 Unported License. To view a copy of this license, visit http://creativecommons.org/licenses/by/3.0/

CC BY-SA 3.0— This work is licensed under the Creative Commons Attribution-ShareAlike 3.0 Unported License. To view a copy of this license, visit http://creativecommons.org/licenses/by-sa/3.0/

CC BY-SA 4.0—This work is licensed under the Creative Commons Attribution-ShareAlike 4.0 International License. To view a copy of this license, visit http://creativecommons.org/licenses/by-sa/4.0/

CC BY-NC 4.0— This work is licensed under the Creative Commons Attribution-NonCommercial 4.0 International License. To view a copy of this license, visit http://creativecommons.org/licenses/by-nc/4.0/

CC BY-NC-SA 4.0— This work is licensed under the Creative Commons Attribution-NonCommercial-ShareAlike 4.0 International License. To view a copy of this license, visit http://creativecommons.org/licenses/by-nc-sa/4.0/

Alternatively send a letter to Creative Commons, PO Box 1866, Mountain View, CA 94042, USA.

Foreword

Clare Island hides beyond a curve of my hillside, but down at the strand it stands guard to the north, at the entrance to Clew Bay, its silhouette and role in our natural history both monumental.

During one visit to the island, I held magic in my hand—a glossy, golden-brown hazelnut conjured up from prehistory and still incredibly whole. Its place in a finger-thick, pollen-rich core, extracted from deep below a little peaty lake, dated it to 8,000 years ago. It began to turn black on my palm, its virtual reality withering in the air.

Paleobotany, the study of fossil plants, was part of the Royal Irish Academy's New Survey of Clare Island, a unique multi-disciplinary endeavour launched in 1991. Its scientists were revisiting, after almost a century, the pioneering survey led by Robert Lloyd Praeger. That great naturalist was energised by Darwin's work on the evolution of island species, but his survey has given today's scientists the chance to explore the measure of change over time in Clare Island's environment, its flora and fauna.

The hazelnut was a token from an island freshly wooded in Ireland's recovery from the last ice age. Above it in the core, charcoal fragments marked the coming of people and their forest fires. On an island now almost treeless, and ribbed with abandoned, grassed-over cultivation ridges, human impact has clearly been profound, both before and since the Praeger survey. The lack of trees and loss of cultivation shows up, for example, in the paucity of landbird species, so that the recent arrival of the chaffinch finds it the island's only finch and the once-common corncrake is no more.

The islanders' lives and culture have been woven through the island's natural world. 'Nowhere else,' as John Feehan writes, 'does the landscape so visibly and dramatically evoke the transformation that occurred in the west of Ireland after the Famine'. His chapter on its past, present and future is informed by his own long study of Irish farming, as well as the natural world. It makes a compelling case for the special commitment and care that should couple both its cultural and natural heritage.

So much of Clare Island's story is remarkable that it might, after all, have offered Robert Lloyd Praeger some small and special island species for discovery. But the triumph of the intense fieldwork of both surveys has been that of comprehensive inclusion, from the role of near-invisible mites on the shore to the history of great erratic boulders on the summit of Knockmore. Few places on Earth, and none elsewhere in Ireland, have yielded such a concentrated inventory of knowledge about the natural world.

As John Feehan suggests, 'It can be hard to believe that Clare Island is home to more than a thousand different animals'. Most are, indeed, very small, and his evocations of their roles in the natural world are vivid and engaging. This is especially true of Clare Island's marine life ('If you are really lucky, you may spot a sea spider'). The Original Survey attempted a census of the creatures of the sea, sampling them by offshore dredging. The New Survey, with the added benefit of scuba diving, could focus more closely on the dense vegetation of seaweeds and the organisms living among them—a far more numerous and varied population than in the vegetation ashore.

Much of the original woodland, including oaks and pines, survives now only as stumps and roots under the island bogs. But in Praeger's time, nearly half the island was heather-covered, a flowering cloak now almost erased through overgrazing by sheep. This has lost the island the woodcock, once plentiful, and reduced cover for the hare. Avian pleasures are now choughs along the clifftops and nesting seabirds crowding the ledges below, one of the largest summer assemblies in Ireland. Among them, the arrival and rapid expansion of a gannet colony has been a dramatic change in the island's avifauna.

Many plant species of the Original Survey have gone, but many others have been added to the record (among them several garden escapes). The prized Arctic species that Praeger took such risks to find on the precipitous cliffs of Knockmore still bloom on ledges beyond the reach of sheep. And the myriad lichens, mosses and liverworts, in fascinating and often beautiful diversity, still flourish in the moist Atlantic air.

The sheer richness of life surviving on Clare Island shines through John Feehan's easy, affectionate prose. His sifting of intricate field reports attends fully to their science, but speaks also of his own lifelong commitment to nature and to the least of its creatures and plants. His wider ecological view helps to give each species its role in the island's life.

The book honours a legion of researchers whose fieldwork, often strenuous or hazardous and carried through in poor weather, has offered an important baseline against which to measure the progress of climate change.

The wider effects in Ireland, as Feehan writes, have already been observed in Ireland's plants, dragonflies, mayflies and moths and more particularly on trees and butterflies. On an island with so many radical changes in habitat and vegetation, the impact on species of a wetter and stormier future remains to be measured.

This book, meanwhile, with its fascinating complexity, is a sharing and celebration of science to complete the Royal Irish Academy's great enterprise.

Michael Viney,
Thallabawn, Co Mayo

March 2019

Opposite

Fig. 1.1 Robert H. Welch's superb photograph of Croaghmore from the north-east: the frontispiece (Plate I) of Part 1 of the Original Survey of Clare Island (1915).

Chapter 1

Introduction: Robert Lloyd Praeger and the Clare Island Survey

I would be measured against you,
and will lose face, pacing out
in the footsteps of the early workers
to where fatigue compels the heart,
and return then, my face aglow
with a booty of old words, and new echoes.

Sean Lysaght,
'The Clare Island Survey' (1991)

The people of Clare Island never sought fame in the annals of natural history; it came to them unasked, in large measure due to a visitor who stepped off the boat at the quay with his wife for the first time in July of 1903. Robert Lloyd Praeger was 38 at the time, the most accomplished Irish field botanist of his generation, at the height of his mental powers and almost legendary physical stamina.

Years later, in *The Way That I Went*, Praeger recalled the spell cast on himself and his wife Hedwig by that first visit to Clare Island. His description of it reads like something in Dante's *Divine Comedy* (to which indeed he himself compared it). They embarked on an utterly calm sea at Roonagh Pier.

> All the hills around were smothered in a white mist, which over the island formed an enormous arch, solid enough seemingly to walk on, and descending nearly to sea level. We lurched across in an ominous stillness, and darkness descended before it was due, as we groped our way to the little quay.[1]

Next day, dense fog and heavy rain kept the couple confined to the little cottage in which they were staying, until evening:

> Late in the day the rain ceased, and a strange red glow, coming from the north-west, spread through the thinning mist. We hurried out to the north point of the island, and there, just sinking into the ocean, was a blood-red sun, lighting up thick inky clouds which brooded low over the black-jagged teeth of Achill Head, rising from a black sea tinged with crimson. It was a scene fitted for Dante's *Inferno*, and if a flight of demons or of angels had passed across in that strange atmosphere it would have seemed quite appropriate, and no cause for wonder.[2]

Praeger was particularly impressed by the magnificent cliffs that fronted the Atlantic waves on the north-west end of the island.

> This great cliff is inconvenient to examine because of the shattered and crumbling nature of the slates of which it is composed; but by dint of patience I explored it from the summit to within reach of the waves, in spite of heavy rain, and the accompanying difficulties of foot-hold. The cliff proved unexpectedly interesting, on account of the alpine nature of its vegetation.[3]

How Hedwig must have felt as she watched her husband of only two years explore these terrifyingly dangerous cliffs we can only imagine: no botanist since has attempted to follow.

Praeger and his wife spent a week on the island, and found the vegetation 'more varied and more interesting than was expected'.[4] It is perhaps not too much to claim that Praeger was enchanted by the memory of this visit, and when the idea of a Great Island Survey was being discussed five years later this was a decisive influence in his selection of Clare Island.

Praeger's early life and influences

Robert Lloyd Praeger was born in Hollywood, Co. Down on 25 August 1865, the second in a family of four boys and one girl, the children of William Emil Praeger and Maria Patterson. He inherited his deep interest in natural history from his mother, who was the daughter of the prominent naturalist Robert Patterson (1802–72). Although Patterson died when Praeger was only seven, he nonetheless was a great influence on the boy's early years, especially through the excursions he organised on Saturday afternoons, 'pointing out

Fig. 1.2 Colour portrait of Robert Lloyd Praeger by Anna O'Leary.

to us delighted children lady-birds and tree-creepers and "devil's coach-horses" and strange chrysalises.'[5] His grandfather also authored several textbooks in zoology that may well have been an early influence on Praeger's botanical writings in this genre.[6] It was, however, to his uncle William Patterson that Praeger himself gave most of the credit for developing his interest in natural history.[7]

During his early years Praeger was deeply influenced by James Lawson Drummond, who was a professor in the Royal Belfast Academical Institution, where Praeger went to school. Drummond's *Letters to a Young Naturalist on the Study of Nature and Natural Theology* (published in 1832) pervaded the philosophy of the Belfast Naturalists' Field Club, in which Praeger was to become so actively involved.[8] Drummond saw natural history as an essential part of general education, and devoutly believed that through it we could trace in detail God's plan in the world.[9] Praeger also acknowledged a great debt to Samuel Alexander Stewart, a founding member of the Belfast Naturalists' Field Club (to whom he dedicated his *Open-air Studies*):

> Ever ready to help, he befriended me when I was still at school, and was tireless in naming specimens and in imparting all that lore relating to botanical and geological field-work that is not to be found in any book, but which passes from mouth to mouth among those who keep the torch of knowledge burning. No young naturalist ever had a better or more patient teacher, or a more delightful friend than I when in after years we tramped the hills of Down together.[10]

When Praeger was eleven his father made him a member of the Field Club, but it was only in 1885 that he became actively involved in its activities. When he moved to Dublin in 1893 to take up the post of Assistant Librarian at the National Library (which he held until his retirement in 1923 at the age of 58), he joined the Dublin Naturalists' Field Club and was elected secretary almost immediately; by 1897 he was vice-president and became president for the first time the following year. By this time, he had developed a particular interest in the origin of island floras, about which much had been written in the later part of the nineteenth century.[11]

Praeger and islands

> Islands are always fascinating—particularly if they are small. Their aloofness makes a curious appeal, to reach them we must cross water—a capricious medium of many moods.
>
> One rambles over them with a not unpleasant sense of restriction; the constant pressure of the circumambient sea is stimulating; and the ships which sail upon it acquire a new significance, from the fact that their ministrations are essential if our sojourn is not to be prolonged indefinitely.
>
> R.L. Praeger (1930). 'Island Memories'. *Beyond Soundings*, 17

On 27 August 1883, out in the deep water between the islands of Sumatra and Java in Indonesia, the little island of Krakatoa all but disappeared as a great surge of volcanic magma exploded out of the ocean floor in the most awesome and cataclysmic volcanic eruption ever recorded, ejecting 21km^3 of rock, ash and pumice with a force 13,000 times that of the atomic bomb that destroyed Hiroshima. At least 14,000 people are known to have died, and the number may have been three times that, mainly as a result of the tsunamis that resulted. This explosion could be heard thousands of miles away, and it affected global weather patterns for years.

All life was extinguished, and for many years it remained utterly deserted. Then, a quarter of a century later, life began to return to Krakatoa, mainly by air. The way in which plant and animal life reappeared on the bare surface of the island attracted the attention of naturalists because of the greatly enhanced interest in colonisation and differentiation generated by the publication of Charles Darwin's *Origin of Species* in 1859. Darwin's discovery of the importance of isolated populations on remote islands for the study of evolution in action generated a new interest in such places among naturalists, and many were scanning the globe for other suitably isolated islands where these phenomena might be studied in the light of the new doctrine of natural selection.

At that time, Robert Lloyd Praeger was eighteen years old, a freshman student at Queen's University in Belfast where he was studying engineering. Although we have no record of this in his own words, it is not difficult to imagine how impressed the young Praeger was first of all on reading the *Origin of Species* and later by the eruption of Krakatoa.

Praeger's first opportunity to look closely at the natural history of an island came in 1904 when Cecil Baring, the wealthy new owner of Lambay Island off the coast of Dublin, approached him with a request for information on the natural history and antiquities of the island; Praeger enthusiastically agreed to the idea of a wide-ranging study of its flora and fauna. He assembled a team of twenty naturalists, and together they studied the island in 1905 and 1906. The results were published the following year in *The Irish Naturalist*, and they exceeded expectations. Although the island covers no more than 240ha and lies only 4km from the mainland, the study added 90 new species to the Irish flora and fauna, seventeen of which had not been recorded in Britain, and discovered five species new to science.[12]

This encouraged Praeger in his idea for a more ambitious and comprehensive survey of a larger, more remote island, where some of the questions that had exercised the island naturalists of the previous half century might be addressed. He accordingly contacted a number of fellow naturalists, inviting them to a meeting in April 1908 to discuss the idea. Out of this the Clare Island Survey was born, though not without opposition from the Blaskets, but Praeger's enthusiasm for Clare Island won the day.[13] The island was a reasonable size, neither too far from the mainland for practical convenience nor too near to be (apparently) unsuited to the first objective of the study (see below). It also had a more diversified surface, 'made up of rocks of different ages, and a hill of 1,520 feet dropping into the Atlantic in a grand cliff covered with alpine plants above and with great bird colonies below'.[14] Importantly, it could offer board and lodging to visiting naturalists, which was not the case for the Blasket Islands. In Praeger's own more elegant words 'the selection of Clare Island was influenced by its suitable size, position, and unusual elevation as compared with most of the western islands; it lay sufficiently far off the coast to raise interesting problems as to the immigration of its fauna and flora, but not so far as to introduce delay and expense to the working parties owing to precarious communication with the mainland; also it was possible on Clare Island to procure accommodation for working parties without outlay on buildings.'[15]

The Clare Island Survey

The Clare Island Survey extended over a three-year period, with fieldwork commencing during Easter of 1909, and ending in November of 1911. The results were published as Volume XXXI of the *Proceedings of the Royal Irish Academy* in a series of 67 parts or fascicles between 1911 and 1915, in which year the Concluding Reports—Praeger's own summary of the results on land and Rowland Southern's discussion of marine ecology—appeared.[16] Many of the contributors to the Original Survey treated Clare Island itself as little more than one of several regions in the 'broader Clare Island district' that included the surrounding seas and the mainland from Achill southwards to Killary and inland eastward as far as Castlebar, 'a district of considerable size and variety of surface'. The experts looking at these groups realised early on that little of exceptional note was to be expected, certainly little that would shed light on the way a distinctive fauna and flora could colonise and then evolve its own distinctive characteristics on an island: Clare Island was too close to, and too recently separated from, the mainland.

The survey was undertaken with two objectives in mind: to see whether 'some beginnings of that differentiation and specialisation which are so remarkable a feature of the natural history of oceanic islands' had taken place since the separation of Clare Island from the mainland; and secondly to see how many new species a concentrated study of this kind would show to be present, especially 'among those more obscure and difficult groups which are apt to be neglected when, on a larger area, a rich fauna and flora prevail'.[17]

In his General Summary published in 1915 Praeger recognised that the first of these objectives had not been realised, partly because Clare Island *is* in fact too close to the mainland: 'A geological study of the area showed that it was not only probable, but almost certain, that a land-connection between the island and the mainland existed long after the Glacial Period, which would have permitted of the immigration of much of the present fauna and flora after normal climatic conditions were resumed.'[18] The second objective, on the other hand, had been achieved with outstanding success: 'the number of interesting additions to the flora and fauna of Ireland and of the British Isles, and the number of species new to science, being far beyond anticipation'.[19] During

CLARE ISLAND

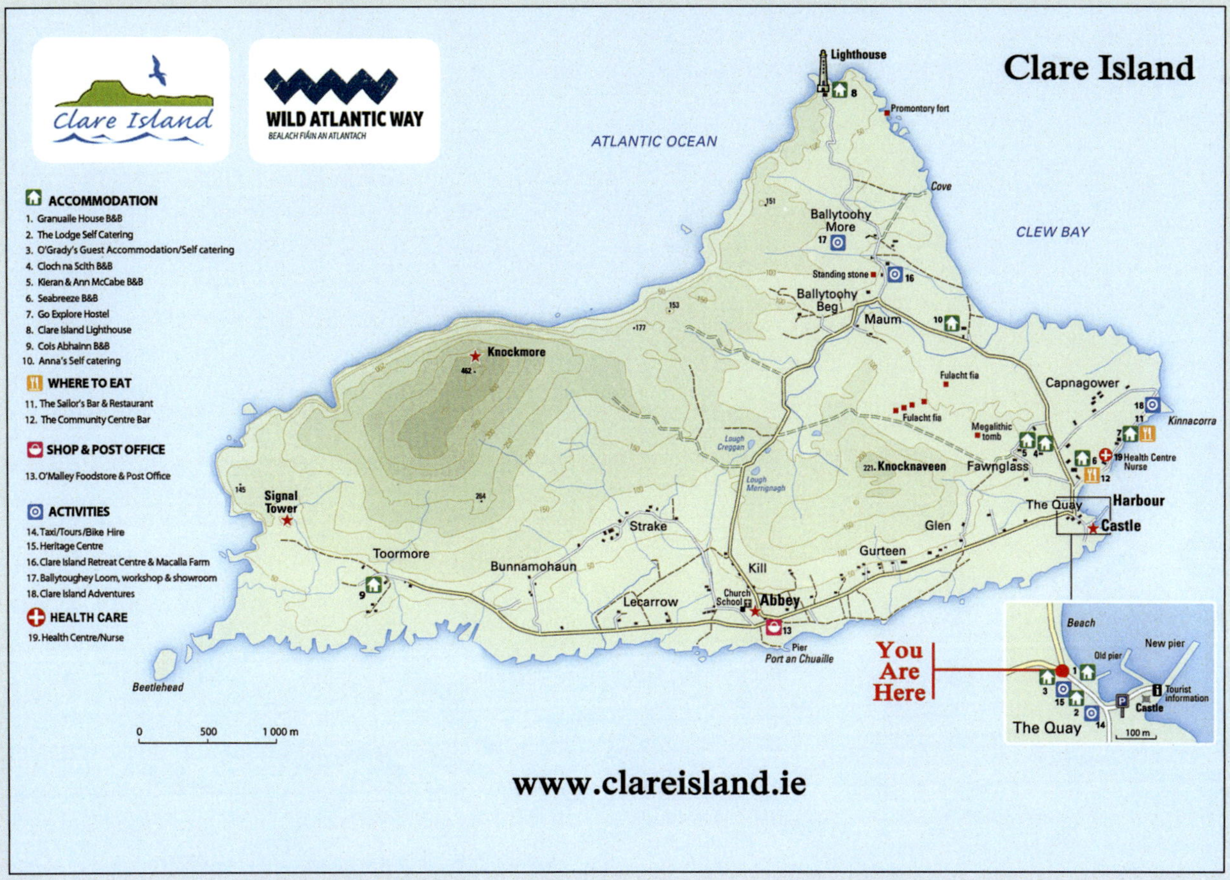

Fig. 1.3 The standard tourist map of Clare Island.

Clare Island stands in the middle of the entrance to Clew Bay, 5km off the west coast of Mayo. It measures some 7km from east to west and 2.5km from north to south, covering an area of 1600ha.

A limestone-filled depression that runs westward into Clew Bay is filled with whale-backed drumlins, which form a remarkable archipelago where they dip beneath the sea: eroded on the ocean-facing western sides where high cliffs front the Atlantic, often protected by a *chevaux-de-frise* of great boulder beaches and submerged banks of enormous stones washed out of the glacial till.[1]

Most of the cultivated land is located towards the eastern end of the island. The exposed western end is storm-swept and 'exceedingly bare'. The coast is almost everywhere surrounded by cliffs varying in height from 15–30m on the east and up to 305m in the north-west. The cliffs on the north and west side are rocky and precipitous, one of them (Allahan, west of Knockmore) rising almost sheer 275m above the sea. On the north-east coast, the cliffs are much lower—between 18 and 21m—and composed mainly of glacial deposits. On the eastern and southern coasts, they are lower, rarely over 15m. [2]

The dominant topographical feature is Knockmore, which is approached on its landward southern side by a steep heather-clad slope, but on the northern seaward side plunges 463m down into the ocean. This terrifying scarp is one of the most interesting places for plants because its inaccessibility to grazing animals has allowed the survival of a remarkable Alpine flora and a rich invertebrate fauna (notably land snails).[3] A high ridge connects the escarpment to Knocknaveen (223m), which is located10km ESE of Knockmore.

[1] Original Survey of Clare Island 1, 11.

[2] Original Survey of Clare Island 7, 2.

[3] Original Survey of Clare Island 1, 12.

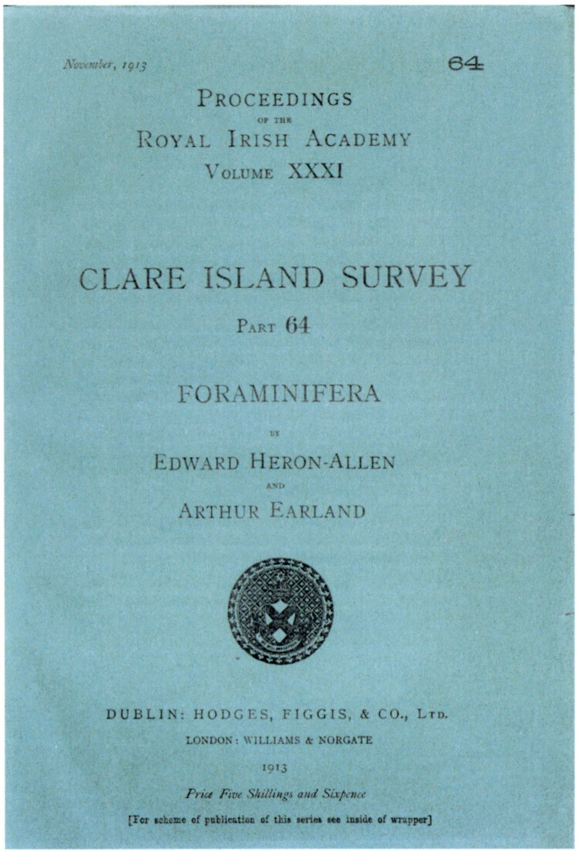

Fig. 1.4 Cover of one of the fascicles of the Original Clare Island Survey.

the course of the survey 3,219 species of fauna were recorded, 1,253 of them new to Ireland, 343 up to that time unrecorded in Great Britain, and 109 new to science. In terms of flora, 3,219 species were recorded, including 585 new to Ireland, 55 unknown in Britain, and 11 new to science.[20] Very few new species were recorded for such well-studied groups as flowering plants, ferns, vertebrates and higher invertebrates. There was a large increase in the species counts for local fungi, Diptera, polychaetes, bdelloid rotifers, nematodes, Flagellata, Ciliata, Rhizopods etc. Some groups—water bears for instance—had never been studied in Ireland before.

Different groups received very different levels of attention. Much time and attention was given to such subjects as archaeology, geology, vascular plants, mosses, fungi, marine algae, birds, molluscs, polychaete worms, sponges and foraminifera, whereas no more than a single visit was accorded other groups—including Arctiscoidea, freshwater entomostraca, rotifers, flagellates, ciliates and slime-moulds. Some visits were spectacularly productive; for example, a single sample dredged in Clew Bay and weighing no more than a few ounces 'furnished an apparently inexhaustible source of new and interesting species'.[21]

The survey was a superb feat of organisation on the part of both Praeger and his wife as much as anything else. Up to a dozen workers at a time visited the island for a week or more over a three-year period between Easter of 1909 and late 1911. The survey on land was supplemented by work carried out at sea by Rowland Southern and others, who had at their disposal the fisheries vessel *Helga II* for dredging and trawling the seas to a depth of 125m, as well as access to earlier surveys carried out at nearby marine research stations (1899–1904).

Praeger managed to recruit a hundred experts in every field of natural history ('the flora and fauna occupied a foremost place'), as well as climatology, geology and agriculture, history, archaeology and other aspects of island culture; many of them paid their own way. Praeger had also been able to enlist the support of the Fisheries Division of the Department of Agriculture and Technical Instruction for Ireland, the Congested Districts Board, the Irish Lights Commissioners, the Geological Survey of Ireland, the National Museum and the British Museum.[22]

> It was a full, stimulating and interesting time. My own work in organizing and conducting, assisting in the field-work, gathering together the sixty-seven separate reports and seeing them through the press, occupied my leisure time for six years.[23]

Although there was general agreement among the participants 'that the narrow strait of sea which separates Clare Island from the mainland represents a very serious barrier to migration, and one across which the existing fauna and flora of the island, taken as a whole, could not have passed',[24] the results disappointed Praeger's hopes of finding a flora and fauna unique to the island. There were no endemic species, although a considerable number of species new to science were discovered. Nevertheless, it was highly successful in overall terms, being the most detailed overview of an island of its size carried out anywhere in the world up to that time, despite the fact that many of the participants contended that it could only be considered a partial survey, and that more time and systematic effort would discover many more species.[25] However, Rowland Southern, who was not terribly enthusiastic about the idea of 'faunal provinces', pointed out that too much should not be read into mere lists of species in terms of geographical significance; 'even in the case of terrestrial plants and animals, the significance of their geographical distribution can only be appreciated by a thorough knowledge of their mode of life and the conditions of their environment, that is to say, of their ecology. The failure of the study of geographical distribution to yield results in any way proportionate to the vast amount of labour which has been devoted to it, has been largely caused by the neglect of the more vital study of ecology'.[26]

The Clare Island Survey of 1909–11 'was in many ways the grand climax in Ireland of the Victorian era of naturalists' field clubs'.[27] In the years following its publication Praeger's botanical interests became more taxonomic, and less exclusively focused on fieldwork in Ireland. His main achievements in later years included monographs on the plant genera *Sedum* (stonecrops) and *Sempervivum* (houseleeks), which widened his horizons to southern Europe and the Canaries.

But the Ireland of those later years had changed in ways that made it less easy to roam the countryside as the earlier naturalists had done, and the study of natural history and antiquities came to be seen as the pursuits of a leisured minority out of step with the confident new march of Irish independence. The destruction of the printing plant and stock of *The Irish Naturalist* during the Easter Rising of 1916 might well have been seen as a portent of things to come.

Through involvement with the Belfast Naturalists' Field Club Praeger developed a religious and political tolerance that served him well in later years. 'The field club functioned as a kind of freemasonry where religious discourse had no place': this was in part a reaction to the Great Evangelical

Fig. 1.5 At work on the Original Clare Island Survey; (a) a group getting ready to pose for photographs in the harbour of Clare Island. Praeger is on the extreme left, sitting on the sand, with his wife Hedwig sitting behind him; (b) Praeger on board the newly-commissioned fisheries research ship Helga; (c) Jane Stephens about to embark on the Helga's boat from the hotel pier at low tide, under Praeger's watchful eye, 1910 (Praeger Collection, Royal Irish Academy).

TABLE 1:
The Original Survey of Clare Island: summary of species

	Total species	New to Ireland	New to Britain and Ireland	New to science
Flora Fauna	3,219 5,269	585 1,253	55 343	11 109
Total	8,488	1,838	398	120

Revival of 1859, four years before the club was set up.[28] However, the attitude to the natural world that pervaded the Belfast club aroused little interest or sympathy in the newly independent and Catholic Free State, and can only be discerned with difficulty in Praeger's writings.

Praeger died on 5 May 1953 in Co. Down, where he went to live following the death of his wife Hedwig the previous year. They are buried together in Dean's Grange Cemetery in Dublin.[29]

The New Survey of Clare Island

The idea of a New Survey of Clare Island was first mooted by the Royal Irish Academy in 1988. This would not only allow comparison and evaluation of change over the century between the two surveys, but constitute a new baseline using modern ecological techniques and approaches that would be invaluable for future environmental monitoring. A feasibility study was carried out in 1990 with the enthusiastic support of the Centre for Island Studies, which had as its mission the exploration, development and promotion of the island's unique natural and cultural heritage, and which undertook to facilitate the integration of the project with the life of the island. This preliminary study concluded that such a New Survey would be both worthwhile and scientifically justified; and it was hoped that it might constitute a resource that would support community development and employment on the island.[30] The New Survey was launched in 1992, and concludes with the publication of this volume.

Like its predecessor, the New Survey can be considered an outstanding achievement, although it has not entirely succeeded in its aim of making and evaluating detailed comparison between the island of a hundred years ago and today. This is due in part to differences in survey methodology

Fig. 1.6 At work for the New Survey: Jenny Bryant and Fabio Rindi collecting freshwater algae from a stream (Michael Guiry).

Fig. 1.7 Cover of one of the volumes of the New Survey of Clare Island.

and intensity of effort between the two. In its study of the island's biota the New Survey has a more ecological focus, and did not always adopt the strictly taxonomic approach of the Original Survey, with a separate fascicle for every taxonomic group or small number of groups, with the result that there are many gaps, particularly of marine taxa (and it should be remembered that only 16 or so of the 36 major plant and animal phyla occur on land). More detailed comment on these limitations will be found when the individual groups are discussed later in this volume.

Mapping the framework of Clare Island's landscape

When you walk around an island
You do not come back
To where you started out
Macdara Woods, *Clare Island (V)* (2006)

Praeger's naturalists were not the first to look closely at the surface of Clare Island. In the 1820s the Ordnance Survey of Ireland had been given the monumental task of mapping the townlands of Ireland (all 60,000 of them) on a scale of six inches to a mile, beginning in 1825 and finishing in 1846. The survey was undertaken as plans were drawn up to reform the country's local taxation system. In addition to providing a land valuation, the survey also provided a base map for the census of the population and a detailed geological survey of the country that could be used for military and other purposes. It was carried out by a department of the army, and the engineers who did the work were actually soldiers in uniform.[31]

Townlands are the fundamental unit of land in Ireland. They probably started out in the distant past as individual holdings, but these were subsequently subdivided and amalgamated in all sorts of ways. The civil parishes that consist of a cluster of townlands of greater or lesser extent were originally established as part of the reform and reorganisation of the Irish church in the thirteenth century, and generally corresponded more or less with civil administrative units (chiefdoms). In this regard, the civil parishes are of great historical significance. Modern Catholic parishes do not correspond to the old civil parishes, which have been divided and amalgamated to take account of demographic change, similar to the way townlands have changed over time.

The six-inch map produced by the Ordnance Survey is an outstanding achievement in cartographic terms, and a work of great beauty, unparalleled anywhere else in the world. Its importance is heightened by the fact that it shows the country in great detail in the years immediately preceding the Great Famine of 1845–48, when it was at its most intensively managed and at the height of population density. It is the single most important document for the detailed study of local history and topography.

The Townland Survey was preceded by a separate and painstaking Boundary Survey, which began in 1826. Its task was to physically mark out or otherwise define the boundaries on the ground in advance of the Townland Survey. These boundary markers were recorded and meticulously described in field books and sketch maps (which can be inspected in the National Archives), and in due course were pointed out to the Ordnance Survey officers.

On completion of the field survey manuscript, Fair Plans were prepared on a parish-by-parish basis at a scale of six inches to a mile. The Fair Plans were then engraved on copper plates for printing, undergoing some small changes in the process. Proof impressions were taken from plates, and these were carefully inspected and checked in the field, and various alterations were suggested or made before the actual printing of the map went ahead. It generally took two or three years to progress from the completed survey to the appearance of the printed map.

It is a matter of deep regret that the Down Survey for Clare Island has not survived. The only detailed map before the Ordnance Survey is William Bald's beautiful Map of the maritime County of Mayo, described as 'the finest Irish cartographic achievement prior to the Ordnance Survey', which surveyed at a scale of four inches to a mile and was printed on 25 sheets at three inches to a mile in 1830. Bald was at work on the map for seven years. He also surveyed 64,750 hectares of bogland in Mayo and Sligo for the great Bogs Survey of 1810–14.[32]

Officers of the Geological Survey used the six-inch map of the Ordnance Survey to record 'by means of marks, writing, and colours, the nature, extent, direction and geological formation of all portions of rock visible at the surface'.[33] Every rock outcrop was coloured and marked in by hand, and notes added on topography and other subjects. The field sheets were the basis for the published one-inch

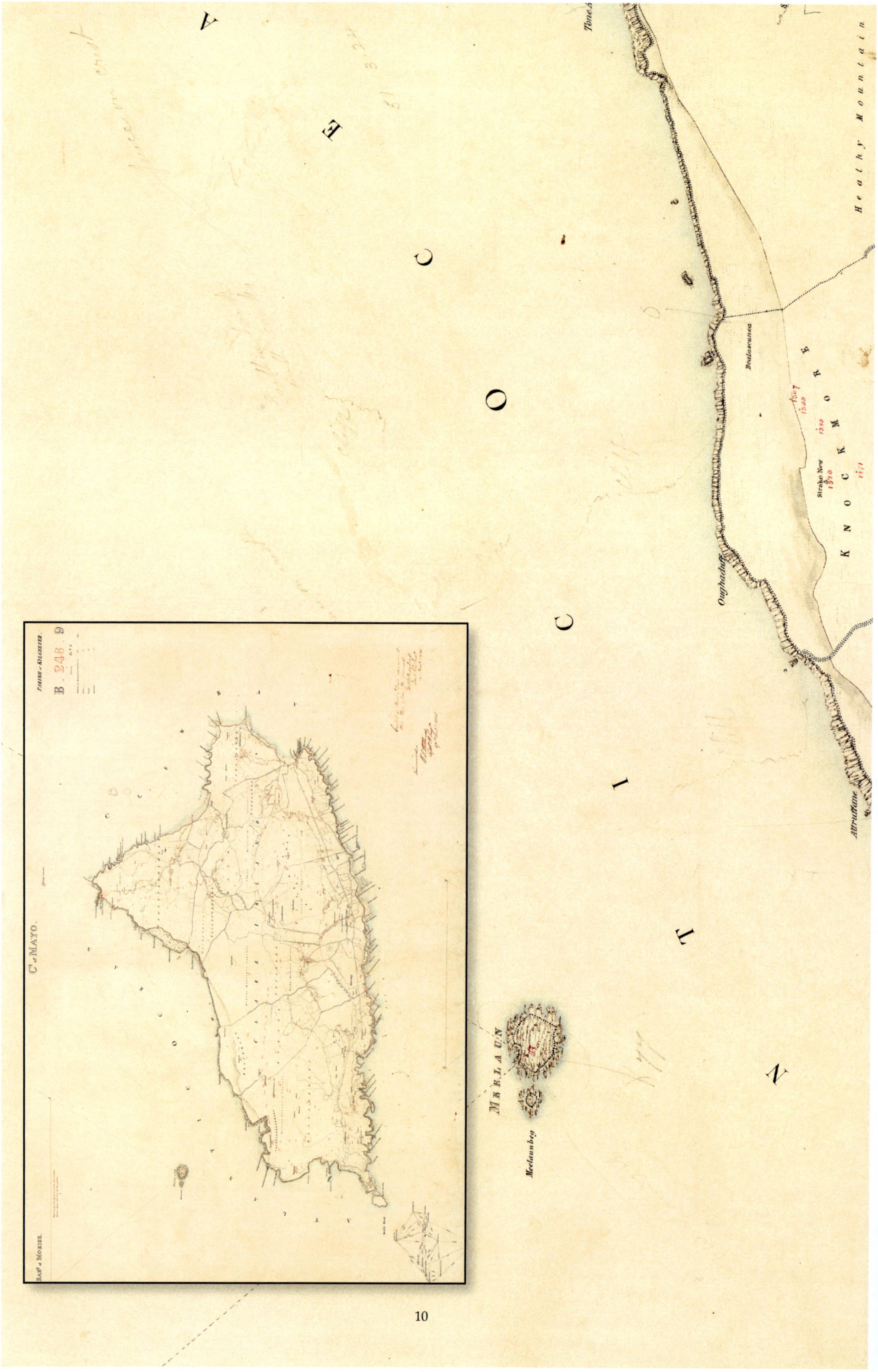

Healthy Mountain
K N O C K M O R E
Strake New
Oughaduff
MEELAUN
Meelaunbeg
C. of MAYO.

Fig. 1.8 The Ordnance Survey Fair Plan of Clare Island (1838).

C L E W
A N
BALLYTOOHYMORE
BALLYTOOHYBEG
CAPNAGOWER
LOUGH-A-VULLIN
Rocky Pasture
Bog
Light House
Ballytoohymore
Rocky Mountain Pasture

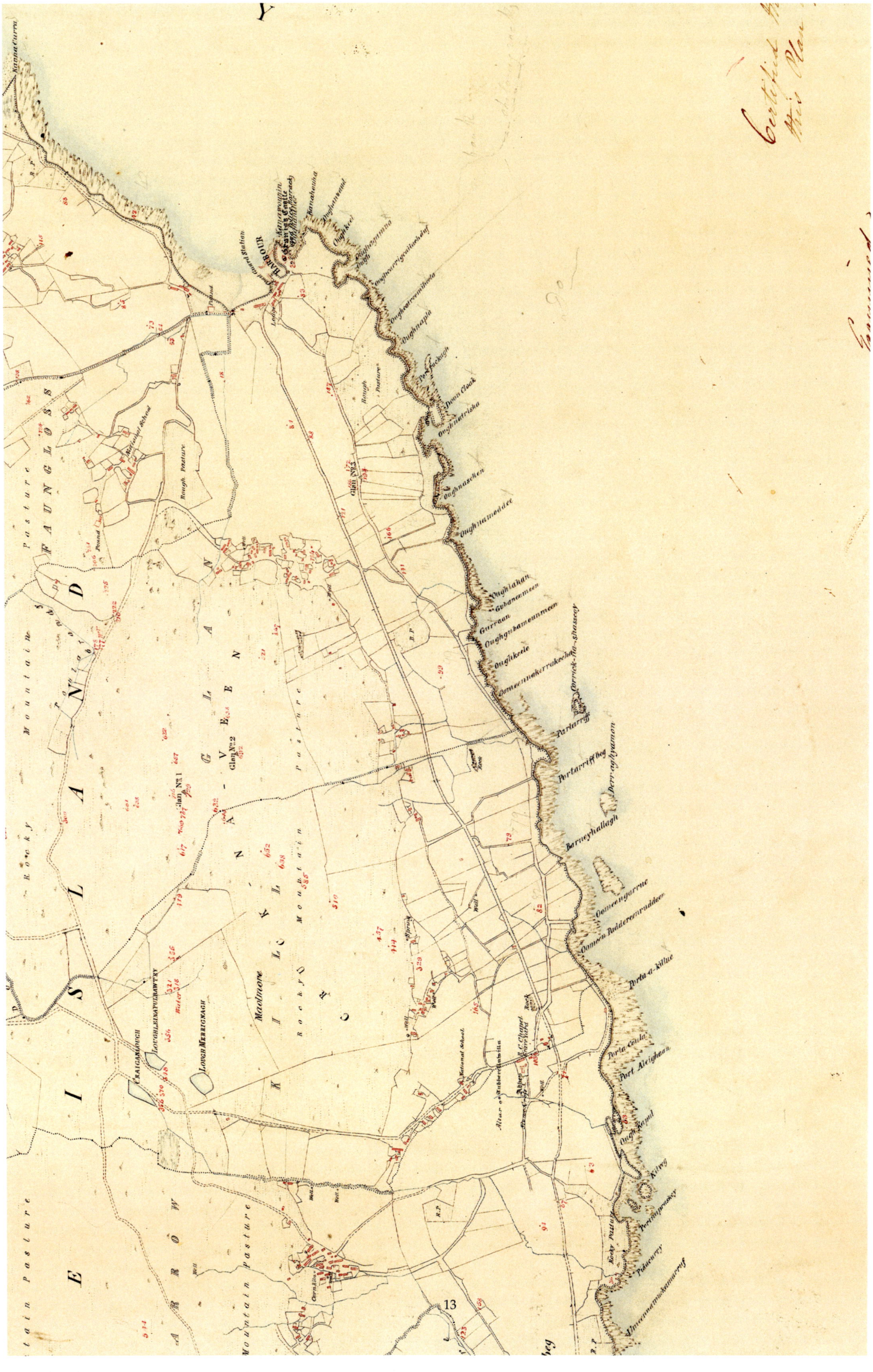
HARBOUR
FAUNGLOSS
Rough Pasture
LOUGH MERRIGNACH
National School

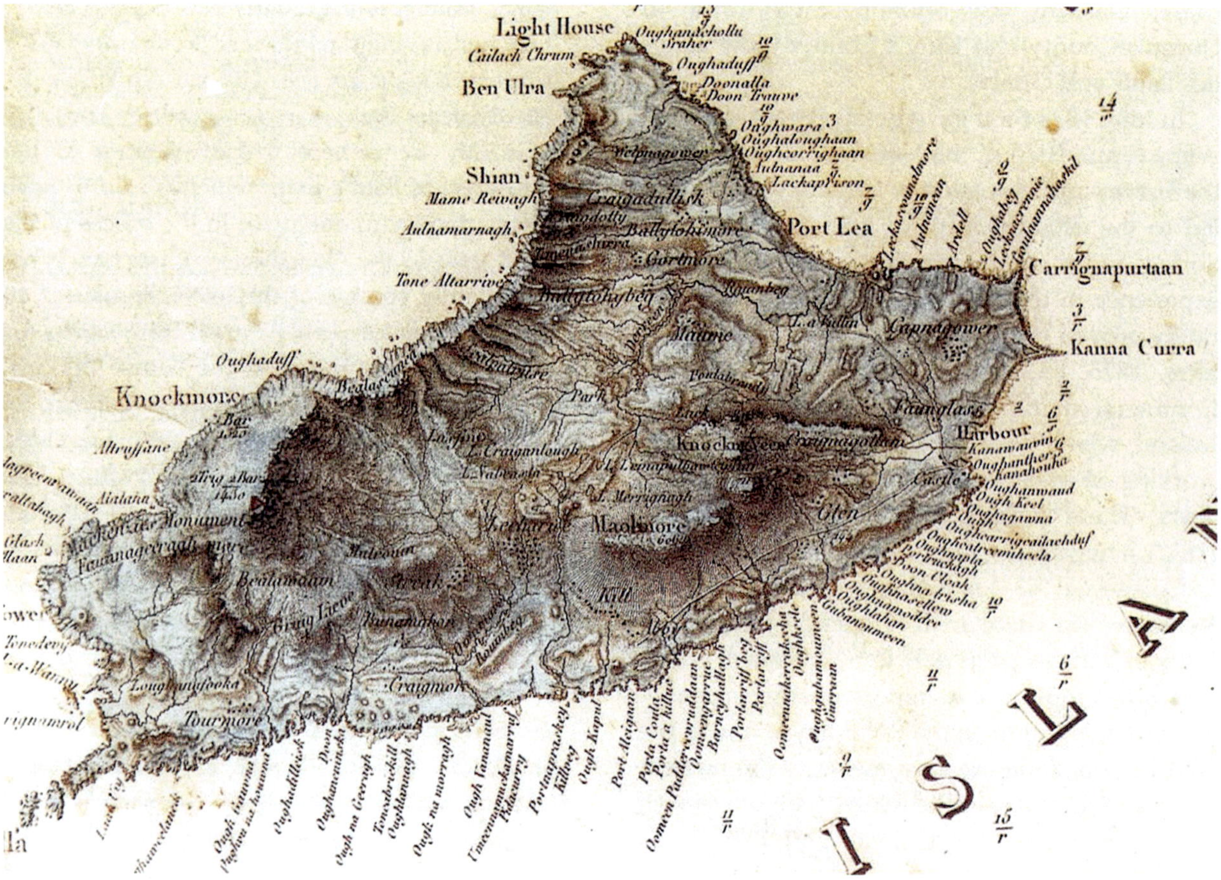

Fig. 1.9 Bald's Map of Clare Island.

geological map of Ireland, and were accompanied by a series of printed Memoirs covering its 205 sheets.[34]

West Mayo was surveyed in the 1860s by a team of geologists from the Geological Survey,[35] but Clare Island received little attention at the time, because the officers were preoccupied with the monumental challenge of deciphering the geology of the region as a whole. The original description of the island's geology, written by R.G. Symes in 1868 (and published in 1876), came to no more than a few paragraphs, but the Clare Island Survey generated new interest, and a much more detailed report was prepared in the years leading up to 1912 by J.R. Kilroe and G.A.J. Cole, while T. Hallissy provided an account of the drift geology and soils.[36] This formed the basis for the publication two years later of Hallissy's geological contribution to the Clare Island Survey.[37]

Placenames

Although the Ordnance Survey and the Geological Survey recorded the topographical and geological character of Clare Island in such detail on their maps, the people who had lived on the island for millennia knew it with a deeper intimacy. Their direct experience hardly extended beyond the seen horizon that was the edge of their world. Nevertheless, they knew every cove and headland, every bush and stone in this small world, every slope and shadow; they knew how its soils varied and what might grow in them; after all, they had to know these things because upon this knowledge their very lives—and their future—depended.

The rich net of placenames that has been woven over the island, across the ridged fields and every patch of pasture, out onto every headland and filling every cove, into every nook and cranny of the land marks out—*chronicles* indeed—the way the growing population exerted an ever more pervasive influence over the island. They are like explorers' flags, signifying some sort of ownership or control—however impossible it often was to replace nature. Every name is to some degree an echo of the human presence. In some sense, the weave of placenames over the land is a record of the advance and retreat of the human tide across

Fig. 1.10 The original Geological Survey field sheet or 'working map' for the west of the island (Mayo six-inch sheet 84).

the island. At the present day for example, the mainly commonage townland of Bunnamohaun consists of little but upland heath and rock. But the name retains the memory of the vanished clachan to which it was once home.[38]

The foundation of our knowledge of the island's namescape was laid by Eoin MacNeill during the first Clare Island Survey in 1913, supplemented by William Bald's survey for his Map of the maritime County of Mayo in the early nineteenth century, and the work of the Ordnance Survey in the 1830s.[39] These three name-gathering ventures collected some 330 names in all. Nollaig Ó Muraíle's study for the New Survey not only meticulously reviewed the pioneering work of the earlier collectors and made it more accessible, but progressed it considerably through his own research on the degree of survival of earlier names at the present day, as well as through the collection of unrecorded placenames (Figure 1.11). Ó Muraíle found that some two-thirds of the names recorded in the early surveys are still remembered on the island to a greater or lesser degree (though not all are in use any more), and he was able to add 270 names not previously recorded, making a total of some six hundred names, remembered or forgotten.[40]

MacNeill noted the way some of the names recorded bygone conditions of agriculture. There are many *gort* (cornfield) placenames, and two names hint at a long-abandoned linen tradition. The most likely meaning of *tuar* in Toormore (*tuar mór*) and *na Tuartha* is 'bleach green', but the word can also mean a home field, often along streams, where cows graze and are fed and milked.[41] Then there was the small group of houses in Maum known as *an Muileann*, the mill, beside the mill stream (*Abhainn an Mhuilinn*) that flows out of Loughavullin (*Loch an Mhuilinn*, the mill pond).

T.J. Westropp drew attention to the 'unusually rich nomenclature of the creeks and caves', showing 'that the life of the people was so much concerned with the shore that every rock and hollow needed a name, unlike the north coast at the mountains, where the vast cliffs discouraged all acquaintance with the shore, save at certain creeks'.[42] MacNeill believed that 'the minute nomenclature of the coast-line is inherited from a time when the business of fishing gave more occupation to the islanders than it does at present'.[43] However, it is more likely perhaps that it is the importance of seaweed which explains the particularity with which every accessible inlet and cove on the coast is named.[44]

Aside from that 'miniature nomenclature of the coast', Ó Muraíle drew attention to the significant number of field and garden names; we can be sure that when population was at its highest and the use of land at its most intense under the prevailing rundale system in the decades before and after the Famine, every plot, however small or remote, will have had its unique Irish descriptor. Names were invented as they were needed, and were 'legal tender' to varying degrees: some known and accepted over the entire island, others much more local.[45] The new map of Clare Island prepared by EastWest Mapping (see page 18) places its main emphasis on names current today.

From theodolite and chain to satellite and plane

In 1992 a special digital map was prepared by the- University of Glasgow for the Royal Irish Academy, and this was used as a base map for the numerous thematic maps that are such a striking

Index of placenames with flora/fauna references

Cows or calves	6, 27, 55	Tree, woods	66, 69, 226, 231, 280
Bull	63, 137–8	Birds	19, 37, 40, 43, 56
Sheep, lambs	68, 104, 106, 141, 267	Crabs	13
Horses	51, 86, 191, 293	Seals	67
Goat	244	Beetle	87–9
Dogs	100, 101	Otter	94
		Limpets	60
Dulse	OS48, 15, 60, 103, 302	Porpoise	115
Wild mustard	118	Cormorant	157
Bramble	149	Pigeon	173
Fern	255	Butterflies	192
Rushes	210		

The numbers refer to Ó Muraíle's placename list (New Survey 1, 115–130).

Fig. 1.11 Nollaig Ó Muraíle's map of the placenames of Clare Island (New Survey, Volume 1). The numbers refer to discussion of the names in his account.

Fig. 1.12 Part of the base map prepared for the New Survey of Clare Island by the Department of Geography and Topographic Science at the University of Glasgow in 1992, based on aerial photography flown by the Ordnance Survey in 1990 (original scale 1:7500, contour interval 5m) (© University of Glasgow).

feature of the published volumes of the New Survey of Clare Island. Many are reproduced in this volume. A proposed series of thematic and dynamic maps 'for educational "odysseys" in the natural and cultural heritage of the island' has not (as yet) materialised.

No modern technique shows topographical detail with such precision as Lidar, which originated in the 1960s. This is a remote sensing technology that combines the light-focusing ability of a laser with radar's ability to measure distance by calculating the time it takes for a signal to travel to an object and return, and then analysing the pattern of the reflected light: visible, ultraviolet or infrared. Its ability to map physical features at very high resolution makes it possible to register subtle topographic features, and its ability to penetrate vegetation allows it to depict land-surface elevation in wonderful detail of these maps enabling us to see many features not evident on the ground or visible with other imaging systems.

In 2007 the Royal Irish Academy commissioned a Lidar survey of Clare Island from the UK Environment Agency's Science Group (Technology) (at a cost of £15,000). This was used in the preparation of the later volumes of the New Survey, and has provided several striking images for the present one.[46] It is perhaps in the depiction of the island's abandoned cultivation ridge systems (discussed in detail later) that its capacities are shown to greatest effect. With the Academy's permission, the Lidar imagery has recently been used in the preparation of a large-scale, highly-detailed map of Clare Island on a scale of 1:7,500 with 5m contour intervals, by East-West Mapping.[47]

Fig. 1.13 Lidar image of the south-west corner of Bunnamohaun. The stream flowing down the centre is the Owenmore River; below: close-up of part of the image.

Notes

[1] Praeger 1939, 185.

[2] *Ibid.*

[3] Praeger 1903, 284.

[4] Praeger 1903; Praeger 1911, OS 10, 2.

[5] Praeger 1949, 16–17.

[6] Especially perhaps *Introduction to zoology* (1846) and *First steps in zoology* (1849), which became a standard text-book for schools.

[7] Collins 1985.

[8] Jones 2009, 12–15.

[9] Drummond 1832.

[10] Praeger 1939, 93-94.

[11] Several detailed island monographs appeared in the first decade of the twentieth century: Christmas Island in the Indian Ocean (1900), Anticosti in the Gulf of St Laurence (1904), the subantarctic islands of New Zealand (1909), the atoll of Funafuti in the Ellice Islands (1896–97), the Faeroes (1901–6) closer to home; and of course, Krakatoa (1908). Praeger's first island visit was to Rathlin in 1889.

[12] Collins 1994, 120–21; Lysaght 1998, 84.

[13] Collins 1994, 121–23.

[14] Praeger 1939, 185.

[15] Praeger 1915, OS 1, 2.

[16] Rowland Southern (1992-1935) was Assistant Inspector of Fisheries with the Department of Agriculture and Technical Instruction, and an expert on marine biology with the Natural History Section of the National Museum. He authored nine of the marine sections of the Clare Island Survey (one of them with G.P. Farran), including the major summary on marine ecology (Part 67), much of the fieldwork for which was conducted from the fisheries research vessel, the *Helga*. He was known for his meticulous and accurate work, and critical of the limits of mere species lists in the absence of a proper ecological framework.

[17] Praeger 1915, OS 1, 1–2.

[18] Praeger 1915, OS 68, 1. Significantly perhaps, the order of priorities is reversed in the General Summary. 'Evidence also went to show that even in the absence of this material [the till choking Clew Bay] the levels at which post-Glacial beds of peat occur on and under the sea-shore of the neighbouring districts pointed to a post-Glacial elevation sufficient to lay bare even the present channel between Clare Island and the Louisburgh area.'

[19] Praeger 1915, OS 68, 1.

[20] Praeger 1915, OS 68, 4.

[21] Southern 1914f, OS 54, 1–2; Praeger 1915, 68, 5.

[22] Collins 1994, 127–28; Lysaght 1998, 82–93.

[23] Praeger 1939, 187.

[24] Praeger 1915, OS 68, 10.

[25] A prediction amply demonstrated in several sectors of the New Survey.

[26] Southern 1915, OS 67, 103.

[27] Viney 2000.

[28] Jones 2009.

[29] Collins 2009, 5–9.

[30] Royal Irish Academy 1991, *Proposal for a New Survey of Clare Island.*

[31] Andrews 1974.

[32] Ó Muraíle (1999), 106–10.

[33] Explanatory note to Geological Survey Memoir (1866).

[34] Herries-Davies 1983.

[35] Cf. Memoir to Sheets 73, 74, 83 and 84 (1876).

[36] Cole *et al.* 1912. There were also contributions on the Carboniferous rocks by Wheelton Hind.

[37] Hallissy, T. (1914). *Geology*. Clare Island Survey Part 7.

[38] The word means 'the low ground of the cabins.'

[39] At the time of MacNeill's work 'the middle-aged and elderly folk in every part of the island could converse in Irish.' (OS 3, 7).

[40] Ó Muraíle 1999.

[41] There is also 'SeanUr' (Ó Muraíle, no. 303). The only literary reference to flax growing on the island 'once upon a time' occurs in the Legend of the Seal Woman told to Nathaniel Colgan (1911) (OS 4, 28).

[42] Westropp 1911, OS 2, 17.

[43] MacNeill 1913, OS 3, 5.

[44] Of the total of 327 names in Ó Muraíle's catalogue, nearly 50 are 'Oo' (cove) names; 14 are 'carraig' names, 9 'port' names, 14 'leach' names and 10 'aill' names.

[45] The placenames of the island, it has to be admitted, are somewhat prosaic as a whole; 'Half these are not names at all' wrote John O'Donovan.

[46] These were prepared by Steve McCarron of National University Ireland, Maynooth.

[47] EastWest Mapping, Enniscorthy 2016. The scale is 1cm = 75m (which takes about 1 minute to walk on level ground). A digital version for smartphone apps is also available.

Opposite

Fig. 2.1 Clare Island viewed from the Corraun Peninsula on the northern shore of Clew Bay. The rocks here are Old Red Sandstone belonging to the Maum Formation.

Chapter 2
Geology: the island's rock framework

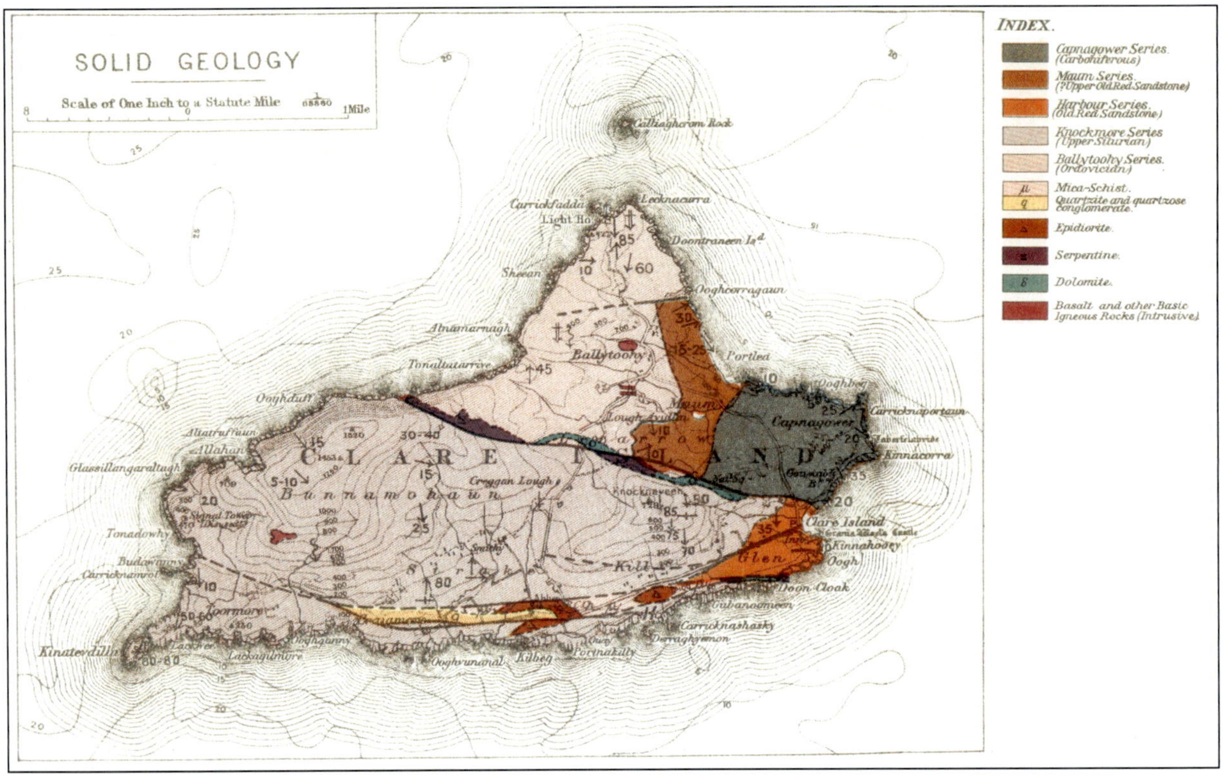

Fig. 2.2 Map of the solid geology prepared for the Original Clare Island Survey, based on J.R. Kilroe's 1910 revision of the rocks, and published in Part 7 of the *Survey of Clare Island* (1914).

The geology of Clare Island is complex and enigmatic, and its challenging make-up has attracted successive generations of geologists to its shores. The earliest geological map of Ireland, Richard Griffith's map of 1838, recognised in broad terms the most immediately obvious geological distinction on Clare Island: the area of older rocks occupying the higher ground of the north and west were described as 'Clay Slate Greywacke and Greywacke Slate', while the area of younger rocks in the east and south-east were described as 'Old Red Sandstone and Sandstone Conglomerate'. The foundations of our understanding of the island's geological make-up were laid by the Geological Survey in the latter part of the nineteenth century. The first survey of the solid geology of the island was carried out by R.G. Symes in 1868, and published in 1876.[1] A new survey was carried out by J.R. Kilroe and Timothy Hallissy, who reported in detail in 1914 on the island's drifts and soils (a detailed account was published in a special Survey Memoir).[2] This formed the basis for Hallissy's geological contribution to the Original Clare Island Survey.[3]

These early studies were largely *descriptive*. They chronicle in some detail the distribution of the different kinds of rocks that make up the island, and their relationships with each other—often using terminology that is now rather outmoded – but their understanding of the *nature* of these rocks is limited. Symes and Kilroe's studies of the geology of the island were informed by the uniformitarian principles first promulgated by James Hutton and Charles Lyle. According to this perspective the processes by which rocks had formed in the past are still at work today, and the story embodied in these silent, challenging rocks could be explained in terms of these processes. However, our understanding of the rock-forming processes was then in its infancy, but it grew steadily as the twentieth century advanced. In particular, the great geological conundrum of how rocks that had manifestly originated in different parts of the globe and in utterly different environments could occur side by side in the same place was not solved until the 1960s, when the development of plate tectonic theory showed that the earth's crust is not a single, solid shell of rock fixed for all time, but is made up of discrete plates that drift slowly on a fluid interior, their movement driven by convection powered by natural radioactivity in the earth's heart.

Armed with this new all-illuminating perspective, Adrian Phillips of Trinity College Dublin re-examined the rocks of Clare Island for

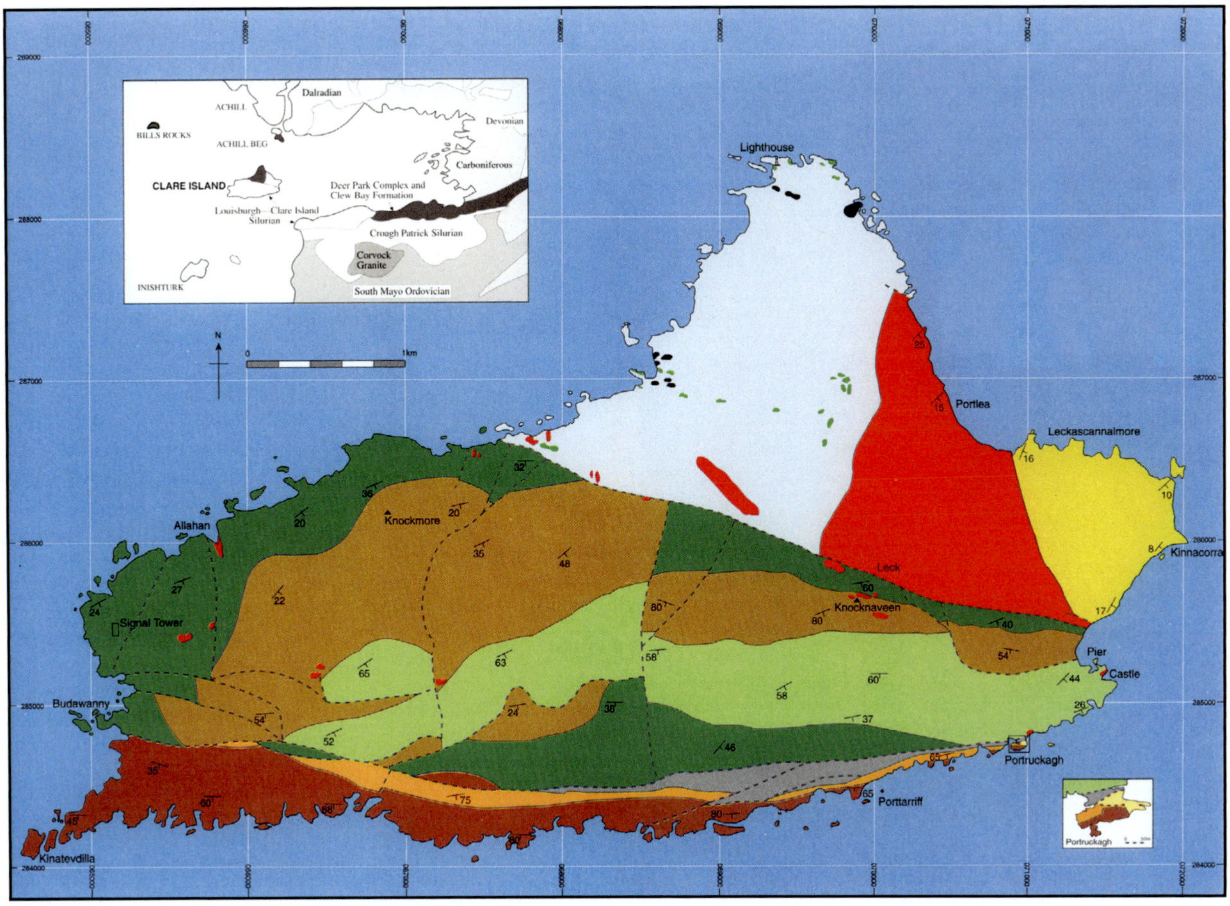

Fig. 2.3 Geological map of Clare Island (Sources: Coastline based on the 1:7,500 map prepared by the University of Glasgow (1992); placenames based on Ordnance Survey 1:50,000 Discovery Series Mayo Sheet 30; remaining detail by J.R. Graham).

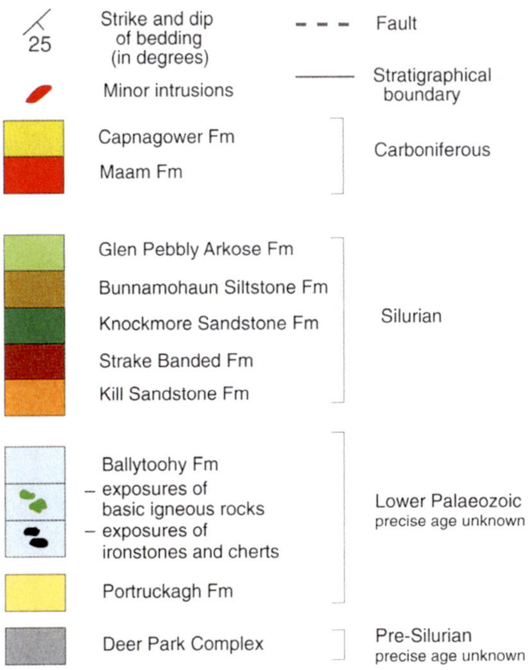

his doctoral thesis in the 1960s, and the account he compiled over many years is the foundation of our modern understanding, though it has been refined and extended by many subsequent studies, and in particular by the new research carried out in the course of the New Survey of Clare Island by John Graham (also at Trinity College).[4] The Quaternary deposits were examined and re-evaluated for the New Survey by Pete Coxon of Trinity College.[5]

***

In geological terminology, a fault is a dislocation across which there has been displacement of the bedrock: one side has moved relative to the other. Such dislocations, once established, are lines of weakness that are often reused, although not always with the same sense of displacement. If you look at the geological map on this page you will see that the Silurian outcrop on Clare Island is abruptly terminated on the north-east by the Leck Fault. The Green Road north of Knocknaveen follows the line of this fault below the prominent topographical cliff known as the Leck, then it continues north-west across the shoulder of Knockmore to the cliffs and beyond. In the last

Fig. 2.4 View of the Leck Fault from the north-west. In the foreground is a peat-drying stand made of sandstone boulders from the Maum Formation.

major movement on this fault the bedrock on the north has been *downthrown* relative to the bedrock on the south by some 600m. South of the fault the rocks are of Silurian age, whereas those north of the fault are of Lower Carboniferous age on the north-east, and of Dalradian age in the triangular promontory that occupies most of the north of the island.[6]

The fault zone is a very ancient and complicated structure: it is not a simple fracture in the bedrock but part of a fault system. Another part of the system is the Kill Fault, which cuts east-west right across the Silurian sequence near the southern end of the island, running through the col that lies between Knockmore and Knocknaveen. This fault system continues along the southern shore of Clew Bay, north-east through Donegal and then on into Scotland as the Great Glen Fault system. The Leck Fault is the major surface expression on Clare Island of what is known as the Fair Head–Clew Bay line, which marks the south-east limit of the Dalradian outcrop in Ireland. To the east of the island it flanks the southern shore of Clew Bay and is generally considered to represent the westernmost extension of the Highland Border Fracture Zone of Scotland, offset (along with the rest of Clare Island) 5km to the north-west by a northerly extension of a major complex of faults on the mainland known as the Doo Lough–Maum Valley tear faults.

The Leck Fault marks the zone of collision between the northern continent of Laurentia and an oceanic volcanic arc that had developed to the south-east. This collision occurred during the Grampian Orogeny (early-middle Ordovician 475–460 mya), which was the earlier phase of the Caledonian Cycle (the later phase was the Acadian that terminated in the early Devonian). There was also extensive horizontal movement along the Leck Fault during the later phase of the Caledonian Orogenic Cycle, and perhaps later during the opening of the North Atlantic, some 60 million years ago.[7] As you walk along the Green Road today it is hard to imagine how profound—in every sense—this fracture is, or how deep a gash into the earth's interior it once was.

The Silurian rocks south of the Leck Fault occupy nearly half of the island, and were deposited during the latter part of the Caledonian Cycle when the major ocean that once separated north-west and south-east Ireland had largely closed. These sedi-

Fig. 2.5 Vertically-bedded rocks of the Strake Banded Formation in cliffs (a) at the south-eastern corner of Bunnamohaun and (b) Fossil mudcracks in rocks of the Strake Banded Formation on the southern coast east of Leckwee.

Fig. 2.6 Cross-bedded sandstones of the Knockmore Sandstone Formation near the Signal Tower in Bunnamohaun.

Fig. 2.7 Current ripple marks on a bedding surface of the Bunnamohaun Siltstone Formation along the Toormore Stream.

mentary rocks—mud and siltstones, sandstones and conglomerates—originated in ancient river systems under a variety of depositional settings, and are sometimes interbedded with layers of volcanic ash. The Silurian rocks rise from the coast northwestwards to the imposing mountain of Knockmore, the angle of whose steep southern slope echoes the dip of the rocks. From the summit the land plunges almost vertically into the sea, splendidly exposing several hundred metres of these Silurian sedimentary rocks in the cliffs.

With our more detailed understanding of the geology of Clare Island we can divide the Silurian rocks south of the Great Fault into five chapters, recorded in five distinct rock sequences in the stratigraphy, each sequence technically described as a 'formation' named after a place in which it is especially well exposed and the predominant rock type in it.[8] The lowest formation is the Kill Sandstone Formation, which varies from 18 to 30m in thickness and comprises mainly cream-coloured sandstones. A coarse conglomerate is often seen at the base, containing cobbles that can be as big as 60cm. These rocks are believed to be of fluvial origin, and the basal conglomerate is probably in the nature of an alluvial fan. They are well exposed along the coast between Portruckagh and Portatarriff.

The Kill Sandstone Formation is followed in the sequence by the Strake Banded Formation, which is 280m thick, made up of red mudrocks interbedded with sandstones. These are the rocks that are so splendidly displayed all along the south coast east of Portatarriff. The formation is made up of four different rock types: tuffs, tuffaceous sandstones, non-tuffaceous sandstones and red mudrocks. A 'tuff' is an igneous rock consisting of ash and other small particles extruded from a volcano. These rocks are ash falls deposited in water; there are five main ones, along with a number of smaller falls. The rocks are pale green or cream in colour and often occur along with tuffaceous sandstones. From current directions and the distribution of the ash we know the volcano responsible was located somewhere to the west.

Thick sandstone sequences separated by purple mudrocks are characteristic of the tuffaceous sandstones. They are often cross-bedded, the cross-beds recording the direction in which the current that deposited the sediment was flowing. These sandstones were probably laid down by floods that followed the volcanic eruptions that produced the tuffs. The interbedded non-tuffaceous sandstones are finer-grained and contain no volcanic ash. They are the deposits of powerful rivers flowing across level terrain and are also commonly cross-bedded. However, these were laid down under a different stream regime from the underlying tuffaceous sandstones, because the current bedding in them shows clearly that they were flowing from east to west—the exact opposite of the river in which the latter were laid down.

Red or purple mudrocks alternate with sandstones throughout the Strake Banded Formation. These finer sediments were laid down for the most part on the floodplain either side of the stream channel, in the distal stretches of laterally extensive sheet floods. Once the transporting water had filtered away or evaporated, the sediment often dried out and the desiccation cracks so characteristic of drying mud developed on the exposed surface.[9]

Overlying the Strake Banded Formation is the Knockmore Sandstone Formation. There is a sudden transition between the two that may represent a break in deposition of unknown duration. Nowhere are the rocks of this new formation more dramatically displayed than in the cliffs west of Knockmore, where they are largely inaccessible to all but guillemots and terns. The best place to get close to them is at the western end of the island, near the Signal Tower. These sandstones are very similar to the non-tuffaceous rocks of the underlying Strake Banded Formation, but they are a bit coarser and mudrock layers are a less important accompaniment. The current bedding in the sandstones shows that these also (like the non-tuffaceous sandstones) were transported from the east and laid down under a broadly similar river regime, though in a more proximal part of the sheet-flood system with a somewhat increased gradient that was not conducive to the deposition or survival of finer grade sediment in the Clare Island part of the basin at this particular time. The green colour suggests a high water table, which may indicate that the transition from the Strake Banded Formation to the Knockmore Sandstone came about through a climatic change.

The Knockmore Sandstone Formation is followed by a fine-grained rock sequence known as the Bunnamohaun Siltstone Formation, which is some 250m in thickness. Much of this consists of green mudstones that often have ripple-marked bedding surfaces and corresponding ripple-scale current bedding in cross-section. Several other rock types occur with these green rocks: red mudrocks (often with polygonal mudcracks), sandstones identical with those of the Knockmore Formation, rocks known as laminites, and a single tuff near the

Amphibolite is a metamorphic rock composed mainly of the mineral amphibole, generally with an oriented fabric. *Amphiboles* are silicates characterised by a double chain of linked SiO_2 tetrahedra. They occur widely in igneous and metamorphic rocks.

Breccias are detrital sedimentary rocks made up of large angular fragments (2mm and bigger).

Chert is a fine-grained rock made of silica, but with crystals too small to be visible with a petrological microscope. The silica in many cherts originated as a gel derived from silica-secreting organisms such as diatoms, radiolarians and sponges.

Greenschist is a low-grade regional metamorphic rock that forms through the alteration of basic igneous rocks such as basalt. It contains a lot of the mineral *chlorite*, the green colour of which gives the rock its name. The chlorite forms through the alteration of ferromagnesian minerals in the parent rock.

Grit is like sandstone, but it is made up of grains of quartz that are more angular or larger than those in ordinary sandstone.

Lamprophyre is a medium-grained igneous rock that contains a predominance of ferromagnesian (FeMg) minerals, accompanied by feldspars.

Metasediment is metamorphosed sediment.

Ophiolites are igneous and metamorphic rocks formed from deep-sea sediment.

Pelites are argillaceous rocks (silts and clays) that have been metamorphosed. (The constituent particles in *silt* are between 1/256 and 1/16mm in size; in *clay* they are less than 1/256mm).

Psammite is sandstone that has been metamorphosed.

Rhyolites are fine-grained to glassy volcanic rocks petrographically similar to granite.

Sandstone is a sedimentary rock in which the constituent particles are between 1/16 and 2mm in size.

Tuff is consolidated volcanic ash, where the fragments are less than 2mm in diameter.

Tuffisites are veins of sintered pyroclastic materials that form as a result of localised fragmentation events during gas-and-ash volcanic explosions.

Turbidites are rocks that originated as slurries of sand-laden water behaving as a dense liquid, usually initiated by slumping off the continental slope.

Untramafic rocks are coarse-grained igneous (i.e. *plutonic*) rocks made up mainly of ferromagnesian rocks with little or no quartz.

Fig. 2.8 A decendant of aquatic green algae, *Peltoclados* is a rare relic of the time plants were beginning to colonise the land surface.

top. These laminites are unusual rocks. As the name suggests, they are delicately laminated siltstones, usually calcareous (although the carbonate weathers easily), and sometimes with carbonaceous films on the bedding surfaces that are believed to be all that remains of algal mats.

The Bunnamohaun Siltstone Formation is interpreted as a non-marine mudflat complex, with repeated exposures allowing the desiccation of the surface that produced the mudcracks and allowed the development of similar algal mats to those in the laminites (though these could also have been laid down in a shallow lake), with the occasional flood bringing sand onto the mudflats. Current directions show that the regime under which the Bunnamohaun Siltstone Formation formed is the same as that responsible for the two earlier formations, but more distal. The rocks of the Bunnamohaun Formation outcrop on the slopes and summit of Knockmore and Knocknaveen, where the steep incline echoes the dip of the bedrock.

The rocks that can be seen all along the coast between Portruckagh and Grace O'Malley's castle belong to the youngest formation in the succession, the Glen Pebbly Arkose Formation, which is more than 500m thick and made up mainly of coarse sandstones and conglomerates.[10] It extends away inland from the ruined tower house at the harbour, underlying the craggy country south of Knocknaveen, and is especially well displayed in the spectacular stretch of coast immediately south of Granuaile's Castle.

This formation is thought to be diachronous, which means that the base is not the same age everywhere: it is oldest in the east and becomes progressively younger as you move westwards across the island. The petrography of these much coarser rocks is quite different to that of the underlying formations, and the direction of flow in the wide, shallow braided river system that brought them here was from the north-east. The pebbles in the conglomerates are especially interesting because they give us a glimpse of the kind of source area the rivers came from. The commonest pebbles are of fine-grained volcanic rocks, but also present are green *serpentinites* that probably came from the ancient Deer Park Complex (see page 37). There are also numerous metamorphic pebbles of different sorts, and—one of the few hints of the age of these rocks—calcareous siltstone pebbles that contain Silurian fossils. Pebbles of locally derived sandstone and mudstone pebbles also occur. All the indications point to a source area of probable Ordovician or Silurian volcanic bedrock somewhere to the north.

We should not rely on today's rivers when we try to envisage what the landscape looked like

Fig. 2.9 Hummock-and-valley topography at the northern tip of the island characterises the landscape underlain by the Ballytoohy Formation (viewed from the south-west).

when these Silurian rivers were in flow. During that time there was no vegetation growing on land; the soils then were very different from those of the present day, and erosion rates were much higher than they are now. Plant life was only beginning to colonise the land, and for a long time the only vegetation was those algal mats now preserved as carbonaceous films on bedding surfaces.

These Silurian rocks can be readily correlated with rocks on the mainland around Louisburgh, but not so easily with rocks further afield. And although they are most probably late Silurian in age, there is still room for doubt because they contain little information we can use to date them. The fossils found by Adrian Phillips in some of the pebbles in the Glen Pebbly Arkose Formation are typical of Silurian shallow-water marine creatures, but not diagnostic of any particular part of the Silurian.

Apart from spores, the only other fossils so far found in these rocks are enigmatic frond-shaped fossils known as *Peltoclados clarus*, which occur very occasionally in red siltstones within the Bunnamohaun Siltstone Formation. Several hundred specimens have been found on the flanks of Knockmore since Adrian Phillips first discovered it in the 1960s.[11] They are thought to be either algae of unknown affinity or an early stonewort relative, but they are no help in pinning down the age of the rocks—or indeed even the habitat in which they grew, though they probably lived in ephemeral lakes that developed from time to time on the floodplain.[12] A fragmentary fish fauna has been identified in the mud-silt laminites at the base of the Bunnamohaun Siltstone formation; a similar assemblage of *Peltoclados* and fish fossils found in non-marine rocks in the Midland Valley of Scotland has been dated as Wenlock (mid-Silurian) in age.[13]

North of the Silurian

The Dalradian rocks with which the probable Silurian rocks come into contact across the Leck Fault in the middle of the island were assigned by Phillips to the Ballytoohy Formation, which can confidently be correlated with the Killadangan Formation found along the south side of Clew Bay.[14] The topography in the area underlain by the Ballytoohy Formation consists of dry hummocks that run in an east-west direction, rising towards the coast, separated by deep valleys in which small bogs are often found. This topography is the result of differential weathering of the underlying rocks; soft shales give rise to deep valleys and depressions, and more resistant grits to the well-defined knolls and ridges.[15]

These Dalradian rocks are much more difficult to interpret than the somewhat younger rocks south of the fault. They were originally a sequence of mostly sedimentary rocks, but have been metamorphosed and now comprise an assemblage of psammites, pelites, cherts, basic volcanic rocks, ironstones, tuffs and limestones. They are of deep water origin, and mainly turbiditic in nature. The Original Survey workers 'failed after a diligent search' to find fossils in them, but a sparse Ordovician fauna has been recovered in the course of even more diligent search over recent decades. A mid-Silurian microflora has recently been described, though there is much argument about this.[16]

There is considerable disagreement over the relationship between the different rocks in the Ballytoohy Formation, because the polyphase deformation to which they were subjected during the Caledonian orogeny and the patchy nature of their exposure make different interpretations possible. Adrian Phillips thought a definite stratigraphy could be recognised in the formation (he believed he could identify five distinct phases of deformation), but others have interpreted the Ballytoohy Formation as a 'chaotic mélange' that formed in mid-Silurian times and consider the areas in which a clear sequence can be recognised to be merely enormous blocks (some of them more than half a kilometre long) in a more jumbled rock matrix.

The basic igneous rocks in the formation resemble rocks typical of mid-ocean ridge basalts, and argue convincingly for the existence of an ocean spreading centre on Clare Island during the Ordovician. The thick pile of coarse turbiditic sandstones that make up so much of the formation are younger. The Portruckagh Formation in the south of the island is similar to the Ballytoohy Formation, but much more limited in exposure, and squashed in between wedges of various other rocks.

Movement on the Leck Fault has brought the metasedimentary rocks of the Ballytoohy Formation into contact with the Silurian. So deeply does this fault extend into the earth that it has facilitated the ascent of magma from deep down: not just deep in the crust, but from the mantle itself. This takes the form of small igneous intrusions—dykes mostly, but there are also some sills and bosses—injected into the Ordovician and Silurian rocks. About sixty of these intrusions can be identified at the surface, most of them in the Knockmore Sandstone and Bunnamohaun Siltstone Formations, though the largest single one is in the area of Ordovician outcrop. A magnetic survey has identified a further fifty igneous intrusions. In the east they are mainly lamprophyres, and in the west mainly ultramafic rocks. Associated with them are gas-emplaced breccias and tuffisites with associated barytes veins.

The Lower Carboniferous

Rocks of Lower Carboniferous age underlie the area of hummocky moraine in the north-east of the island. They rarely appear at the surface inland, because the lower ground in which they outcrop is largely blanketed by glacial sediment, but they are spectacularly displayed all along the east coast between Ooghcorragaun and the beach at the harbour. They are divided into two formations: a lower Maum Formation (*c.*200m thick) that outcrops in the western part of the area, and the Capnagower Formation (*c.*100m thick) in the eastern part.[17] The two formations constitute a single cycle of sedimentation that took place in an intra-montane river basin at the beginning of the Lower Carboniferous period, and they can be correlated with rocks of the same age that outcrop around the north shore of Clew Bay and in Glen Hest and Glen Nephin.[18]

The rocks of the Maum Formation are largely red in colour and attain a thickness of perhaps 200m. Their contact with the Ballytoohy Formation is an unconformity, but the contact is never seen; rather, it is inferred on the basis of topography.

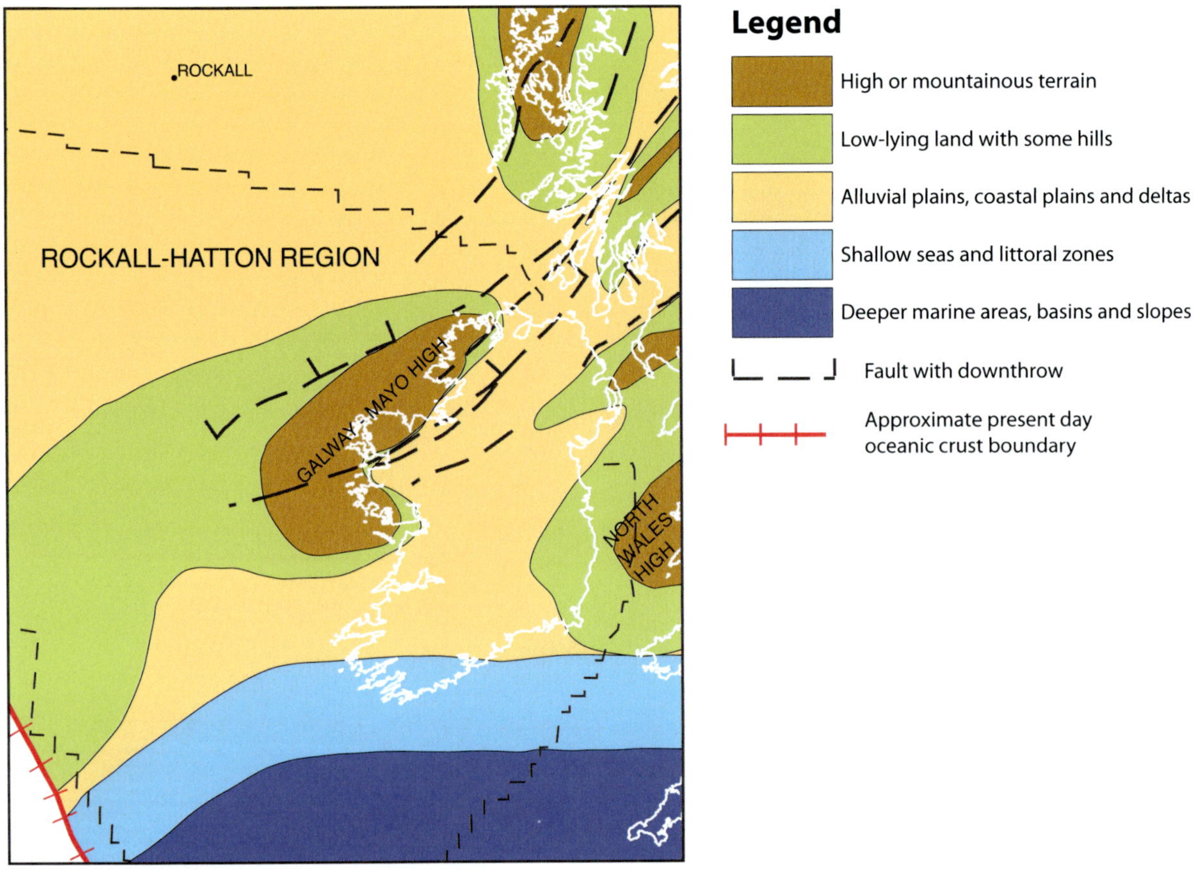

Fig. 2.10a Ireland's Late Devonian geography (Dave Naylor).

Fig. 2.10b The distinctive red rocks of the Maum Formation are beautifully exposed along the shoreline at Portlea.

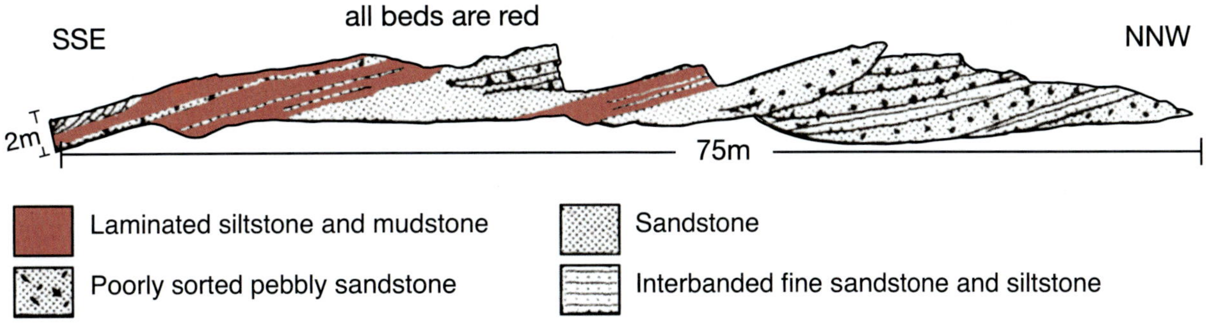

Fig. 2.11 Simplified cross-section of the Maum Formation rocks at Portlea (redrawn from a sketch by Adrian Phillips).

The lowest part of the Maum Formation consists of almost unbedded red breccias and conglomerates, part of a southerly-derived aerial scree or alluvial fan banked up against the Leck Fault. The lowest beds are best seen along the coast south of Ooghcorragaun.[19] The composition of the suite of pebbles in them is important because it can give us some idea of where the rivers in which the sediments originated came from. The most common pebbles are vein quartz, psammite, quartzite, mica schist and jasper, with lesser numbers of rhyolite, basalt and tuff: an assemblage that suggests derivation from a Dalradian hinterland that lay somewhere to the west or south-west. Evidence of current direction in the same rocks also shows that the ancient river in which the conglomerates were laid down was flowing to the east or north-east. Comparison with modern river systems makes it likely this was a bedload-dominated river that flowed intermittently, with frequent flash floods.

These lowest Maum Formation rocks are followed by coarse red conglomerates, well-exposed along the northern part of the coastal

Fig. 2.12 The massive conglomerate blocks developed at the base of the Maum Formation are best seen in the area north of the Green Road.

Fig. 2.13 *Taenidium* fossil burrows in sandstones of the Maum Formation at Curraun, on the northern coast of Clew Bay between Mulranny and Achill Island.

section. They contain pebbles and cobbles up to 20cm in diameter of vein quartz, meta-psammite, quartzite, mica schist and jasper, with smaller amounts of volcanic pebbles, and comprise a series of fining-upwards alluvial cycles comparable to the deposits of modern braided rivers.[20] A great place to see these fantastic conglomerates further inland, and a place where they are readily accessible and can be examined at really close quarters because you can climb all over them, is north of the Green Road below the Leck—the inland cliff after which the Great Fault that runs across the island is named—where gigantic blocks of the conglomerate excavated by frost shatter are strewn across the landscape.

The greater part of the Maum Formation (*c*.145m) consists of red siltstone and mudstone interbedded with approximately similar propor-

Fig. 2.14 Rocks of the Capnagower Formation on the shore at Leckascannalmore east of Portlea.

tions of red sandstone and conglomerate in cycles several metres thick. These were laid down as sand and gravel of northerly origin in a system of braided channels, signaling a dramatic change in the responsible river regime. The pebbles in the conglomerate indicate a Dalradian metamorphic source to the north-west.[21]

No fossils have been reported from the Maum Formation on Clare Island, but on the coast of the mainland west of Mulranny where it is much more extensively exposed, abundant trace fossils occur at certain horizons. These are interpreted as the excavations of large millipede-like creatures known as eoarthropleurids, and thought to be aestivating burrows excavated in cool sand closer to the water table where the animals could shelter during periods of prolonged heat. The landmass of which Ireland was a part was close to the equator at this time and the climate was semi-arid, with seasonal rainfall.

The transition to the overlying Capnagower Formation is marked by a pronounced shift in the river regime and a change in colour. This new formation is generally better exposed than the Maum Formation, especially along the coast, where there are some spectacular (but mostly inaccessible) cliff sections. Most of the Formation consists of a succession of sedimentary cycles up to 15m thick. The coarse or pebbly sandstones at the base of each cycle often cut channels into the underlying fine sediments that range in width from 1 to 5m and in depth from 30cm to 4m. Sandstone dykes injected into underlying mudstone are very well developed, and fragmentary plant material is common, along with simple vertical invertebrate burrows. The meandering river in which these sediments were laid down was flowing north-eastwards from the Dalradian country that lay to the south-west.

A distinctive and widely-developed characteristic of the formation is the way particular strata often cut into earlier sediments. This applies not only to the coarse sandstones laid down in the main channel of the river (which was probably in the order of tens of metres wide), but on a smaller scale to the finer sands and mudrocks laid down in splay channels cut into the floodplain. Another distinctive feature is the widespread development of fossil soils (palaeosols) within the formation. These resemble modern vertisols, which are characteristic of a range of sub-humid to semi-arid climates, with a pronounced dry season. The total assemblage of features preserved in these palaeosol horizons suggests a monsoonal climate with a mean annual precipitation of 800–1000mm.[22]

The sedimentary characteristics of the Capnagower Formation suggest that it formed in a series of sinuous, laterally migrating rivers that periodically flooded and distributed sediment onto the floodplains by means of crevasse systems. Such behaviour is typical of low gradient rivers found on fluvial coastal plains.[23] This would be entirely consistent with hints that brackish water conditions were developing near the top of the formation.

Abundant traces of plant life have been found in these Capnagower rocks: twigs, bits of branches and rootlets, preserved either as coaly impressions

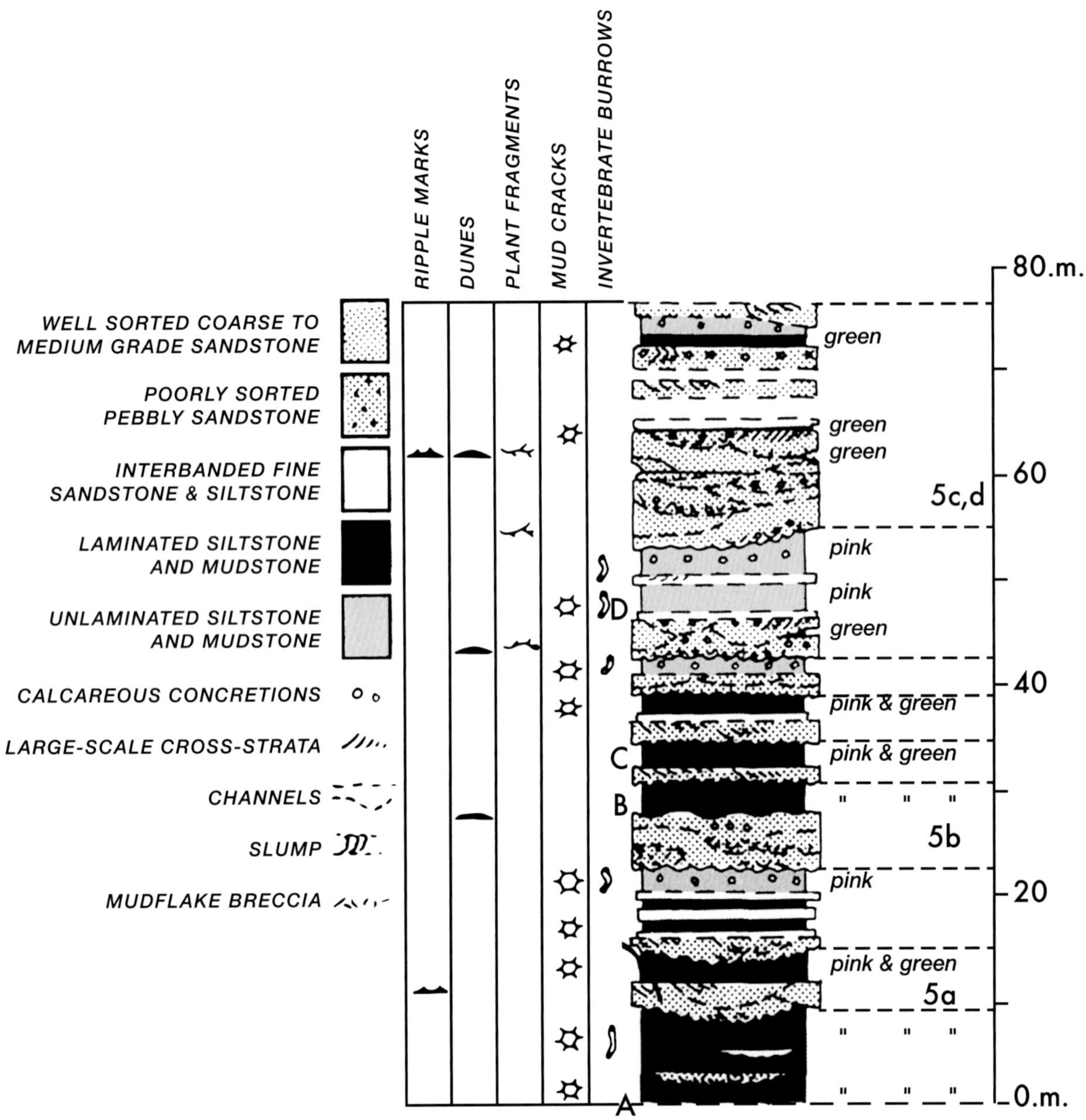

Fig. 2.15 Stratigraphical log of the Leckascannalmore Member of the Capnagower Formation measured along the cliff sections east of Portlea. A-D miospore horizons (Phillips and Clayton 1979).

or sandstone casts too poorly preserved for identification. The top 20m or so of the formation are distinguished from the rocks beneath by the presence of green and black siltstones and shales instead of red beds at the top of each cycle. Trace fossils (*Rhizocorralium* and *Chondrites*) in the black shales suggest a marginal marine environment for these particular lithologies. A poorly preserved fauna of bivalves, gastropods, ostracods has been found in a 4m exposure of black siltstone just below the top of the formation, which also contains a diverse miospore flora. This enables the formation to be dated to late in the Tournaisian or early in the Visean. The Maum Formation is also probably Tournaisian, or possibly latest Devonian.[24]

These Carboniferous rocks—like the Silurian rocks we met earlier—are fluviatile sediments, but the rivers in which they formed were more like today's rivers, because the surface of the land was largely forested. The coarser sediments in the succession represent sand and gravel deposits laid down in river channels, and the finer red lithologies are the silt and mud carried out of the main river channel and deposited on the floodplain.

Fig. 2.16 Serpentinite hand specimen.

Serpentinite is a metamorphic rock formed by late-stage hydrothermal action on ultrabasic igneous rocks. The characteristic constituent mineral is *serpentine*, which is a layer-lattice silicate that is produced through the metamorphism of ferromagnesian minerals in the parent rock. It is extremely common in the ultrabasic rocks of which ophiolites are composed. In chemical terms serpentines are mainly hydroxyl-bearing magnesium silicates formed from original olivine and pyroxenes. Serpentinite is a characteristic lithology among the intrusions associated with movement along the Leck and Kill Faults. The process of serpentinisation involves the incorporation of large amounts of water, destroying the original fabric so that the rock often has a shattered-looking texture.

Grenville Cole remarked that 'were visitors frequent in the island, ornaments and fancy articles might be made from the western serpentine, as is done at Lizard Head in Cornwall'. (Cole 1914, 26).

Put in simple terms then, it seems that there was a valley in the Clew Bay area at the beginning of Lower Carboniferous time. What is now Clare Island lay somewhere in this valley, into which sediment was being fed by rivers from both sides to form what became the Maum Formation. The sequence of sediments that constitutes the Capnagower Formation formed as a more mature alluvial fan built out northeastwards into the intramontane basin. This river basin extended eastwards across Clew Bay into the country between Castlebar in the south and Glen Nephin on the north, and far beyond.[25]

The oldest part of Clare Island

The oldest rocks on Clare Island appear in a fault-bounded wedge of inliers between Portruckagh and Kill in the south of the island, upon which the Silurian succession rests unconformably. These rocks, which are believed to be identical to rocks on the southern shore of Clew Bay known as the Deer Park Complex, comprise high-grade metamorphic rocks: amphibolites, serpentinites and metasediments with a mesmerisingly complex, polyphase deformational history. Adrian Phillips and others made a first attempt at deciphering this in the 1960s and early 70s, but there is still considerable uncertainty, disagreement and argument surrounding the history of the complex. These rocks are also generally assigned to the Dalradian Supergroup of late Precambrian or early Palaeozoic age. The most likely interpretation for the serpentinites in the complex is that they represent oceanic material younger than the metasediments.

In the south-east corner of the island, also near Portruckagh, is another small area of metamorphic rocks, but of much lower metamorphic grade than the Deer Park rocks, with which they are in tectonic contact. These are assigned to the Portruckagh Formation, and comprise green-

Portruckagh

Portruckagh is the name given to the small bay 700m south-west of the harbour on the north-east side of the promontory of Doon Cloak. It is one of the most scenic spots on the entire coastline, and of particular geological and archaeological interest. There was once a promontory fort on the headland, which is connected by a narrow neck of till that allows passage for sheep. When Thomas Westropp first described the fort in 1911 this was 'a broad grassy neck 27′ wide, with steep slopes at either end, and low precipices on each side, the western at a curving creek with fine high natural stone arches; the eastern at a crescent of shingle beyond which are abrupt cliffs sheeted with masses of Osmunda wherever a stream descends their faces'.[1] It was fenced off by a bank of earth and stone, and there is a hint that this was once reinforced with a drystone wall.

Here the Deer Park Complex reaches the coastline (See Figure 2.3), which is one of the few places this complex composition and convoluted fabric can be puzzled over: on the north side of the little storm beach (Westropp's 'crescent of shingle') where a small area of the rock is kept washed and polished by wave action. This is also the only place where the eponymous Portruckagh Formation outcrops; faulted against the Deer Park rocks on the north-west, and with the basal conglomerate of the Kill Sandstone Formation resting unconformably on it to the south. Numerous small shears within the Portruckagh Formation show the intense crushing pressure to which the rocks were subjected over time, resulting in what one geologist described as 'a mashed scum of graphitic slates'.

The Deer Park rocks are faulted against the Glen Pebbly Arkose Formation on the north. At the south end of the promontory the Kill Sandstone is conformably overlain by the lowest rocks of the Strake Banded Formation, which has prominent tuff bands. Both the Kill Sandstone and Strake Banded Formations were significantly affected by movement along the Kill Fault, but this was much later than the deformation of the Portruckagh Formation.

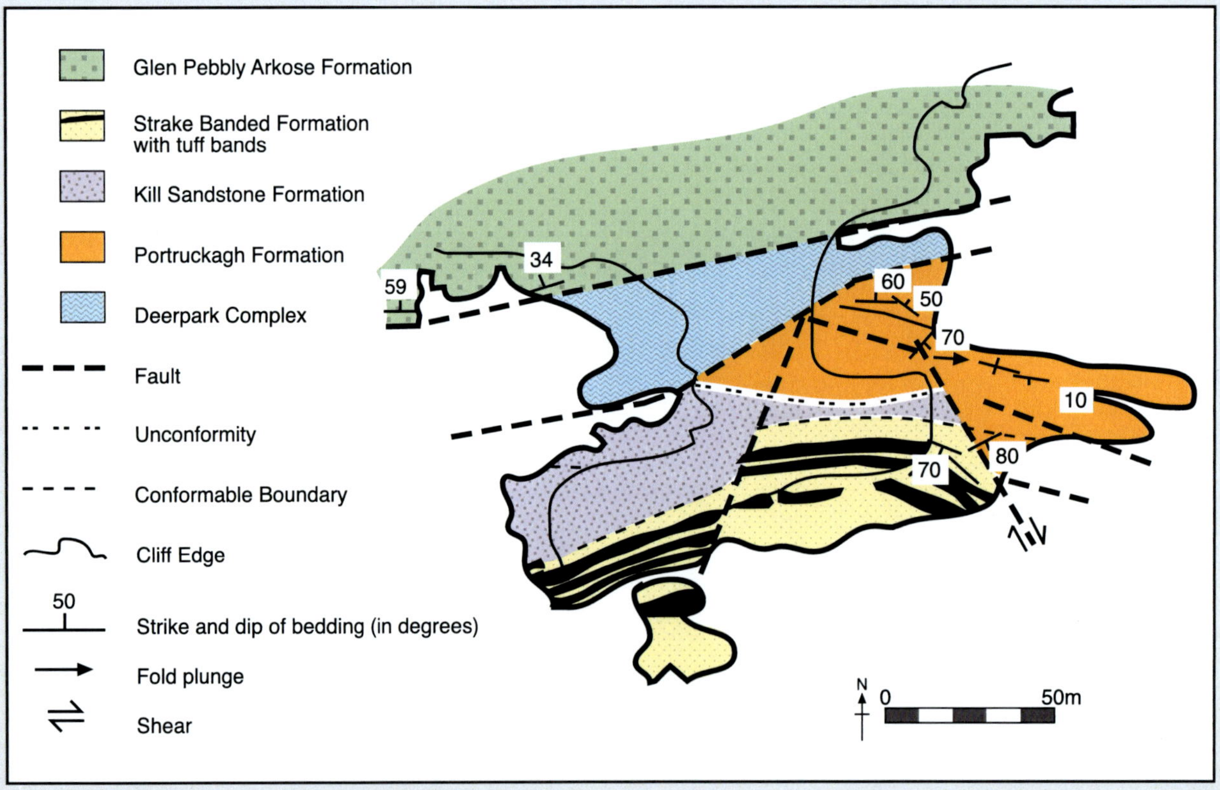

Fig. 2.17A Geological map of the headland (After Phillips 1966).

[1] Westropp, Original Survey, Archaeology, 22.

Fig. 2.17B (a) Fabric of the Deer Park rocks, (b) shear plane with gouged quartz cobble.

The sheltered cove on the western side (Westropp's 'curving creek with fine high natural stone arches') is known as Ooghnadrishoge. This formed through preferential erosion of softer rocks, leaving a barrier of coarse white sandstone. Such differential erosion is a characteristic feature of the southern coastline in general. The offshore rock to the south is Carraig na mBealaí.

Fig. 2.17C Portruckagh from the east.

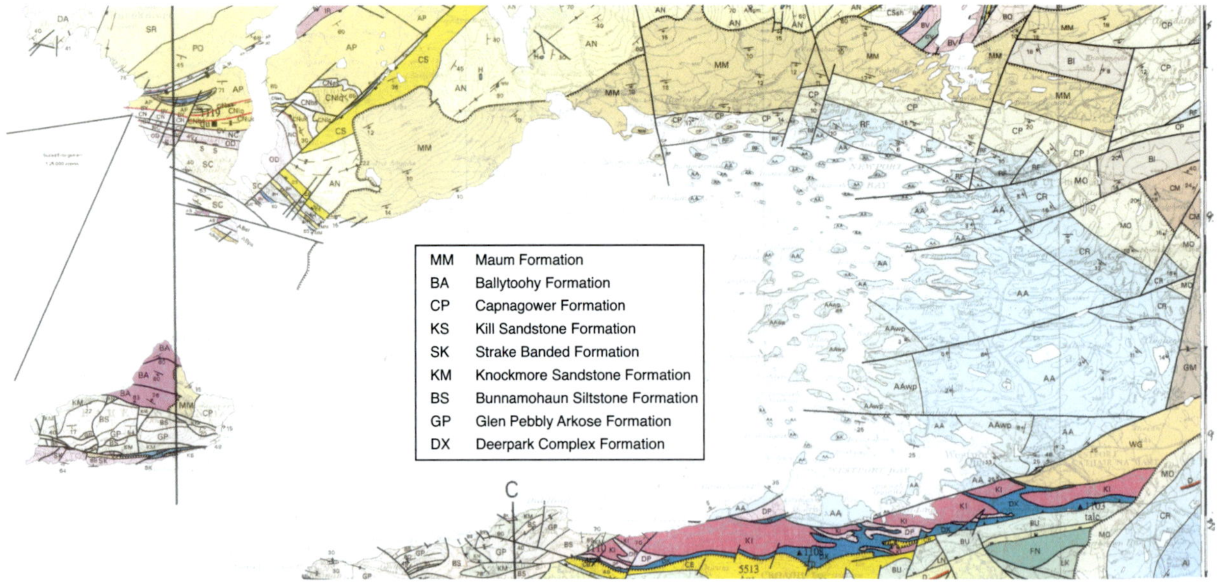

Fig. 2.18 Clare Island in the broader geological context. All the rock formations on the island can be matched with formations on the mainland, of which it is a recently detached island ('recent' in terms of geological time) on the continental shelf (Long *et. al.* 1992).

schists, with graphite and pyrite-rich pelites: the metamorphic product of rocks that were originally sandstones and dark mudrocks. The Portruckagh Formation is taken to be equivalent to the Ballytoohy Formation.

Our piece of the jigsaw

Clare Island is a detached piece of the geological jigsaw that is Ireland. The pieces that once joined it to the mainland have been lost, torn away or drowned by wind and wave and successive glaciers, but the different colours on this remaining piece—the individual subjects in the jigsaw picture—can all be matched sufficiently closely with those on the mainland for the overall picture to be seen.[26]

The entire assemblage of Clare Island rock formations can be picked out in the geological map of the mainland from which it has so recently been detached. Clare Island was formed by northwest displacement of about 5km of the east-west axis of the rocks seen around Louisburgh and Old Head. The bedrock east of Roonagh Quay is a similar patchwork of the Knockmore Sandstone, Bunnamohaun Siltstone and Glen Pebbly Arkose Formations of Clare Island. There is also a tiny slice of Strake Banded Formation immediately north-east of Emlagh Point. The Killadangan Formation north of Croagh Patrick and running eastwards towards Westport is equivalent to the Ballytoohy Formation of north-west Clare Island. The little sliver of the Deer Park Complex on the island is a shard detached from the main outcrop between Croagh Patrick and Westport. The Maum Formation is spectacularly exposed along the coast of the Wild Atlantic Way between Achill and Mulranny, and the Capnagower Formation outcrops extensively around Newport.

All these different rock units, each with their own unique history and characteristics, have been welded together tectonically. As we have already seen, all the geological pieces of which Clare Island is made can be matched with others on the mainland, but the forces of weathering and erosion over uncountable ages have reduced all that lies between to sand and silt: the materials from which the sedimentary rocks of the future will be shaped.

This immense, scarcely imaginable process of erosion has been at work on Clare Island in two senses. First, in geographical terms, the process wears away the rock that once joined the island to the mainland—and extensions of Clare Island that extended for unknown lengths in other directions. Second, in a stratigraphical sense, the process is stripping from the Carboniferous rocks that are the youngest 'hard' rock on the island the enormous pile of younger rocks that accumulated on top of them in later geological epochs, only to be worn down again in the time between the Carboniferous and the Ice Age. The Carboniferous rocks that once covered

Fig. 2.19 Tellus imagery for Clew Bay (DPC is the Deer Park Complex) (Geological Survey of Ireland).

the Silurian rocks south of the Leck Fault have also been stripped away, except for a thin veneer that survives on top of Knockmore.

More than a third of Clare Island consists of a 'marginal platform' from which the hills of Knockmore and Knocknaveen rise 'with remarkable abruptness'. Hallissy identified this as 'part of the great post-Carboniferous peneplain that extends over the central plain of Ireland, while the hills of Knockmore and Knocknaveen are remnants of a higher and older plateau from which the mountains of the western highlands have been carved'.[27] The platform which dominates the southern part of the island rises from 8m above sea level at the harbour to 40m at Toormore, where the steep slope of Knockmore, rising above it is probably an ancient sea cliff of late Tertiary times, some 3 million years ago. The tilt is probably due to isostatic rebound after the Ice Age.[28]

Data from the Geological Survey's Tellus Programme has just become available for Clare Island, and the magnetic data in particular shows some impressive and intriguing features.[29] The magnetic signature picks up an echo of deeper geology than is visible at the surface; the persistently high magnetic anomaly for the southern part of Clare Island (red) is thought to be picking out the deeply-buried Deerpark Complex and its extensions to the east (Figure 2.19).

The northeast-southwest linear feature is a previously unmapped igneous dyke. Such dykes are a characteristic feature of the north-east of Ireland, but they also extend much further south (for example, there is a great swarm of them in the Killala Bay area). These were emplaced some 58 million years ago (in the early Tertiary period of earth history) at a time when the earth's crust was being stretched as a result of plate movement, and injected with magma from a 'hot spot' within the underlying mantle: events preceding formation of the great rift that opened in the crust and then widened to become the North Atlantic ocean. The Giant's Causeway and the other basalt formations of Antrim formed at the same time.

Notes

1 Kinahan *et al.* 1876 This devotes only three paragraphs to Clare Island.

2 Cole *et al.* (1914).

3 Hallissy 1914 OS 7.

4 Phillips (1966, 1973, 1974); Graham (1981, 2001a, 2001b). The thorough nature of Adrian Phillips' survey became something of a legend on the island. Not content with leaving no stone unturned on land, be insisted on exploring every cove by currach.

5 Coxon (2001); Coxon and O'Connell (eds) (1994).

6 A cap of fine-grained sandstone of Carboniferous age survives on the summit of Knockmore. Dalradian is the name given to the youngest stratigraphic division of the Precambrian in Ireland and Scotland. It comprises a vast belt of metamorphic rocks that extends from western Ireland through Tyrone and north-east Antrim into Scotland. These rocks originated in the Iapetus Ocean prior to collision of the two continents.

7 The most recent analysis disputes this later phase of movement (Worthington and Walsh 2011), as does John Graham (pers. comm.).

8 The Silurian formations correspond broadly with the Knockmore Series of the early Survey geologists.

9 Graham (Silurian), 38–39.

10 Arkose is a quartz sandstone with more than 25% of the mineral feldspar, characteristic of deposition in an arid environment.

11 Graham (2002).

12 Higgs and Williams, 19–34.

13 Palmer *et al.* (1989).

14 The Ballytoohy Formation correlates broadly with the Ballytoohy Series of the early geologists.

15 OS 7, 1; 'The northern portion of the island presents a very rugged appearance due to the unequal denudation of strata formed of different materials. The strata being composed of grits alternating with shales, in a nearly vertical position and having an east and west strike. The grits stand out in bold outline, whereas the shales have been denuded more deeply, leaving a series of small east and west valleys.' (GSI Memoir, 11).

16 Brück, P.M. and Rushton, W.A. (2011). Further careful research in recent years has shown that the Dalradian and the Clew Bay Complex have shared the same deformation history during the Grampian Orogeny (caused by the attempted subduction of the Laurentian margin with an infant oceanic arc) during mid-Ordovician times, *c.* 470Ma. This casts doubt on the Silurian age attributed to the Clew Bay

Complex on the basis of its presumed Silurian microfossil content (Chew *et al.* (2003, 2010).

[17] Phillips and Clayton (1980).

[18] More specifically, during the later part of Tournaisian into Visean time.

[19] Adrian Phillips recognised this as a distinct 'member' of the Maum Formation, which he called the Lecarrow Member. These are the first rocks you see after you step off the boat at the harbour. They are at least 36m thick, but very poorly exposed.

[20] Phillips distinguished this as the Allanahy Member of his Maum Formation. It is up to 24m thick. Distinctive though these rocks are, they are possibly a lateral equivalent of the Lecarrow Member. Meta-psammite is metamorphosed sandstone.

[21] These rocks are Phillips' Leckaprison Member.

[22] Graham (2001), 82, 84.

[23] Graham (2001), 84.

[24] Phillips and Clayton (1980), 125-133; Graham (2001), 85. The Tournaisian is the earliest stage of the Lower Carboniferous, between 345.3 and 359.2 mya. It is followed by the Visean stage, between 326.4 and 345.3 mya.

[25] Phillips *et al.* (1969). The dramatic difference in the kind of early Carboniferous rocks seen in these two areas is thought to mark the eastern continuation of the Leck Fault escarpment.

[26] It is much more complex than the familiar jigsaws where all the pieces contribute to a single picture. The detached piece of Ireland that is Clare Island consists of several pieces from different pictures all welded together: and each has its own geological story to tell.

[27] Hallissy (1914), OS 7, 3.

[28] Cullen and Gill (1992b), 8, quoting Phillips.

[29] Tellus is an airborne survey to gather geochemical and geophysical data of rocks, soils and water.

Overleaf

Fig. 3.1 This spectacular LIDAR image by Steve McCarron shows the extent of the hummocky moraine deposited during the last advance of ice across Clare Island, and in particular the abrupt nature of its lobate terminus. Notice the way the moraine stops short of the col between Knockmore and Knocknaveen. Pete Coxon describes this 'the best glacial margin in the country that isn't a corrie moraine'. The 'blisters' on the summits of the moraine hummocks are peat-drying stands.

Chapter 3
The legacy of the glaciers

The rocky skeleton that forms the solid framework of Clare Island is a product of rock-forming processes and tectonic events that occurred hundreds of millions of years ago. The geological character of the island inherited from that remote past has been profoundly altered by events over the last 2.6 million years. During this time the story of the earth—and of the northern hemisphere in particular—has been dominated by a succession of cold, glacial events separated by interglacial interludes when conditions returned to what we probably tend to think of as 'normal'. There have been approximately 20 of these glacial cycles, each one lasting some 100,000 years and separated by intervals lasting on average for 10,000 years. The most recent glacial epoch in Ireland began about 118,000 years ago and ended some 11,700 years ago, reaching its greatest extent between 23,000 and 18,000 years ago during what is known in Ireland as the Glenavy Stadial or Late Glacial Maximum (LGM), when much of the country was under a blanket of ice that extended far offshore, smothering all of what is now Clare Island.

Our understanding of the Ice Age has developed enormously in the 175 years since Louis Agassiz first recognised its existence. Few can match Agassiz's words that conjure up the sheer wonder of it all:

> The surface of Europe, adorned before by a tropical vegetation and inhabitated by troops of large elephants, enormous hippopotami, and gigantic carnivora, was suddenly buried under a vast mantle of ice, covering alike plains, lakes, seas and plateaus. Upon the life and movement of a powerful creation fell the silence of death. Springs paused, rivers ceased to flow, the rays of the sun rising upon this frozen shore (if, indeed, it was reached by them), were met only by the breath of the winter from the north and the thunder of the crevasses as they opened across the surface of this ice sea.[1]

Nowhere in Ireland can you get a better sense of the sheer scale and wonder of the Pleistocene ice sheets and the role they have played in the shaping of today's landscape than in this part of western Ireland.

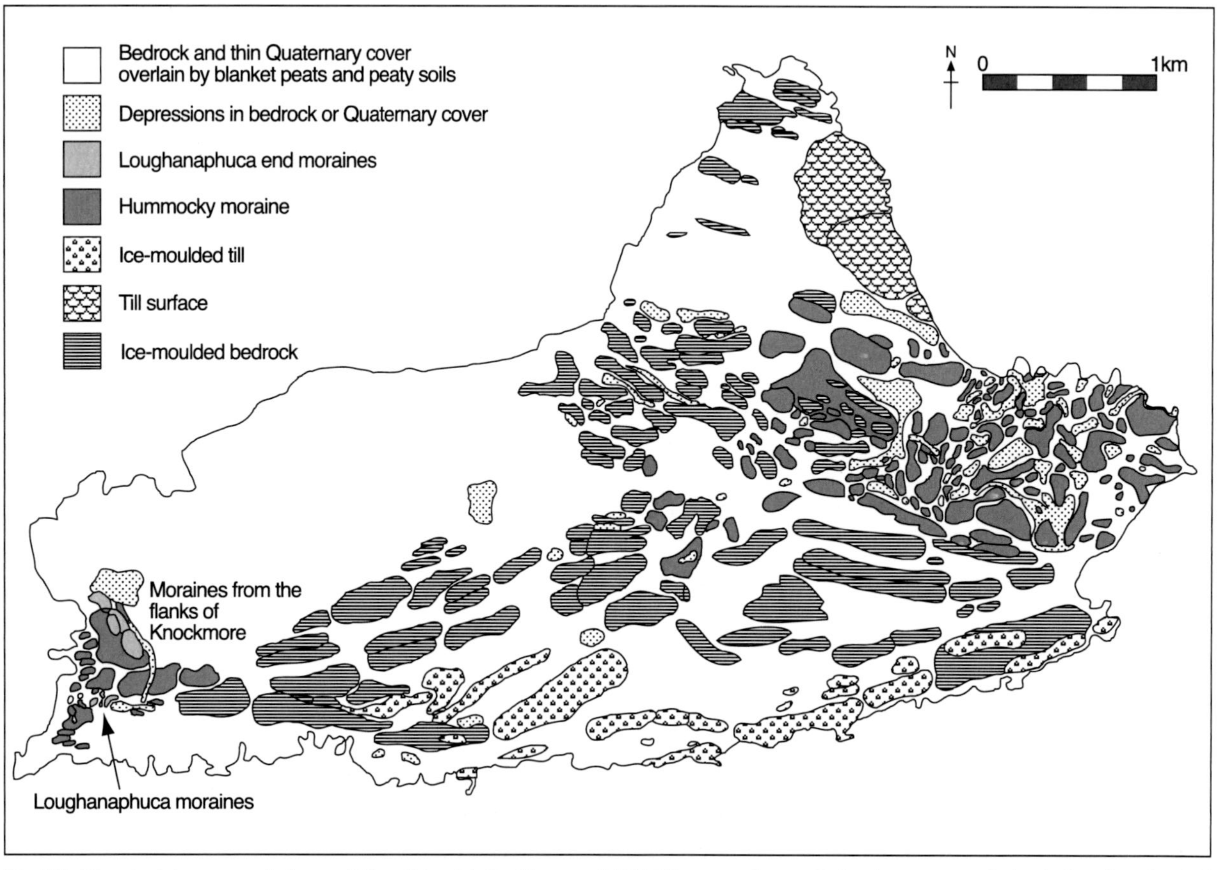

Fig. 3.2 The glacial geomorphology of Clare Island (after Browne, 1991). The map shows distinctive geomorphological and geological units only. Much of the area left unshaded is covered, as is so much of western Ireland, by a thin drape of Quaternary sediment, principally a soliflucted mantle of frost-shattered debris and/or remobilised glacigenic material (*New Survey* volume 2, 91).

Fig. 3.3 Hummocky moraine in Lecarrow viewed from the north-east end of Knocknaveen.

Two 'events' are recognised within this Glenavy Stadial: a 'Main Event' that took place between 24,000 and 19,000 years ago, and a 'Drumlin Event' (18–16,000 years ago), which saw an extensive readvance of glaciers before rapid deglaciation set in 14,000 years ago. The Midlandian Late-glacial—the period between 13,000 and 10,000 years ago—consisted of the Woodgrange or Late-glacial Interglacial (13,000 to 11,000 years ago) and the brief final cold snap of the Nahanagan or Late-glacial Stadial (12,600–11,700 years ago).[2] Almost all the glacial features seen in Ireland developed during the Midlandian and the period of deglaciation that followed.[3]

There were two successive phases of glacial advance on Clare Island, resulting in two distinct till lithologies, which probably represent the Main Event followed by the Drumlin Event. Clare Island lay at the western margin of the glacier that produced the drumlin swarms in Clew Bay. Ice surged over the entire island all the way to the outer edge of the continental shelf, some 50–100km offshore, where moraines have been found deep beneath the sea. Ice movement across the island had a profound effect on the landscape, moulding the underlying bedrock and depositing sediments of glacial origin across the eastern and southern parts of the island.

Glacial deposits mantle the rock platform that occupies most of the lower ground of the island. They are seen in outcrop along the north-eastern, eastern and southern coasts, and extend inland for well over a kilometre, giving rise to a characteristically hummocky terrain, the hollows between the 'drift knolls' occupied by little pockets of peat or alluvium. The best sections are on the north-east coast, where the sea has cut into the till. The cliffs here are often tens of metres high; at Leckaprison, for example (south of the cove at Ooghcorragaun), a vertical section of more than 20m of till is exposed.[4] If you look closely at these sections you can see that there are two till lithologies: a lower till (the Roscahill Till), which is up to 20m thick and contains numerous scratched limestone boulders (along with sandstone, quartzite, siltstone) in a fine-grained dark-grey matrix, and an upper till (the Newport Till), which is coarser, looser and lighter in colour and contains no limestone blocks.[5]

Fig. 3.4 The two tills: coastline north of Capnagower.

This limestone till appears at the surface only along the south coast, where it has been ice-moulded and drumlinised. Inland it can be traced from the drumlin at Rooanbeg on the coast in Bunnamohaun northwards along the Owenmore stream up to the 100m contour.[6] Towards the higher ground it is often overlain by a drape of soliflucted material. In Ballytoohy townland, the Roscahill Till consists of a thick layer of subglacial lodgement till laid down on a 'till plain'.

The Roscahill Till is overlain in the east of the island by a reddish-brown sandstone-rich till—thinner and less extensive than the Roscahill Till—whose composition is similar to the Newport Till seen on the mainland. This is responsible for the hummocky landscape of the north-eastern part of the island, the classic kame and kettle topography characteristic of an area where ice has recently stagnated and wasted *in situ*, and for the drumlins in Glen, which often have a core of Roscahill Till. On the north and north-east of Knocknaveen the ice was impeded by high ground; the sloping hummocky surface that abuts the northern end of Knocknaveen at around 90m is a lateral moraine of this glacier. It is this upper, hummocky till that represents the re-advance of the ice stagnating in the col between Knockmore and Knocknaveen. This time ice did not flow through the col between the two hills but stagnated against its north-east slope.

The lower till is predominantly a lodgement till, whereas the upper till is predominantly a melt-out till. The two may have originated from different parts of the same glacial system, or they could represent successive phases of ice movement that would have involved the earlier one moving westwards across Clew Bay and the later one in a more north-westerly direction. The fact that the two tills are not always distinct often makes interpretation difficult. The sequence is often capped by angular boulders in a coarse matrix. This is probably sediment from the upper part of the melting ice, let down as the glacier stagnated to form the hummocky topography we see in the north-east.

As you travel westwards across the island the till becomes thinner and eventually peters out altogether. The surface becomes increasingly rocky, with more evidence of frost shatter and ice plucking, and soil and vegetation change in parallel. Peat replaces the more fertile till soils further east; this has largely been replaced by centuries of cultivation and the addition of manure, so that the humic gleys of the east are replaced by scaplic podzols and scalpic lithic podzols. With the reversal of this former arable to grassland, bog and wet heath vegetation characterised by such species as mat grass, heather, rushes and bog asphodel, and with patches of plantain sward, are now widespread.

However, it is clear from glacial striae and the distribution of erratics (see below) that much of the island was overrun by ice before the deposition of these tills, during the LGM when the ice stream ended way out on the shelf edge. There are erratics high up on Knocknaveen at 200m, and ice-moulded rock occurs in many places at higher levels than the limits of the till. All this indicates that there were earlier glacial events than those recorded by the tills on today's surface.

The rocks of the low platform along the southern shore are deeply furrowed and polished by ice that moved out to sea 10 to 15′ south of west. The orientation of the conspicuous drumlin that occurs on the south coast at Rooanbeg, south of Knockmore—as well as the Clew Bay drumlins—

Fig. 3.5 Prominent 'bullet clast' in till west of Portnakilly.

Fig. 3.6 Ice moulding of Glen Pebbly Arkose at the south-eastern corner of Knocknaveen (Glen townland).

Fig. 3.7 Multi-phase till on the south coast west of Portruckagh.

and the *roches moutonnées* that are a prominent feature of the island up to 200m, all present their ice-dressed surfaces towards the east.[7] Good examples of these landforms can be seen south of the lighthouse, near the signal tower, and on the east side of Knocknaveen.

Fine examples of glacial striae occur on the rocks 'along the whole of the southern coast from Ooghnamaddy to Kinatevdilla', mainly pointing a little south of west.[8] In some places a second set of striae can be seen cutting across the earlier ones, indicating a later but less intense episode of glaciation. These later striae indicate ice movement from the south-east (pointing 30 to 35′ north of west).

Erratics

Boulders of scratched limestone that must have come from the limestone area east of Clew Bay are common in the older till, as are boulders of granite identical with that of Corvockbrack south of Clew Bay in the southern part of the island. Great blocks of this granite occur along the course of the Owenmore stream on the south-eastern slopes of Knockmore, and smaller fragments occur elsewhere, even on the summit of Knocknaveen at 200m above sea level. Scratched serpentinite boulders identical with that of Croagh Patrick are common along the southern coast. Granite and serpentinite occur both on the surface or embedded in the more superficial till. These boulders were clearly brought by the more recent glacier from the south-east. One of the most impressive erratics on the island is a great boulder of red sandstone conglomerate that must be over a tonne in weight, perched almost on the very summit of Knockmore.

The last glacier

There are several moraines that date to the last phase of local glaciation, when there were small local glaciers on the island. At Loughanaphuca on the western flanks of Knocknaveen there is a fascinating complex of parallel arcuate end moraines between the corrie and the sea, 'that mark pauses in the retreat of a shrunken glacier that debouched through a small rock-bound valley to the east'. There is another small terminal moraine across the mouth of a former cirque about half a mile farther north. These moraines once banked up small lakes now drained dry by a streamlet which has cut its way through the morainic barriers that confined them.[9] These corrie glaciers seem to have been active during the deposition of the Drumlin Event tills, and again during the short-lived Nahanagan Stadial.

To sum up: a large outlet glacier streamed westwards across Clew Bay onto the Continental shelf to the west. Knocknaveen stood in the path of the ice that deposited the Roscahill Till, and so it streamed around its northern and southern flanks, flowing nearly all the way across the south of the island, a lobe of it extending up the col between Knockmore and Knocknaveen. Ice-moulded rock south-west of the col shows that at some time ice crossed between the two hills; this could have been at this time, or it could have been earlier. The evidence is not clear on this point.

More recent geological events

A small area of machair (sea-derived sand) occurs along the margin of the storm-swept western coast. This is stratified, often interbedded with peat, hill-wash or local detritus. Another wind-borne deposit forms a barrier of blown sand between the bay north of the harbour and the low-lying marshy flat immediately to the east of this.[10] Some of these deposits support the unique plantain sward vegetation in the west of the island. The most easterly point of Clare Island is occupied by a v-shaped storm beach or 'spit', which encloses the island's only salt marsh.

Peat occupies—or did so at one time—the hollows of the glacial drift, but the high ground is also covered by peat. The area of peat is much less today than it was in earlier times as a result of peat-cutting and erosion. Under the bogs of the lower ground are the remains of ancient pine forests (see pages 65–67).[11] Solifluction deposits are widely developed elsewhere on the island.

Fig. 3.8 (a) Section through the hummocky moraine; (b) Section of till on the east coast showing lower limestone-rich till overlain by a till richer in sandstone (Pete Coxon).

Fig. 3.9 (a) Ice-moulded bedrock (Strake Banded Formation) forming 'whaleback ridges' west of Portnakilly. (b) Striated exposure of Ballytoohy Formation rocks at Port.

Fig. 3.10 (a) Maum Formation erratic boulder high on Ballytoohy; (b) erratic boulder of Corvockbrack granite beside the Owenmore River in Strake; (c) Maum Formation erratic boulder on the flanks of Knockmore (Pete Coxon).

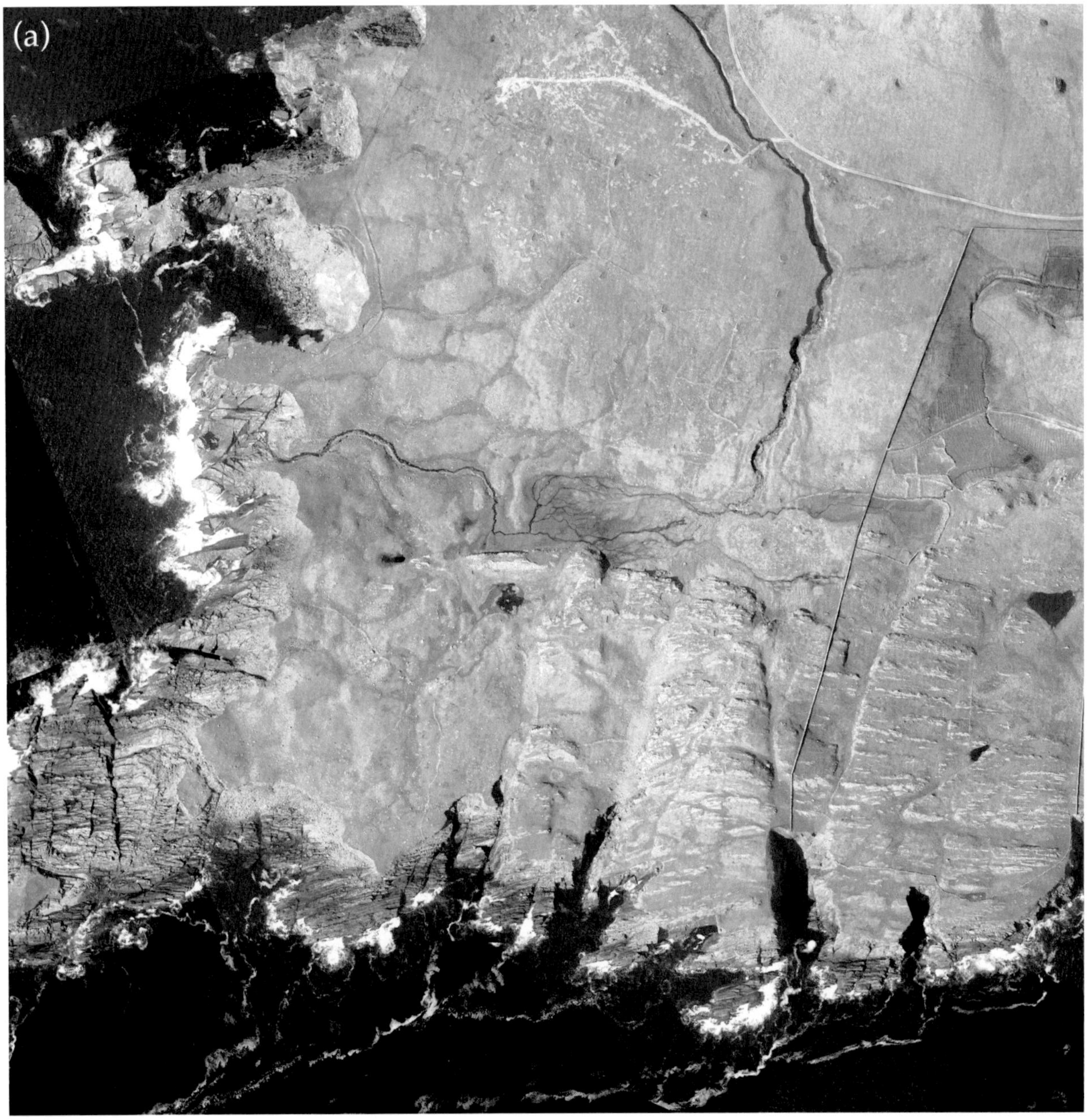

Fig. 3.11 **The Loughanaphuca moraines**

(a) Topographical setting of the moraines. The West Road through Bunnamohaun curving around the south-west slopes of Knockmore is at the top of the picture, where it comes close to the River of Rooing. Loughanaphuca itself is at the centre of the picture (Cambridge University Collection of Air Photographs: copyright reserved).

(b) Close-up image of (a) showing the arcuate end moraines (Pete Coxon).

(c) Lidar image of the south-west corner of the island.

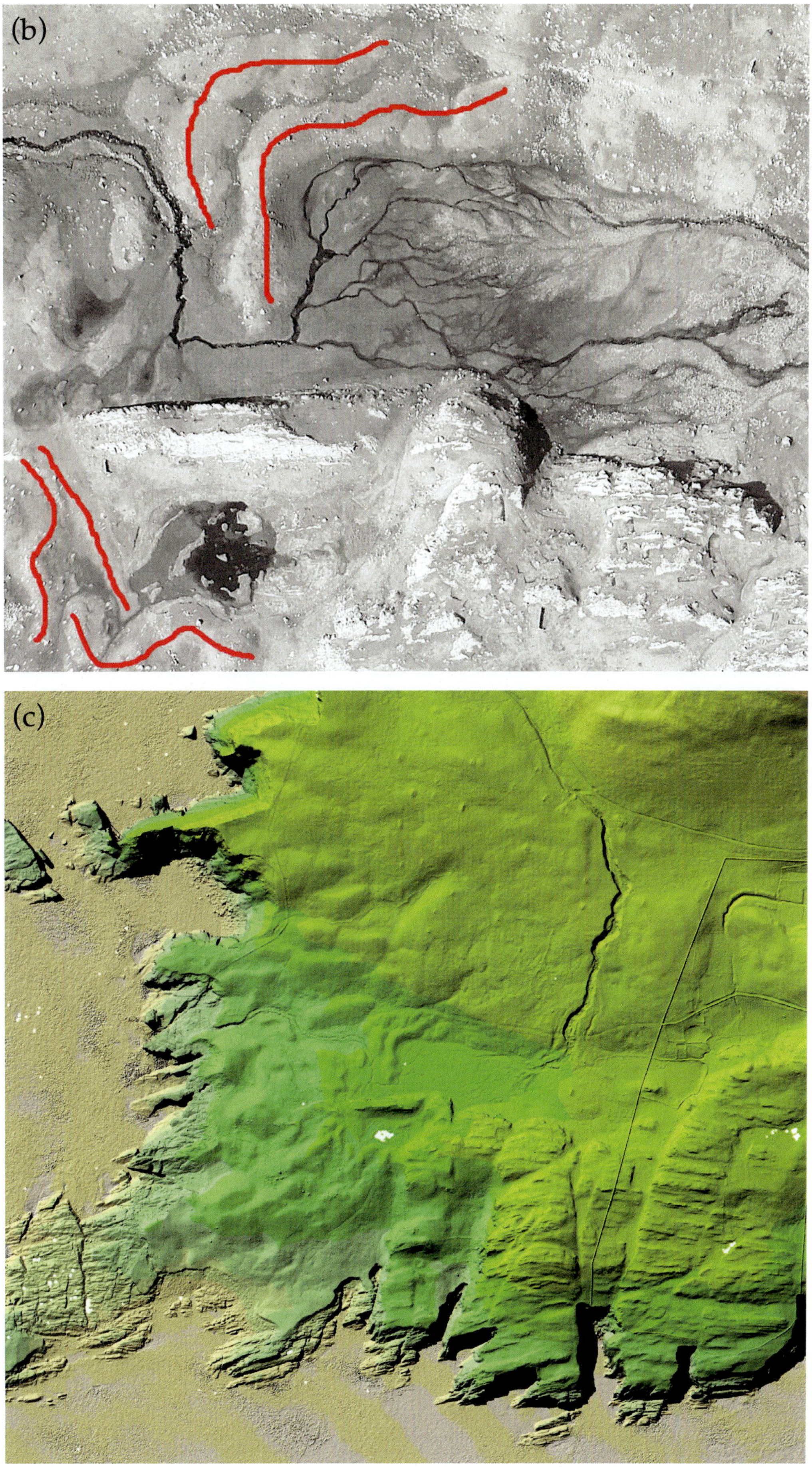
(b)
(c)

Fig. 3.11 **The Loughanaphuca moraines**

(d) Colour Lidar image of the corrie at the south-western end of Knockmore from the coast to the south-west, showing the headwaters of the River of Guaile Lich that flows into the bog of Currach. From here the river flows southwards, until it meets the higher ground of the Strake Banded Formation, which steers it into a westerly course before entering the sea at Leckacanny. There is some evidence for what remains of a moraine of Younger Dryas age (Pete Coxon *pers. comm.*), and the pattern of the landslides that cover buried prehistoric farmland at the foot of Knockmore can be clearly seen.

A core taken from inside the corrie shows that there was no ice here during the Younger Dryas, but there are ramparts that suggest it contained a lobate rock glacier at the time (Pete Coxon, *pers. comm.*) A rock glacier is made up of angular rock debris frozen in interstitial ice, the debris in this case derived from the steep slopes of Knockmore above.

The course of the river through the bog of Currach has been artificially straightened (notice the short feeder channels in the image). This hints at the existence of a mill in its lower course, of which nothing now remains, but which may have been connected with the scutching of flax. There is the supposed site of a mill also on the Toormore river to the west.

(e) (On facing page) The moraines viewed from higher ground to the northeast.

Notes

1 Agassiz, 1840.

2 Originally referred to as the 'Younger Dryas'. The Holocene starts 11,700 years ago.

3 Coxon 2001, 88–9.

4 Hallissy 1914, OS 7, 14.

5 Coxon 2001, 92–104.

6 Hallissy 1914, OS 7, 15.

7 Hallissy 1914, OS 7, 10.

8 Hallissy 1914, OS 7, 13.

9 Hallissy 1914, OS 7, 17. In Vullings *et al.* (2013) the small glacier that developed on the south-west flank of Knockmore was fed by the smaller corrie higher up.

10 Hallissy 1914, OS 7, 17.

11 Hallissy 1914, OS 7, 18.

Overleaf

Fig. 4.1 LIDAR image of the hummocky moraine east of the Leck Fault, at the south-east corner of Clare Island. Note how the prominence of the Leck Fault is emphasised by the vertical exaggeration of the image. The harbour is at the bottom left corner. (Steve McCarron).

Chapter 4

After the Ice Age

Fig. 4.2 Pete Coxon and students coring at Poirtín Fuinse (Pete Coxon).

For a long time after the retreat of the ice, Clare Island was a forlorn, cold and windswept place, its landscape of ice-planed rock blanketed with the glacial debris of the moraines. Without vegetation to protect it, till materials on higher ground moved downslope under the influence of freeze-thaw processes to form the soliflucted material seen in many parts of the island. Wind and rain swept finer material out of the till into numerous hollows between the hummocks, in which small lakes were impounded. We must now turn our attention to these hollows, because it is in the sediment that accumulated in them after the Ice Age that the next chapter of the story of Clare Island is recorded.

The pollen record preserved in the sediment of these water-logged hollows provides us with an outline description of how the surrounding vegetation has developed and changed since the Ice Age. Pollen grains have a very tough outer skin or integument, so that the annual rain of pollen falling on lakes or bogs is preserved in the peat or bottom sediment. And since every species or group of species has its own distinctive pollen grain, the pollen falling on the surface—mainly from wind-pollinated plants, which produce enormous quantities—gives us a measure of the surrounding vegetation at any particular time. Furthermore, the pollen record preserved in the peat profile (or the sediment at the bottom of the lake) provides us with a record of how the vegetation has changed over time.

Pete Coxon and his colleagues have analysed two pollen profiles from Clare Island, one from a 7m peat core in a deep hollow at Lecarrow in the north-east of the island, the other from an 11m core from Lough Avullin (Figures 4.4, 4.7).[1] A number of other sites on the island have also been examined in lesser detail; doubtless these will be followed in time to come by other sites on the island. A core from Poirtín Fuinse proved to be exceptionally rich in plant macrofossil remains.

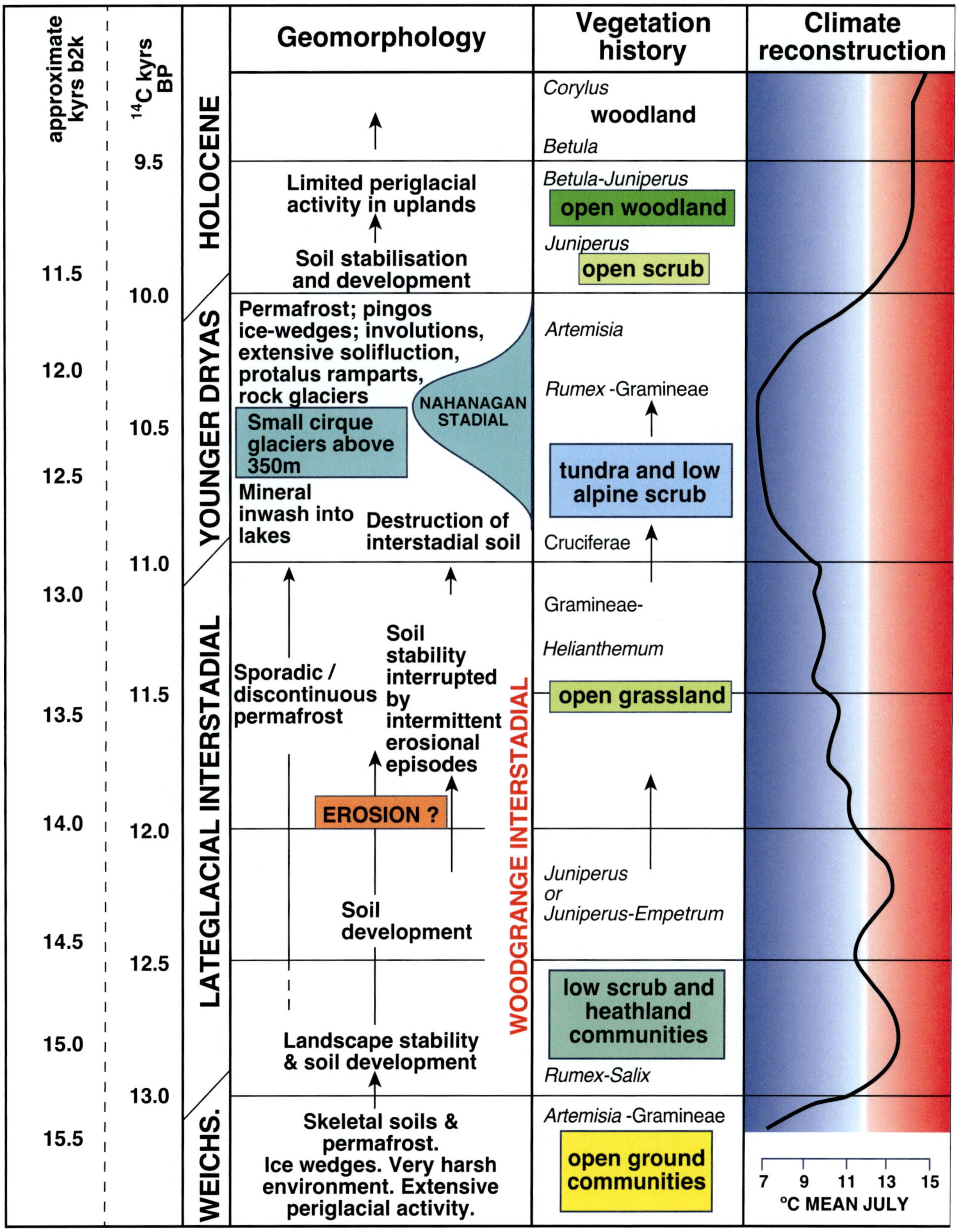

Fig. 4.3 The late-glacial timescale, showing geomorphology, vegetation history and climate reconstruction (kyrs BP = thousands of years before present) (Coxon and McCarron (2009).

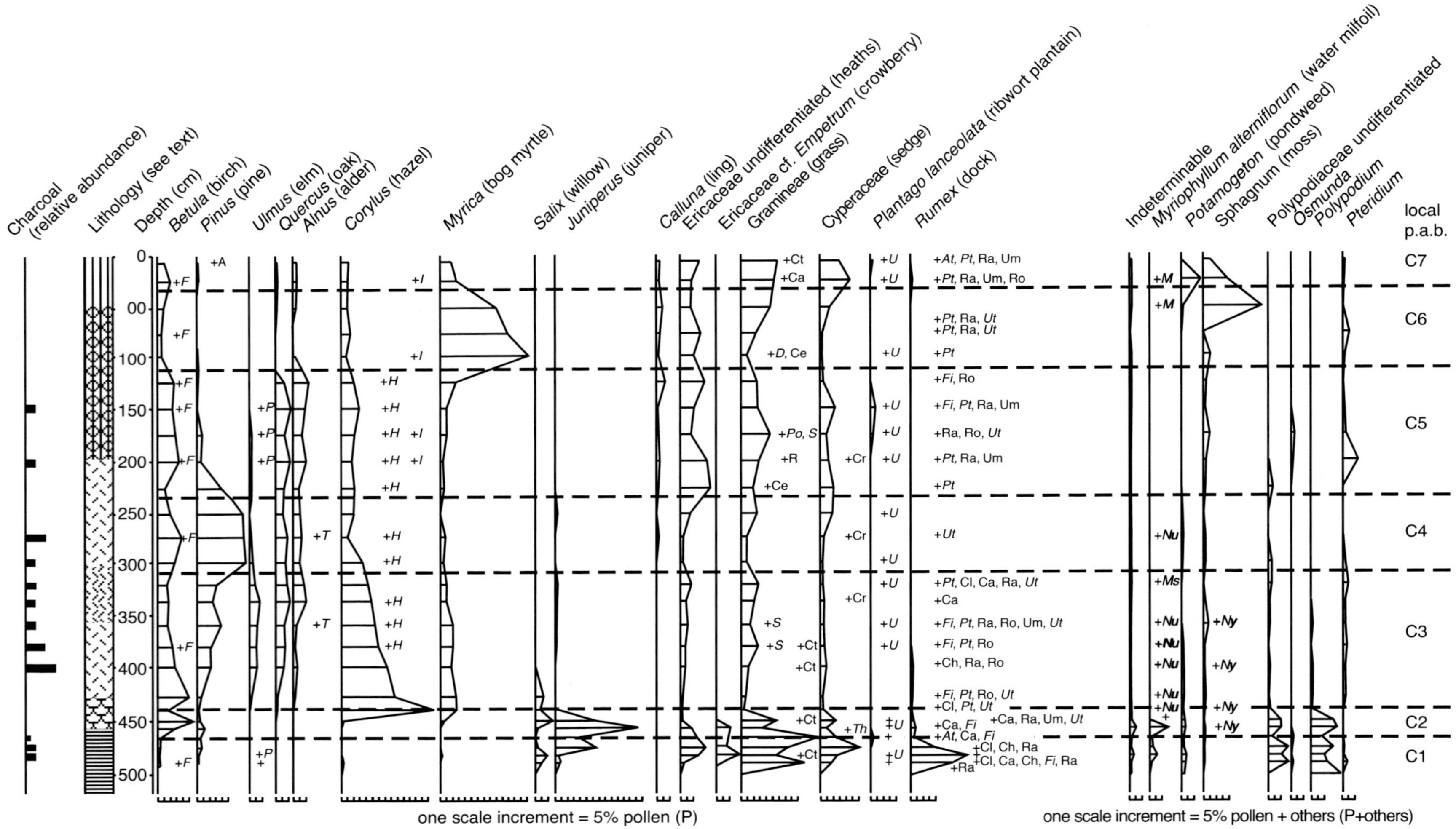

key to abbreviated taxa

F = *Fraxinus* (ash)
A = *Acer* (maple)
P = *Populus* (poplar)
T = *Taxus* (yew)
H = *Hedera* (ivy)
I = *Ilex* (holly)

C = *Centaurea*
Po = *Polygonum*
Ct = Compositae tubuliflorae
D = *Drosera*
S = *Succisa*
R = *Rubiaceae*
Fi = *Filipendula*
Ce = Cereals

U = *Plantago undiff.*
Th = *Thalictrum*
Cr = Cruciferae
At = *Artemisia*
Cl = Compositae lig.
Ca = Caryophyllaceae
Ch = Chenopodiaceae

Pt = *Potentilla*
Ra = Ranunculaceae
Ro = Rosaceae
Um = Umbelliferae
Ut = *Urtica*

M = *Menyanthes*
Ms = *Myriophyllum spicatum*
Nu = *Nuphar*
Ny = *Nymphaea*

Fig. 4.4 Percentage pollen diagram, Lecarrow townland, Clare Island. A set of pollen assemblage biozones is determined for each site by examining the pollens and spores found at the site. The succession of one characteristic group of pollens and spores by another such group marks the boundary between different pollen assemblage biozones. These pollen assemblage biozones were determined by Coxon (1987).

Lecarrow

At the bottom of the profile from Lecarrow there is lifeless sand and gravel washed in from the adjacent moraine slopes in the aftermath of the glacial epoch. Above this are laminated sediments that contain the first pollen assemblage biozone (pab) in the profile, referred to by the people who study these things as zone C1. These sediments contain the pollen rain of a plant assemblage that is very characteristic of Nahanagan Stadial sites elsewhere: hardy grasses and sedges, crowberry and docks, dating to around 10,200–10,000 years ago. This is followed, as it is all over Ireland, by a plant assemblage of tundra, dominated by juniper, which reached its peak between 10,000 and 9,500 years ago. Between 10,000 and 8,000 years ago the juniper tundra gradually gave way to pioneer birch and hazel woods. This is the second pab, C2.

In due course taller canopy woodland with oak, alder and pine took the place of these early woods (C3). The alder expansion has been dated to between 7,500 and 6,500 years ago. Very significantly from our human point of view, charcoal fragments become common in the profile during C3: this is significant because it is our first sign of the arrival of people on Clare Island, reflecting as it does frequent burning of the woodland. At about 330cm in the sediment core the proportion of elm in the profile declines dramatically. This is the famous elm decline recorded everywhere in the pollen record around this time, and it probably occurred on Clare Island between 5,000 and 4,200 years ago. Following the decline of the elm, pine came to dominate the forests, along with birch (C4).

Apart from the pollen record of these pine forests, they also survive as fossil trees buried under the bog. The forests of fossil pine are mainly concentrated in the north-eastern part of the island, particularly in and around the townlands of Maum and Lecarrow (Figure 4.6). The fossil stumps coincide with a dramatic decline in the cover of pine on the island as C4 gives way to C5 around 4,000 years ago, when the pinewoods were replaced by woods of birch, oak and alder, and plant communities characteristic of open heath. Oak is more common in the fossil woods that are found around the edges of the small bogs at higher level, especially between the 200 and 400 foot contours in the circular depression between Curraghmore and Lough Avullin and due north of Creggan (together with birch and willow, but very little pine). Pine is the predominant species at lower levels (within 100 foot of sea level). The pines are often up to a metre in diameter at ground level, which would mean a breast height of as much as 60cm. The oak trees are often 45cm in diameter at breast height, which would mean trees were between one and two centuries old.

Further incursions into the woodland are reflected in an increasing presence of grasses and ribwort plantain (and more charcoal). Around 4,000 years ago birch, alder and oak replaced pine as the dominant trees. The large increase in the amount of grass and plantain pollen in the profile from this time on reflects human incursion into the forest cover to make fields. An increase in the amount of charcoal in the sediment, and the charred appearance of some of the pine stumps seen under the bog at Maum also reflect the clearance activity of early farmers. The much more detailed pollen analyses from Church Lough on nearby Inishbofin record considerable and sustained levels of human activity around the same time (between 3,970 and 2,345 years ago).

The two final pabs (C6 and C7) are marked by a dramatic increase in bog myrtle (locally at least) and the spread of bog across the island. Tree cover continued to decrease and an increasingly open vegetation pattern developed.

The only woodland of any extent remaining today is in a sheltered valley at Lassau, west of Portlea. This is dominated by birch and grey willow, never more than 5m high, and is more in the nature of scrub. Canon Lett (who studied the bryophytes of Clare Island for the original survey) thought this little scrap of woodland maintained ecological continuity with the forest that had once covered the Island, and that the mosses and liverworts occurring here were 'the survivors of the Mosses and Hepatics that abounded on the trees and ground of the primeval forest, the bleached stumps of which stud the ground near Maum and other places, and conjure up the idea of what a congenial habitat their presence formerly provided for our little plants'.[2] As we will see later, the more likely explanation is somewhat different (pages 272–274).

Fig. 4.5 The fossil pine forest in Maum and Lecarrow.

Timeline for Irish Archaeology

Early Christian	*c.* 400–1000AD
Iron Age	*c.* 500BC–400AD
Bronze Age	*c.* 2500–500BC
Neolithic	*c.* 4000–2500BC
Mesolithic	*c.* 8000–4000BC

Lough Avullin

The bog at Lough Avullin at one time covered 10-20 acres of the valley in which the lake lies—or rather lay—because there is almost no open water today. It is said to have been drained by Sir Samuel O'Malley in the early 19th century, when a 3m-deep drainage cut was made on the eastern side to provide a headrace for the mill.

The core from Lough Avullin tells much the same story as that from Lecarrow.[3] The pollen found at the base came from postglacial tundra dominated by scattered pine, willow and juniper, and eventually with abundant crowberry. The start of the Holocene is marked by a rapid increase in organic deposition and abundant pollen. The open tundra gave way in time to birch and hazel, pine, elm and oak, which dominated the landscape between 10,000 and 700 years ago. Alder appeared just before 6510±40 years ago, around which time there was a marked decrease in hazel.

The elm decline that is such a feature of pollen diagrams around 5,000 years ago was accompanied by a decline in the representation of other trees and shrubs, and a rise in the amount of charcoal. This is generally taken to indicate the arrival of the first farmers, who cleared away forest to provide open land for crops and grazing animals. Around this time too we find an increase in the pollen characteristic of the farmed landscape consisting of abundant grasses and plantain.

There was a dramatic decline in pine around 4,000 years ago. Arthur Forbes thought pine lost its dominance as improved climatic and/or soil conditions gave greater competitive advantage to broad-leaved trees; he also speculated that widespread honey fungus might have been a factor in the decline of pine.[4] From that time on ash became more prominent, and there was a noticeable increase in heather and heath, as well as of royal fern, polypody and sphagnum moss. There

PROC. R.I. ACAD., VOL. XXXI. PART 9, PLATE I.

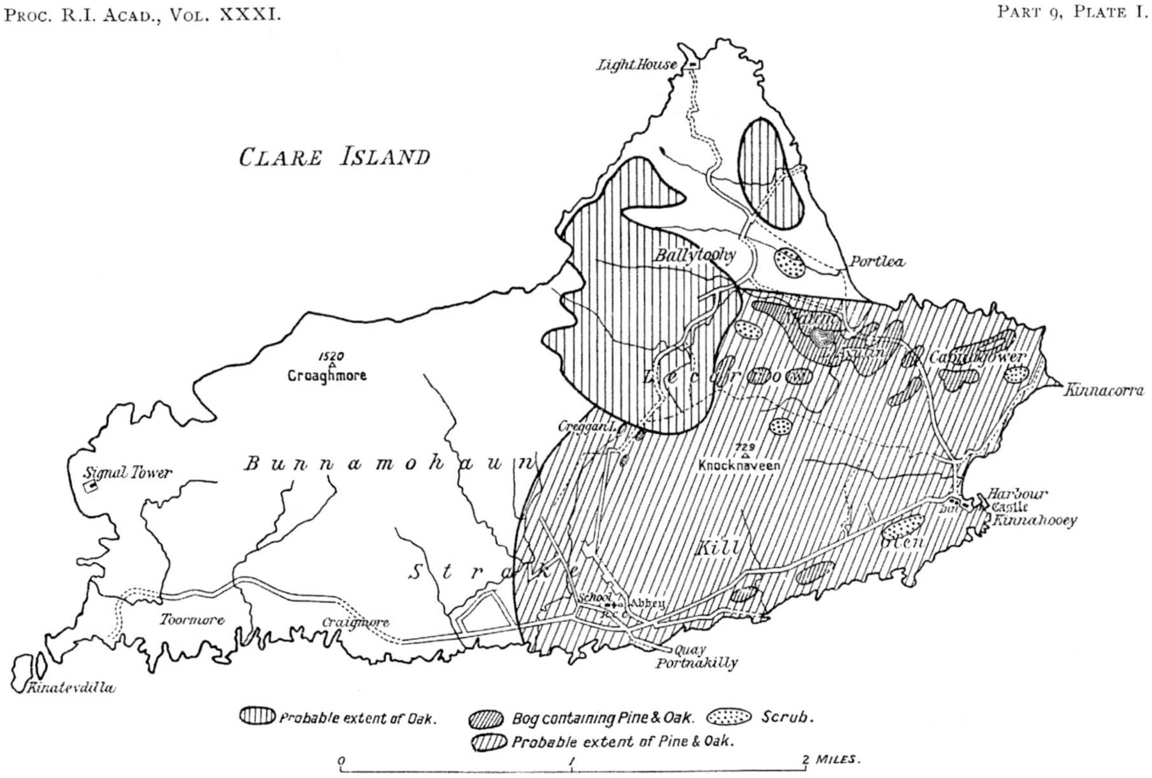

Clare Island, showing areas with scrub and bog timber.

CLARE ISLAND SURVEY.—FORBES: TREE-GROWTH.

Fig. 4.6 The distribution of fossil forests on Clare Island: Plate 1 in Forbes, Original Survey 9.

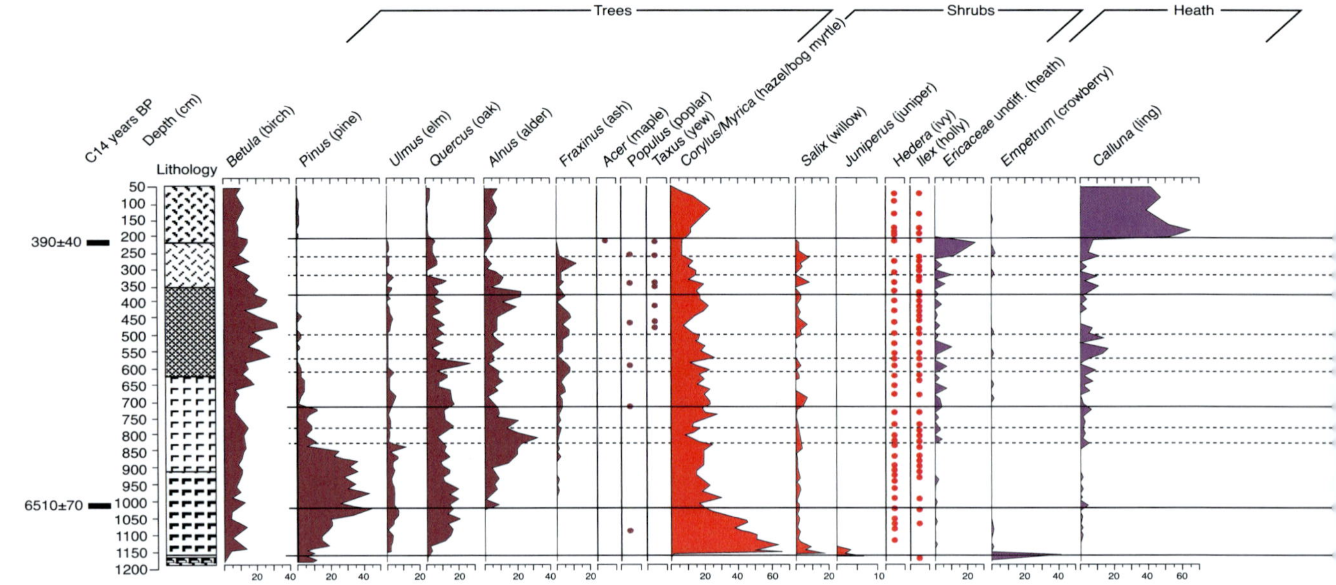

Fig. 4.7 The Lough Avullin pollen diagram (*New Survey* Volume 7).

was a pronounced increase in open ground around 1,300 years ago, reflecting a more intensive agricultural use of the island during the Early Christian period. There was also a striking increase in white water-lily and pondweed, painting a picture of a lake where these open-water species were abundant. We can picture a time when the open water of Lough Avullin covered 20 acres, its surface starred with white water-lilies, its depths green with pondweed, fringed with open forest dominated by pine and oak. Echoing the evidence of the pollen record, the spacing of the fossil trees found under the bogs is also suggestive of open parkland rather than dense forest, the trees varying in age from 50 to 300 years old.[5]

However, it was only around 350 to 400 years ago that woodland more or less disappeared, and heather began to assume the dominance it had at the time of the original survey. Alder, birch and bog myrtle were still common on wetter ground, where sphagnum became very abundant. Cereal pollen and the signature weeds of farming—daisy / dandelion, plantain and goosefoot—are well represented from this time on. There is a higher percentage of charcoal than ever before in the profile, and an increase in minerogenic inwash that suggests increased bare ground subject to erosion.

The overall picture is of fluctuating tree cover, reflecting phases of forest clearance and agriculture along with periodic burning alternating with periods when there is less evidence of human interference with the forest cover. The rich archaeological record of Clare Island chronicles a human presence for the last 5,000 years or more, but it is only in the last 400 years that people have been sufficiently numerous or technically advanced to significantly alter woodland dominance of the vegetation.

A map included in the Geological Survey's Special Memoir for Clare Island shows the localities where submerged peats and forests occur around the coast of Ireland (Figure 4.8). Since they are found nearly all around the island, it can be inferred that sea level has risen considerably since the time of these submerged forests and bogs.

Nowhere does the depth in Clew Bay exceed 20 fathoms, and the narrow strait that separates the island from Louisburgh is less than 60 feet deep, which means that the depth of till was sufficiently thick to bridge the narrow strait between Clare Island and the mainland.

But as well as this, even though the bog-entombed pine forest on the island is well above sea level today, the submerged peats and forests that occur all round the Irish coast show that at one time sea level was several tens of metres lower than it is today. This means that even without the vast accumulations of till left in Clew Bay by the retreating ice, Clare Island would have formed part of the mainland at the period in which the bogs and pine forests flourished.[6]

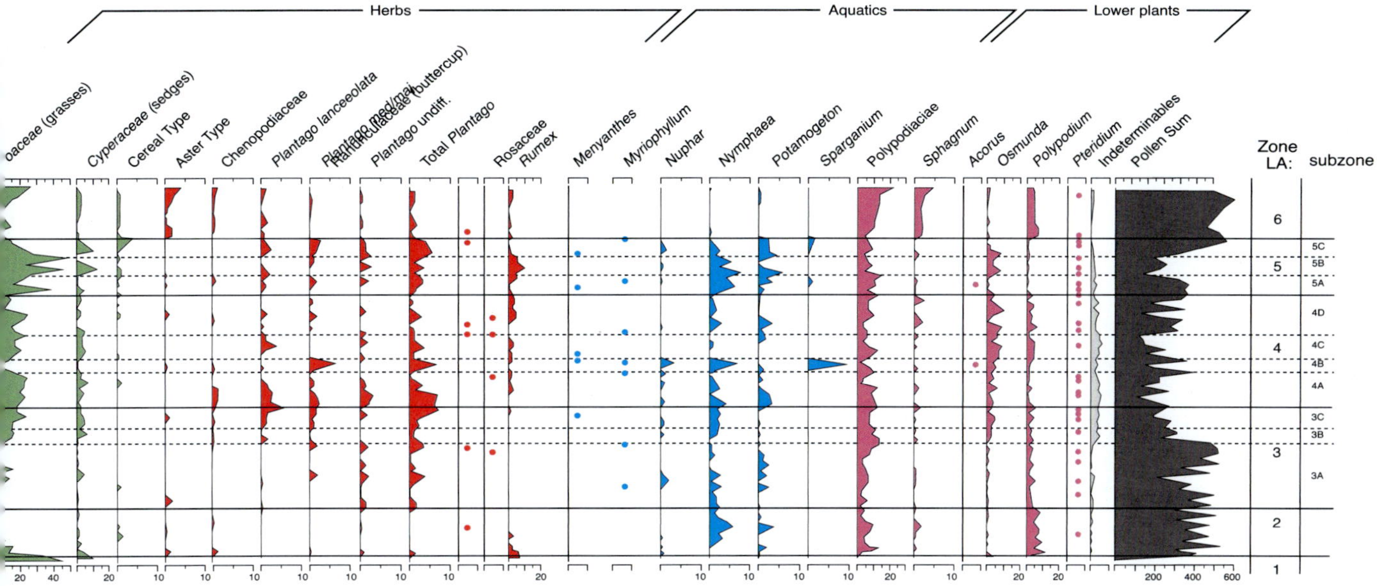

Fig. 4.8

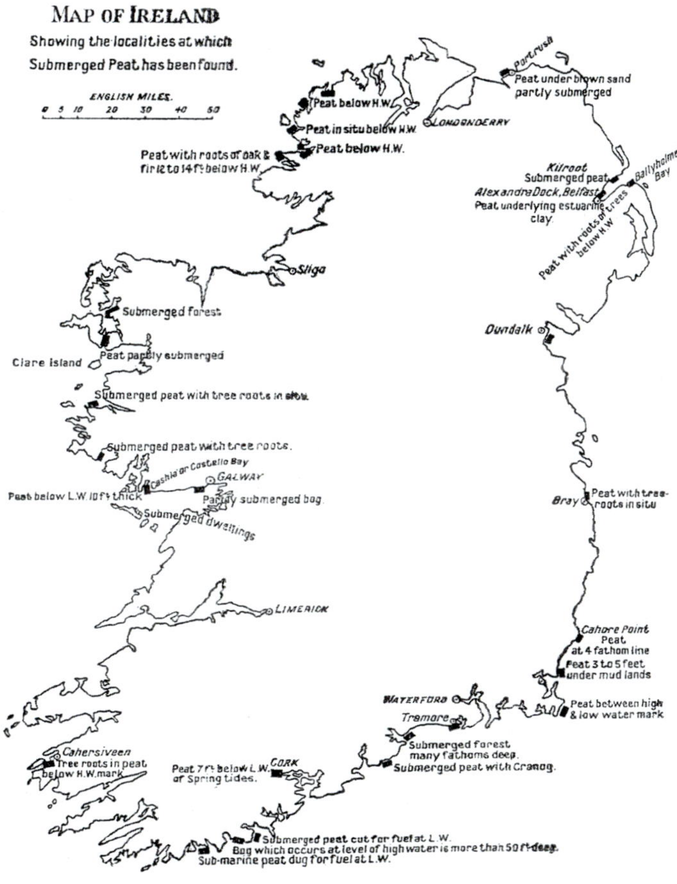

Notes

[1] Coxon 1987, 2001; Coxon *et al.* 2013.

[2] Lett 1912, OS11, 2, 3–4.

[3] Coxon *et al.* 2013.

[4] Forbes 1914, OS 9, 27–30.

[5] Forbes 1914, OS 9, 16.

[6] Hallissy 1914, OS 7, 21; Cole *et al.* 1914, 43–4, 40.

[7] Cole *et al.* 1914, 43–4.

Overleaf

Fig. 5.1 Bunnamohaun turfbanks.

Chapter 5
Soils on Clare Island

The kinds of soil found in any particular place depends first of all on the nature of the parent rock material. Another factor we must consider is topography, which is greatly influenced by geological factors and itself influences temperature and rainfall. The third factor is climate: specifically the temperature and rainfall to which the rock material has been subjected as well as the length of time over which they have operated. Temperature and rainfall determine the possibilities for farming on Clare Island, as they do everywhere. These climatic conditions are affected by variation in altitude and aspect: what can be achieved in terms of farming in different parts of the island is profoundly influenced by the degree of exposure to Atlantic storms; the west and north of the island is extremely vulnerable in this respect, the south and the east less so.

Soil types are also influenced by the nature and degree of biological activity, that is, the vegetation at the surface and soil biota that produce the organic matter in the soil. Finally, the nature and degree of human interference can radically alter the character and potential of the soil in any particular place.

At its simplest, the soil profile will have three horizons: a topsoil designated as the A horizon, a subsoil designated as B, and underlying parent materials designated as C; additional layers rich in organic matter are designated as O. Where rainfall is high, there may be an E (*eluvial*) horizon from which material has been leached out and deposited in an I (*illuvial*) horizon between A and B. The underlying rock is designated as R (Figure 5.2). Taking all of these factors into consideration, soils are divided into a number of Great Soil Groups.[1] The various horizons can be further qualified by such qualifier terms as the following (distinguished in the legend by lower case letters).

b (buried)	s (spodic)*
g (gleying)	t (illuvial clay)
m (pan formation)	w (weathering)
p (ploughing/disturbance)	x (fragic)**
r (rock structure)	

* Spodic soils are defined by the illuvial accumulation of organic matter and/or sesquioxides, and are generally derived from a coarse-textured parent material.

** Fragic denotes very hard and compact subsurface horizons, with a blocky or prismatic structure such that roots or percolating water can only penetrate along cracks between the blocks.

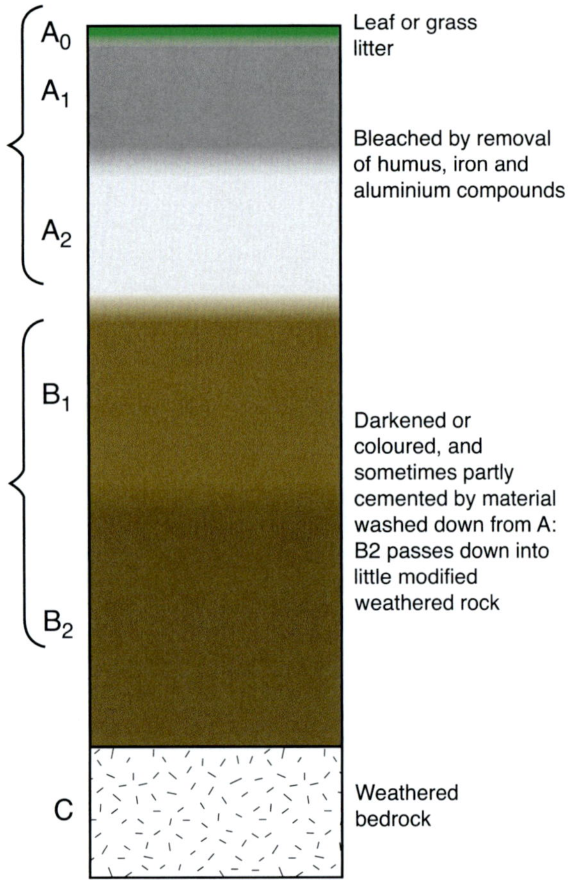

Fig. 5.2 Generalised diagram of a soil profile, showing the different horizons.

Soils were not treated separately in the Original Survey of Clare Island, although it may have been assumed that they would form part of James Wilson's contribution on 'Agriculture and its History'. In any event, the word 'soil' is hardly mentioned. Timothy Hallissy contributed two chapters on soil in the GSI Memoir on Clare Island,[2] although little of this appeared in the Original Survey geology fascicle.[3]

Hallissy's account included consideration of several aspects that are central to modern soil survey and could be considered progressive for its time, but his report is more in the nature of a survey of land potential than a soil survey as such, just as the Land Valuation Survey that preceded it by some 70 years had been.[4] This was rather an estimation of the value of the presumed output of each particular parcel of land, based on land use, soil, vegetation and topography, modified by consideration of such special circumstances as access. Although Richard Griffith emphasised the importance of soil and insisted that there be spadesmen in the survey teams, 'the evidence suggests that Griffith's spadesmen did not dig deeply very often'.[5]

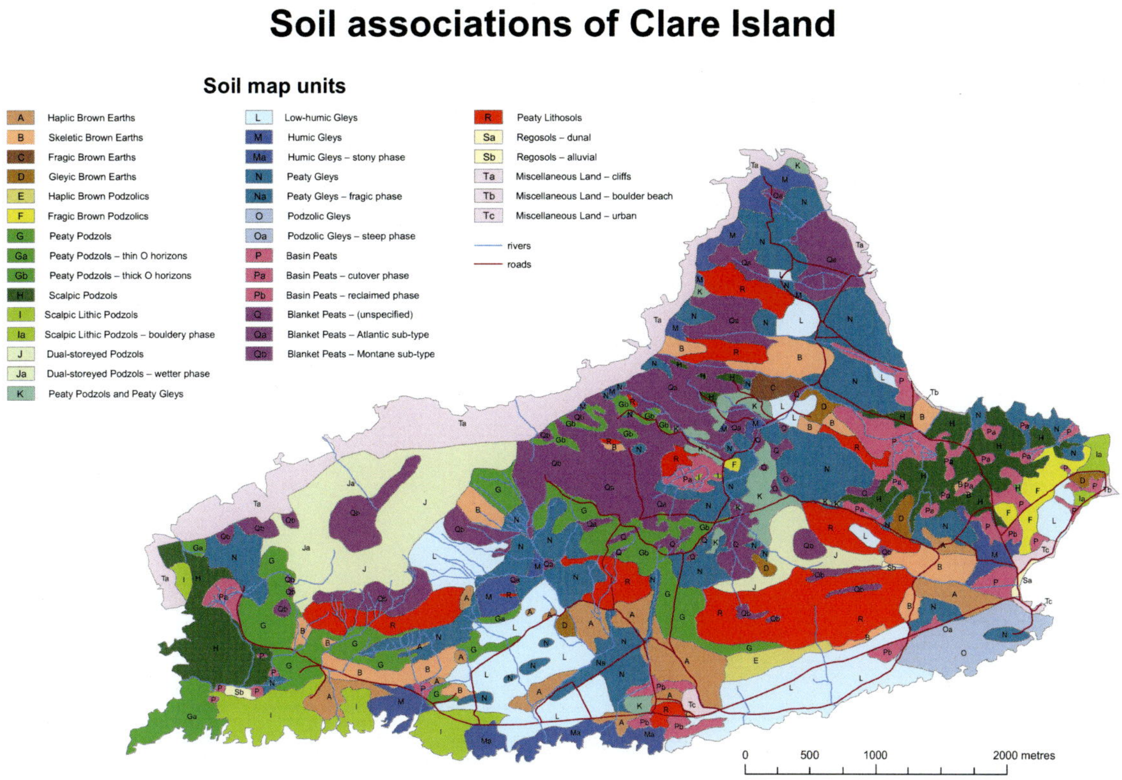

Fig. 5.3 The soils map shows the dominant soil types at any one location. The pattern is so complex that smaller patches of other soils are always intermixed with the dominant soil, but this cannot be shown on the small scale of the map. Chapter 3 in the *New Survey* Soils Volume (8, 97–109) outlines the distribution of the map units in detail, and gives their agricultural use and potential.

Hallissy drew attention to the great variety of soils on Clare Island, which were so complex he thought a complete description of them would be impossible. On the other hand, the great topographic variety of the island and its soil portfolio, daunting as it is, can be seen as the ultimate pedological challenge. All five of the major 'physiographic divisions' that accommodate the 44 soil associations depicted in Gardiner and Radford's 1980 General Soils Map of Ireland are represented within the compass of a small island of 1,600 ha. Eight of the ten Great Soil Groups found in Ireland occur on the island, mostly in small parcels that constitute an intimate and complex intermixture, the mapping of which was the formidable task undertaken by Wies Vullings, with the help and guidance of Jim Collins and George Smillie of University College Dublin between 1997 and 2000. Most of the eight soil groups are further divided into two or more subgroups to give a total of twenty soil types (presented as 'associations' rather than as a 'series'), the majority of which occur in small parcels intimately interwoven to form the complex tapestry shown in Figure 5.3. The new study includes descriptions of 24 soil profiles, supported by tables listing the physical and chemical characteristics of each, at a level of detail greater than that in any earlier published report on Irish soils.

Patches of Brown Earths—the most productive soils on the island—are scattered throughout the lower ground of the island's townlands, but these account for only 10% of its soil resources. The best of these soils are well-drained and well-aerated, with a fertile loamy crumb structure and extensive biological activity.[6] The Abbey is situated close to the largest parcel of Brown Earths on the island, and two of the largest pre-Famine clachans (Lecarrow and Strake) were located beside two other patches a little way north-west of it.

Grey-brown Podzolic Soils are of limited occurrence. These are soils with a well-developed A/Bt/C profile where 't' represents a level of clay enrichment. Brown Podzolic Soils are characterised by a distinct podzolic (spodic) B horizon, but they lack a distinctive eluvial E horizon.

Where a subsurface leached (E) layer is developed overlying a zone enriched in humus and iron oxides the soil is a podzol. The Peaty Podzol subgroup is widespread and common, especially in three areas: there is an almost unbroken strip that extends from the southern slope of Knocknaveen (Kill townland) to the high ground on which the feeder streams for the Owenmore river rise; on the western and south-western flanks of Knockmore (with a 'satellite' parcel on the Island's most westerly peninsula at Loughanaphuca); and on some of the drier parts of the hummocky landscape between Knocknaveen and the northern coastal cliffs.[7] Peaty podzols that long ago had their surface organic layer removed to burn as fuel are assigned to a distinct subgroup of Scalpic Peaty Podzols. The widespread occurrence of peat-drying stands (thuiles) and the survival of peat in the deeper hollows in 'scalped' areas are tell-tale evidence of how common this practice was in the past.

The most common soils are assigned to the Gley Great Group; they cover some 500ha, nearly a third of the island. Their most distinctive characteristic is their wetness, due to a high water table and/or high surface runoff. Three subgroups are distinguished on the basis of the amount of surface organic matter—low-humic, humic and peaty gleys—merging at the upper end into soils of the Peat Great Group. (A fourth subgroup of Podzolic Gleys is characterised by the presence of some organic matter dispersed in the profile).[8] The geographical extent of both humic and peaty gleys was considerably greater in the past, but this is less today because the removal of the surface organic O (peat) horizon has shifted them across the soil spectrum into different classes. It is significant that most of the island's archaeological features are situated on land that was ranked second and third from the bottom in Griffith's Valuation (ranging from nothing up to just under two shillings an acre), suggesting that there has been significant deterioration over the intervening millennia.[9] Thus, although the soils of the island's prehistoric fields are generally of poor quality now, they may have been less so when under cultivation in ages past.

The drumlins

The soils on the drumlins are mainly podzolic gleys, poorly and imperfectly drained because of their sandy clay texture, and the massive, low-porosity structure of the compacted, ice-moulded till beneath. The derelict cultivation ridges on the drumlins are often remarkably well preserved and of considerable height and amplitude, the whaleback ridges generally separated by waterlogged depressions, some of which are fen-like areas of peat. Some of these ridge systems may be little more than a century old, because the wet, stubborn soils made cultivation difficult in earlier times. The map prepared by the Land Judges Court shows few field boundaries on the summit of the Glen drumlin in the 1890s.

Peaty Gleys are found on the crest and eastern (stoss) half of the Strake Drumlin, in stark contrast to the well-drained Brown Earths that surround them on three sides. What seems to have happened is that most of the O horizon was removed originally, and lazy beds dug into the exposed Ag horizon (g=gleyed). The material dug from the furrows and transferred to the ridges was mostly Ag and Eg material, both dense and massive in consistency.[10]

Peat soils

Climate in the west of Ireland is highly conducive to peat formation, which occurs where surface conditions are so wet that the remains of plants cannot decay properly because the fungi and other microorganisms responsible for their decomposition do not thrive in waterlogged conditions. This can be due to the nature of the topography (which gives rise to topogenic mires) or because there is so much rainfall (climatic mires), or from a combination of the two.[11] The hollows that are such a characteristic feature of the hummocky terrain in the east of the island were ideally suited to act as templates for peat formation. Some of the smaller areas of these basin peats are relatively intact (Currach an Locha in the Lough Avullin depression is the best example). The peat reserves in many other areas have been cut as household fuel, and where adequate drainage was possible other peat reserves were drained and converted to hay meadows.[12]

Blanket bog accounts for 14% of the land area of Clare Island, making it the second commonest soil (after peaty gleys at 16.2%).[13] Most of this occurs on the higher ground, but smaller patches of low-level bog of 'Atlantic' type, which has a higher nutrient status and supports different vegetation, occur in flush areas and along some streams. Repeatedly over centuries there has been slippage of saturated peat from the higher ground of Knockmore. Such bog bursts greatly alter the montane bog stratigraphy and vegetation, and give rise to a characteristic multi-phase or dual-story podzolic stratigraphy on lower slopes (Ja).

Shallow, poorly developed soils on resistant parent material such as quartzites and sandstones are described as 'lithosols'. Peaty lithosols are

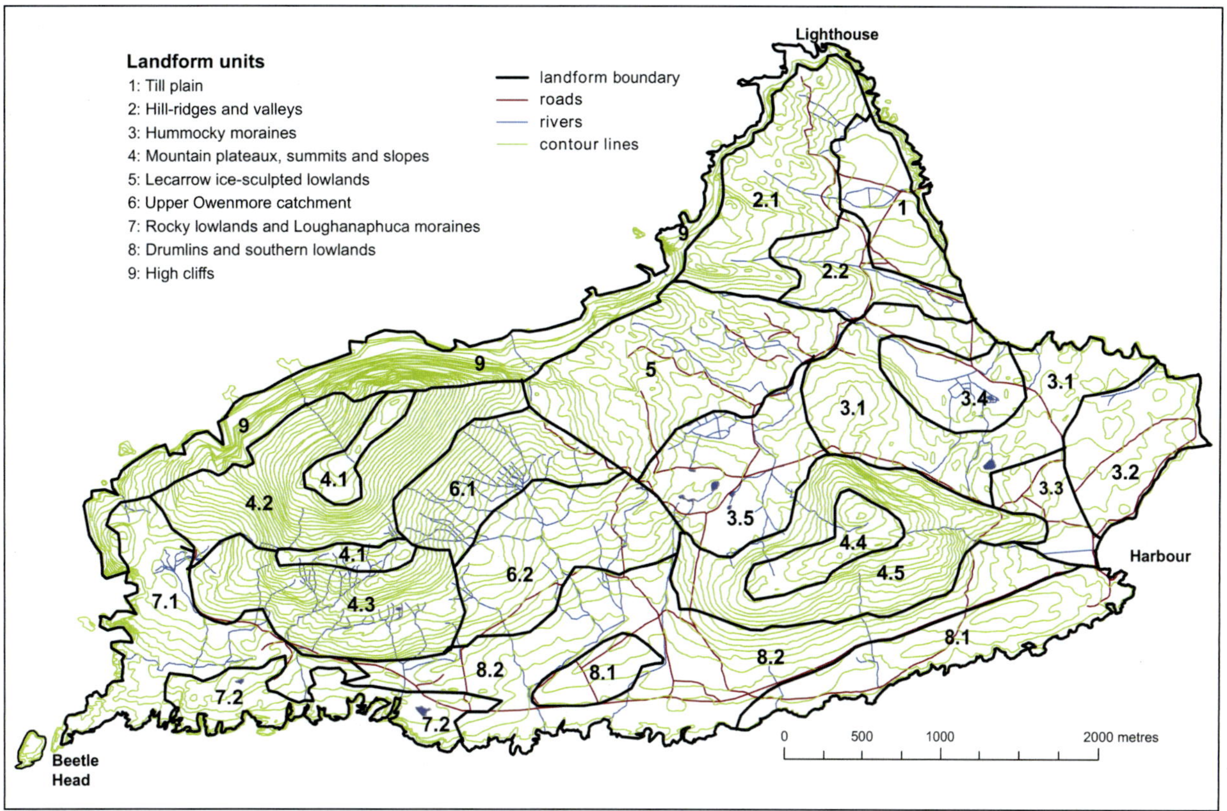

Fig. 5.4 Landform units map (in table below).

Table 1
Landform units and subunits on Clare Island

Landform units and subunits	Landform units and subunits
1: Till plain	**5: Lecarrow ice-sculpted lowlands**
2: Hill-ridges and valleys	**6: Upper Owenmore catchment**
2.1: Peaks and valleys	6.1: Owenmore headwaters
2.2: Marginal, morainic foot-slopes	6.2: Foot-slopes and pediments
3: **Hummocky moraines**	**7: Rocky lowlands and Loughanaphuca moraines**
3.1: Bouldery moraines	7.1: Loughanaphuca moraines
3.2: Capnagower peninsula	7.2: Rocky lowlands
3.3: Underneath Craignagollum	**8: Drumlins and southern lowlands**
3.4: Lough Avullin depression	8.1: Drumlins and Drumlin-like ridges
3.5: Upper Dorree valley and lakeland	8.2: Southern lowlands
4: **Mountain plateaux, summits and slopes**	**9: High cliffs**
4.1: Knockmore summit	
4.2: Knockmore side- and shoulder-slopes	
4.3: Knockmore benches and ledges	
4.4: Knocknaveen plateau and summits	
4.5: Knocknaveen slopes	

widely distributed on the island, most notably on the south and north-east slopes of Knocknaveen, on the hill ridges of Ballytoohy, across the col between Knocknaveen and Knockmore and on parts of the south face of Knockmore.

Areas of dune sand and alluvium have no more than a thin and poorly developed soil because soil-forming processes have not been active for long enough. These immature soils (which are of very limited significance in terms of the area they cover) are regosols.

There is more of a pattern to the patchwork of soil colours than is at first apparent because different soil combinations characterise distinct landforms. Vullings, Collins and Smillie have divided the island into nine Landform Units (many with subdivisions), each with its geomorphological characteristics and distinct array of soils.[14]

Cultivation ridges

'This island has been pleated like a piece of paper that cannot be smoothed out.'

Brian Lynch, 'Old Furrows II' (1993). *Easter Snow. An Island off Ireland*

Although the main instrument of soil change over historical time has been the spade, which 'has left its imprint in the form of thousands of ridges and associated furrows, in acre upon acre of peat-stripped knolls, in over-deepened drains and in unmeasured lines of sod fences', it is likely that the farming activities associated with this use of the spade 'are likely to be insignificant when compared with accretions to the whole island from the atmosphere and ocean spray over millennia'.[15]

Agricultural potential and productivity are controlled in the first instance by the nature of the soil, but the maintenance of its fertility depends on the farmer's ability to manipulate it by drainage, and by augmenting fertility in order to make up for nutrients lost to crops and grazing animals. The key factor in maintaining soil fertility on Clare Island was the ready availability of an endlessly renewable supply of seaweed fertilizer, linking farming on the island with the surrounding seas (discussed in some detail on pages 268–270).

Fig. 5.5 Cultivation ridges.
(a) (Cultivation ridges on the Glen drumlin south of the nineteenth century cluster settlement.
(b) Ridges on the slope north-east of the cluster settlement in Glen.
(c) Truncated ridges on the south coast.
(d) Lidar images of ridges in Strake.
(e) Lidar image of ridges in the area of Capnagower.

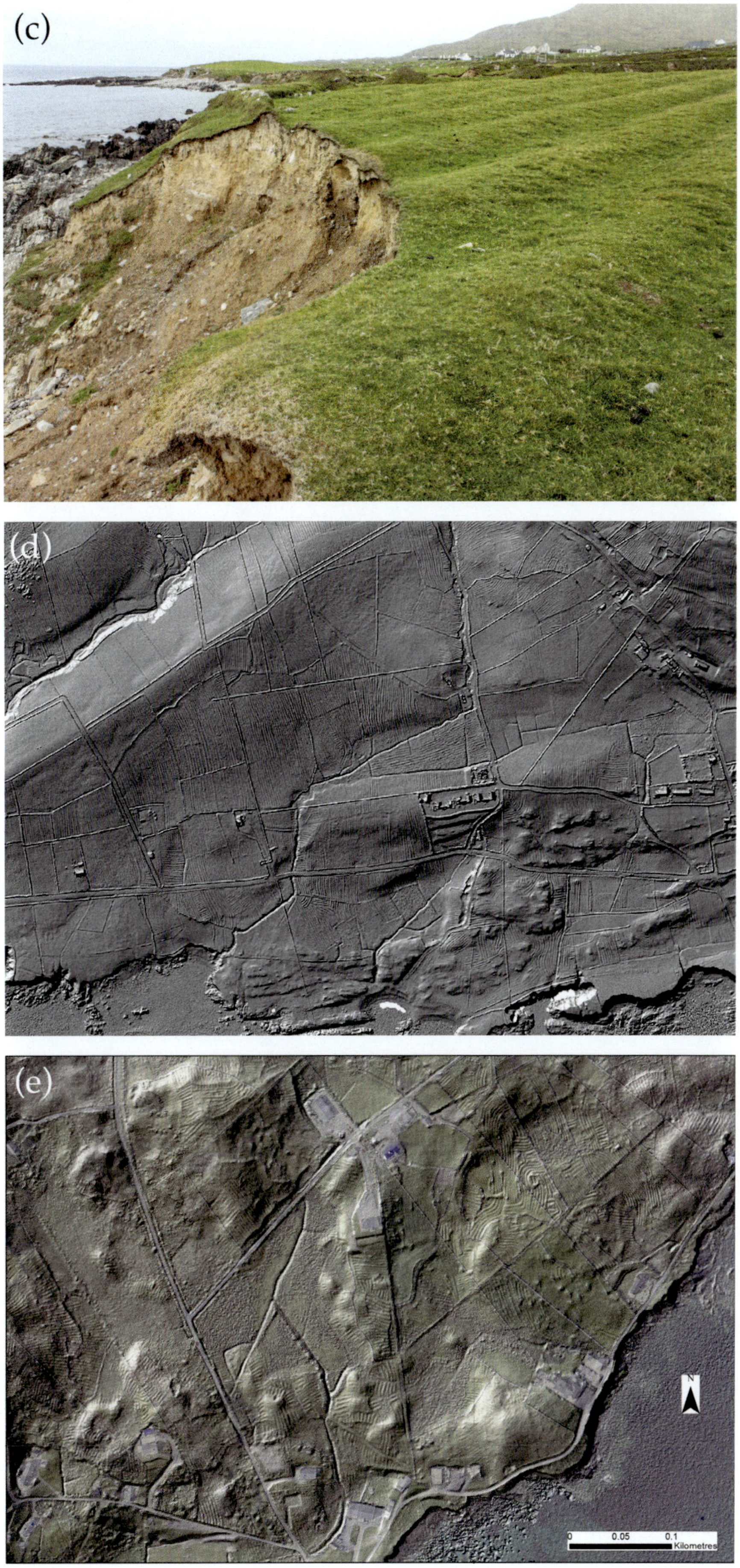

Cultivation ridges (lazy beds) are found on most types of soil and on all but the steepest slopes. They were a simple and effective way to carry away surface water in areas of high rainfall or impeded drainage, and helped to counter the restrictions in this regard of Low-humic Gleys; indeed, the infield of the late 1890s more or less corresponds to this soil type.[16] Not surprisingly then, these soils were given the highest values by the Griffith valuators. The amplitude of the vast majority is between 1.3m and 3.5m.

It is probable that ridge and furrow cultivation has been practised on the island for centuries—perhaps for millennia—but most of the fossil ridges seen today were constructed as the population began to rise after the middle of the eighteenth century to its peak in the years before the Famine.

The sight of these ridges is like the light in a film projector, projecting through the lens of informed imagination the human drama of their making on the screen of the mind. There is nobody about, but if you listen carefully enough, and if you look hard enough, from this simple imprint on landscape you can hear voices and you can see faces. They are as pregnant with memory and meaning as the relics in earth and stone we classify as archaeological artefacts, more richly so because we can fill in much of the detail of the unchronicled lives of their makers in a way that is seldom possible for archaeological features. Their historical significance is another reason why these ridged soils should be accorded the specific cartographic and taxonomic status advocated by Vullings, Collins and Smillie, and formally referred to as *rugic* soils.[17]

Other landscape features that survive from that time can be seen as extensions of this ghostly underlay of ridges. Nowhere else in Ireland does the fabric of that layer of the landscape palimpsest, which lies immediately below our own, survive so clearly and in such detail. Where the archaeological heritage of an area is particularly rich it is sometimes protected and managed in its entirety as an 'Archaeological Complex' rather than as a series of isolated features. In a similar way, Clare Island should be designated as a 'Landscape Complex' on the grounds of its own unique natural and cultural heritage, and it too should be similarly protected and managed.

A note on climate

The confident rampart erected by the Congested Districts Board around the island's perimeter—ometimes strengthened with iron posts set in concrete to support wire fencing—has become something of an archaeological monument, breached in many places ages since by the advancing sea. Several subsequent generations of post and wire fences have fallen to the same assault of the winter seas. Even on the more sheltered northern shore the sea has in many places eaten into the ridged fields of a century ago. It is not really possible in word or image to convey any real sense of the sea's power during storms, because it is simply not possible at such times to come anywhere near it to observe, but the size of the boulders picked up and hurled effortlessly and with unrelenting force against the land at such times gives us some sense of it.

The Original Survey included a fascicle on climatology by W.J. Lyons, but it is little more than a snapshot of the weather conditions prevailing for a short run of years (1894 to 1910) leading up to the survey.[18] Casual inspection of this suggests that the pattern of weather differed little from today; even the months of May and June had on average more wet than dry days, and the average number of wet days in the wetter months of November and December was 26 in each. In the year 1898 there were only 60 days when no rain at all fell.[19]

Only those who live on the island all the year round can really understand its weather. Within the community a body of weather lore grows over the years and is passed on from one generation to the next. Changes in the atmosphere, in the pattern of wind and cloud, in the appearance of sun, moon and stars, and the behaviour of wild and farm animals observed over generations were used to predict the weather. This was more central to island living in an earlier age before broadcast weather reports began to make it largely redundant.

It is easy to dismiss the weather lore gathering dust in folklore libraries as outmoded by the standards of modern weather forecasting. But it was the product of generations of careful observation of cause and effect, and is often remarkably accurate for short-term forecasting for a particular place, sometimes more so than modern forecasts. However, these forecasts are strictly local, relying on signs given by particular landscape features and the behaviour of animals in very specific environmental contexts that are not transferable.

Glimpses of this earlier lore are preserved in the folklore and folk life records collected as part of the Schools Collection in the 1930s:

Swallows fly low in pursuit of insect prey descending in response to changes in atmospheric pressure in advance of rain. The sheep begin to bleat and come down from the mountain.

When the cock crows merrily and loud all understand that a change for the better is drawing near.[20]

It is a sign of rain when the cat stands with her back to the fire.

A storm is imminent if the cat scratches the bark of a tree.[21]

Rain is imminent when the fire burns with a blue flame.[22]

Snow on Nephin in October means there will be some snow every month until the following April.

When the hens gather in a sheltered corner and preen themselves rain is on the way.

When the cricket sings loudly rain is on the way.

When the rooks become agitated in the air rough and stormy weather is coming.

A fog with the new moon means dry weather; a fog with the old moon means bad weather directly.
If there is a ring around the moon at night it is a sign of rain. The wider out the ring extends the closer the rain.[23]
Bad weather is on its way when the ducks are agitated.

Much of the weather lore of Clare Island is common to the weather lore of the west of Ireland generally. Although for some reason no returns were made from Clare Island for the Schools Project, traces of traditional weather lore that pertain specifically to the island are still remembered locally:

- Mare's tails clouds in the sky are a sign that windy weather is on the way.
- If horses are seen rising to the top of the hill/mountain it is a sign of good weather to come, but if they come down from higher ground and keep close to the boundary, bad weather is on the way.
- A weathergall west of Roonagh is a sign of bad weather in the coming weeks.
- The Maurfhadai (waterdog) up on the land in Curraghmore is a sign of bad weather to come.
- It is a sign of good weather you hear the boulder beach at Kinnacorra rolling on a calm evening without swell.
- Dolphins playing in the sea close to the shore signal a change in the weather.
- Willy wagtails are on the street outside the house mean rain is on the way.
- Sounding on Carraig na Meara means the wind is going south, while sounding on Carraig na Ruain means the wind will be from the north-east.
- Froth coming up over Breacail signals three weeks of bad weather.
- A low white mist coming in over the land from the sea (cuisne) is a sign of hot weather.
- If the mainland appears to be close to the island it is a sign of bad weather, while the mainland seeming to be far away signals good weather.[24]

Between now and mid-century Ireland's mean temperature is likely to increase by between 1–4°C, continuing the trend observed over the last 35 years; highest daytime temperatures are projected to rise by up to 2° in summer, and lowest night-time temperatures by up to 2–3° in winter.[25] Mayo, like the rest of Ireland, can expect warming of 3–4° by century's end, with mean July temperatures rising to around 18°C and mean January temperatures to the level currently experienced throughout the county in an average April.[26] Winters are expected to become wetter (across Europe on average by up to 20%) and summers drier (up to 20%). There is likely to be an increase in outbreaks of cold continental air over Ireland in winter.

Although summers will be significantly drier, the more frequent drought periods are likely to punctuated by higher-intensity rain events.[27] Winters are likely to be as much as 25% wetter, with a notable increase in periods of heavy rainfall during the winter.[28] The annual value of wind-driven rain is projected to increase from somewhere in the region of 600–700 litres/m^2 of precipitation per annum on vertical surfaces to 700–800 litres/m^2.

Sea level along the coast is expected to rise by 0.5–1m over the course of the century.

Storm wave heights are likely to increase in winter and spring, and model results suggest an increase in both the frequency and height of extreme storm surges in winter. The combination of these changes is likely to bring about significantly greater coastal erosion.

It is next to impossible to predict how these changes will affect the flora and fauna, not just of Clare Island, but of Ireland generally. The effects of climate change in Ireland have been observed on plants, dragonflies, mayflies and moths: and more particularly on trees and butterflies. Spring warming has had a significant impact by advancing the timing of key phenological phases, but it is often difficult to assess to what extent habitat change or loss are factors in local or more general decline.

One of the more startling predictions of the work carried out by Irish researchers using the EC-Earth global climate model is that there will be a century-long period of abrupt Northern Hemisphere cooling 450 years into the future, with temperatures over Northern Europe on average up to 8 degrees cooler between November and March, and in Ireland up to 3 degrees cooler, signaling a temperature anomaly similar to that of the Little Ice Age (*c.*1250–1850 AD).[29]

Notes

[1] Vullings, Collins and Smillie 2013, 36.

[2] Cole, Kilroe *et al.* 1914.

[3] Hallissy 1914, OS 7.

[4] Vullings, Collins and Smillie 2013, 2.

[5] *Ibid.*, 6.

[6] *Ibid.*, 38.

[7] *Ibid.*, 46.

[8] *Ibid.*, 54–64.

[9] *Ibid.*, 8–9.

[10] Taken more or less verbatim from Vullings *et al.*, 60–61.

[11] Mires are peat landforms.

[12] Vullings, Collins and Smillie 2013, 64–66.

[13] The total area of peat soils is 19%.

[14] These are described in detail in Chapter 4 of Vullings *et al.*, 111–39.

[15] Vullings, Collins and Smillie 2013, 29–30.

[16] *Ibid.*. 142–43.

[17] *Ibid.*. 159–60.

[18] Lyons 1914, OS 6.

[19] Lyons 1914, OS 6, 17.

[20] The Schools' Collection Vol. 0115, 54–56.

[21] The Schools' Collection Vol. 0094, 263.

[22] The Schools' Collection Vol. 0090, 188 and widely elsewhere.

[23] The Schools' Collection Vol. 0149, 159–160.

[24] Michael McNamara, Michael O'Grady, Martin O'-Malley (pers. comm.).

[25] Gleeson *et al.* (eds) 2013.

[26] John Sweeney is 'confident enough in an estimate of a rise of 0.8–1.0 degree in both winter and summer mean temperatures for Clare Island by mid-century as compared with the 1961–90 averages' (pers. comm. 18 October 2016).

[27] Sweeney *et al.* 2013.

[28] Gleeson *et al.* (eds.) 2013.

[29] Gleeson *et al.* 2013; in Gleeson *et al.* (eds), 91–94.

Opposite

Fig. 6.1 The kelp zone on rocks of the Ballytoohy Formation north of Portlea.

Chapter 6
Sea and shore

In this difficult world of the shore, life displays its enormous toughness and vitality by occupying every conceivable niche. Visibly, it carpets the intertidal rocks; or half hidden, it descends into fissures and crevices, or hides under boulders, or lurks in the wet gloom of sea caves. Invisibly, where the casual observer would say there is no life, it lies deep in the sand, in burrows and tubes and passageways. It tunnels into solid rock and bores into peat and clay. It encrusts weeds or drifting spars or the hard, chitinous shell of a lobster. It exists minutely, as the film of bacteria that spreads over a rock surface or a wharf piling; as spheres of protozoa, small as pinpricks, sparkling at the surface of the sea; and as Lilliputian beings swimming through dark pools that lie between the grains of sand.

Rachel Carson, *The Edge of the Sea* (1955)

Whatever about the great diversity of life on land, it is in the seas that the greater profusion of life occurs; of the 36 major *phyla* of living things, only 16 occur on land. Although much is hidden from all but those who can don wetsuit or snorkel, there is so, so much to see, especially along the shore, between the lines of high and low tide. The creatures that live here must cope with all the challenges posed by an environment that is land for half the time, a water world for the other half. The life cycle of many of these creatures is an alternation of generations more fantastic than that of insects on land, the brief early days or weeks of their lives spent as creatures of the plankton utterly different from the adults they will become.

Each is as intimate to its place in the ocean as every plant and animal is fitted for the unique niche it occupies on land. The animals we see exposed on the shore when the tide is out—mussels, barnacles, periwinkles, limpets—are the ones evolution has fitted to withstand the full force of the waves, but these familiar animals are a tiny minority of the creatures that live here. A particularly favourable stretch of rocky shore can harbour several hundred thousands of animals per square metre.

Living conditions at the upper and lower ends of the shore vary greatly: the sea reaches the upper extreme only at the highest level of spring tides, whereas the other end is under water except at the lowest level of spring tides. Plants and animals optimally adapted to life at one extreme are likely to be ill-equipped to live at the other extreme. However, every interval between the two extremes is defined by its own unique suite of physical and chemical characteristics, and each has its own assemblage of plants and animals adapted for life at that interval. Some range across a wide zone, while others are very particular about the interval they occupy. For some groups (e.g. periwinkles and the larger seaweeds), we have a fairly good understanding of the preferences of particular species within the group, but for many others we do not.

Every fragment of rock and shell on the shore has a story to tell, although we have to use the eye and ear of the mind to translate what we see at the tide line into the ecology of the submarine worlds out of which they have come, similar to the way the palaeontologist must translate the fragmentary remains of vanished life entombed in rock into a world of living things.

The biggest difference of all is that between eroding and depositing shores; the former characterised by bare rock and pounding waves, the latter by abundant sediment and lesser tidal forces. Very few depositing shores are found on Clare Island, exposed as it is to the full force of the ocean on all but its more sheltered lee side. Such is the degree of exposure to the waves that there is nowhere for sand to settle; the intertidal zone is predominantly rock. The only beach (apart from a few patches of sand on the south coast) is the one at the harbour. There are numerous large rock pools, each an entire aquarium with widespread development of encrusting calcareous algae. The extensive pinkish-purple blankets of this *Lithothamnion association* that develop on moderately exposed shores support a wonderfully diverse fauna. Especially noteworthy features of this association are the abundance of sponges on rocks, of sea anemones and purple urchins in the pools, and the general richness of molluscan life.[1] It is beautifully and luxuriantly seen in many places along the south and east coasts, where the gently sloping rock surfaces are well-suited to its development. Coralline algae (mainly *Lithophyllum incrustans)* dominate this community, often growing in large, thick sheets and masses in the interstices of which a rich fauna finds food and shelter. Although they often grow in rock pools, coralline algae also live out of sight on the bare rock below the lowest of the spring tides. Their coralline form is due to the lime they incorporate in their tissues. The main population centres of sea urchins, starfishes and the common whelk are in deeper waters offshore, whence they advance into the intertidal zone.

The Original Survey stretched its net far beyond the borders of Clare Island itself; indeed some marine contributors did not pay that much attention to the island. This, together with uncertainty as to the precise locations at which many species were found, and different survey approaches, makes any successful comparison between the New and Original Surveys of limited value. And whereas the Original Survey attempted a census of all the creatures of the sea, the New Survey peers more intimately into the lives of just a few of the key organisms. Accordingly, the New Survey decided to concentrate on detailed study at two representative rocky shores: the extremely exposed shore at Leckacanny, and the moderately exposed

The fauna of rocky shores on Clare Island (Original Survey 67, 52–56)

50 species of Foraminifera; especially abundant in tufts of *Corallina officinalis* in rock pools
31 sponges
30 bryozoans
22 coelenterates
6 echinoderms
8 free-living flatworms (Turbellaria)
5 ribbon worms (Nemertea)
10 nematodes
5 oligochaetes
2 siphulculid worms (Gephyrea)
20 copepods
3 isopods
8 amphipods
12 decapods
1 pseudoscorpion
1 sea spider
5 mites
50 molluscs
4 fishes
2 centipedes (both seaside specialists, hunters of the land between the tides
– and a handful of insects specialised for living in this salty world

shore at Portnakilly. A detailed survey of seaweeds and molluscs was however carried out, and in the case of these two key groups comparison between the Original and New Surveys has been possible.[2]

In general, the rocks of the intertidal area can be divided into several zones, each with a rich and characteristic faunal assemblage:

1) Exposed on rocks or weeds
2) Under boulders on bedrock
3) In crevices
4) Associated with *Lithophyllum*
5) Rock pools
6) The *Laminaria* zone
7) The zone of red seaweeds (down to 8–15 fathoms)

The cliffs of the west and north-west coasts of the island are practically inaccessible; here the intertidal zone is a vertical face. Much of the south coast is also very exposed, but the steep dip of the rocks prevents the development of gently sloping surfaces. There are few boulders about, and so there is little shelter for creatures of the intertidal world. This great exposure limits the range of habitats possible, and only those animals (and plants) that are more specially adapted to withstand the tremendous buffeting of winter waves can survive.

The fauna of the rocky shores of Clare Island was investigated 'with a fair degree of thoroughness in many groups', and a detailed summary was prepared by Rowland Southern.[3] Most of the animals that make their homes along the shores are to be found in tidal pools and in crannies and crevices in the bedrock. Nearly twenty sponge species are associated with the *Lithophyllum incrustans* that is so characteristic of the southern and eastern shores, along with seventeen species of polychaetes, but this list is very incomplete.[4]

Life on exposed shores

Because they are so much poorer in terms of biodiversity, exposed coasts have attracted much less attention from ecologists than the species-rich habitats of more sheltered situations. Indeed, Leckacanny, studied by Ruth O'Riordan and her team for the New Survey, is one of few exposed coasts in the world that have been looked at in some detail.[5] During their study they recorded only fourteen species of seaweeds, fourteen macrofaunal species and four lichens, distributed in four interfingering zones across the intertidal area:

1) The black lichen zone (see pages 151–52 in the section on lichens; see also pages 189–91 on algae)
2) A zone dominated by the red seaweed *Porphyra umbilicalis*
3) A zone dominated by the red seaweed *Mastocarpus*
4) The *Alaria* zone

The black lichen zone is often covered with a slippery cover of small algae, sometimes accompanied by *Porphyra linearis* and tattered sheets of the brown seaweed *Ralfsia*; this is especially noticeable during wetter seasons. The few animals found in this zone have to be able to withstand long periods out of the water (such as the rough periwinkles, *Littorella saxatilis* and *L. arcana*, and another small periwinkle called *Melarhaphe*).

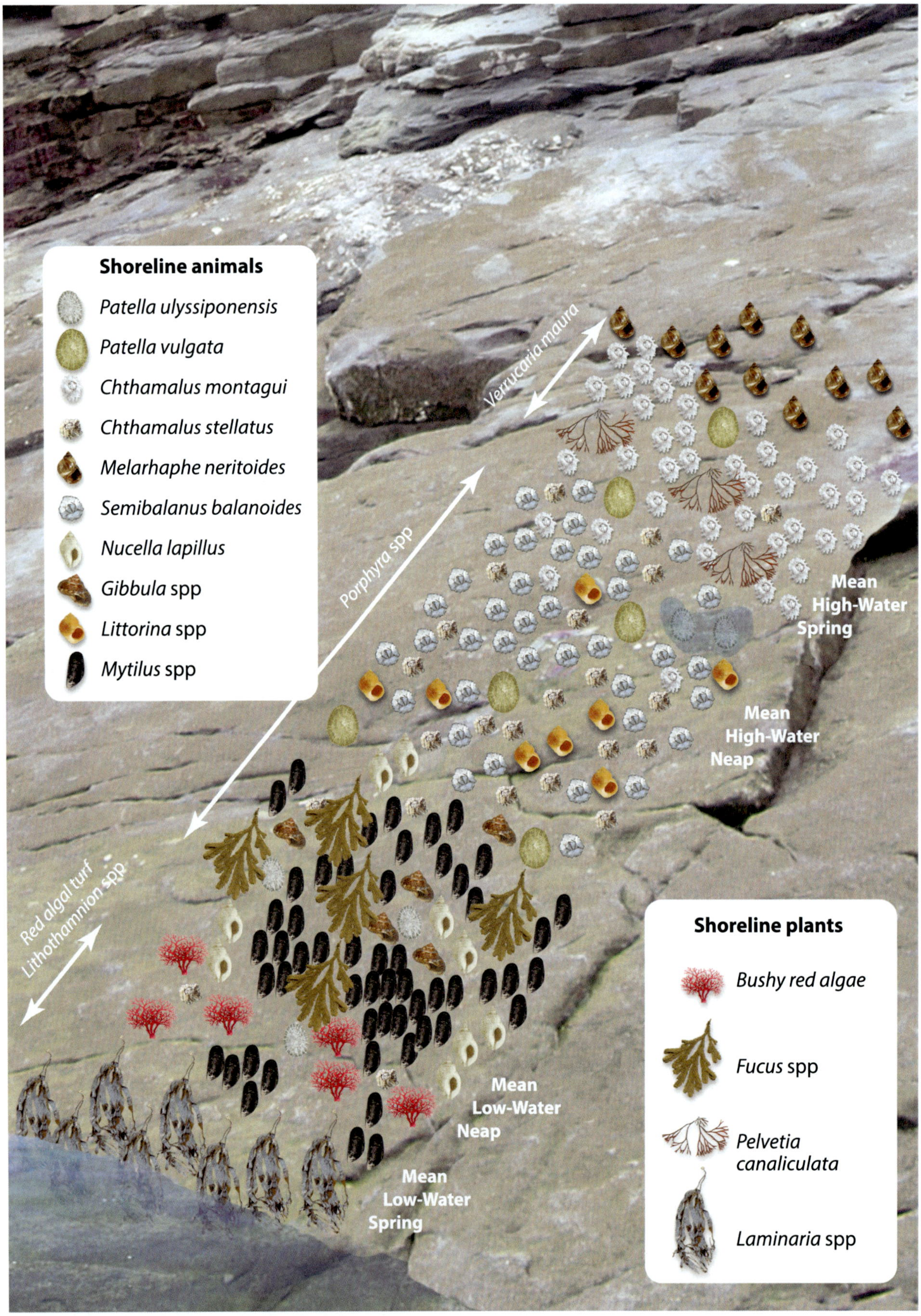

Fig. 6.2 Zonation of the characteristic plants and animals on moderately exposed shores on Clare Island. Redrawn from Volume 3 of the *New Survey of Clare Island* (facing page 1).

Almost all of the algae are red algae, the only brown seaweed (apart from *Ralfsia*) being *Alaria* (at first glance it looks like a floppy marine version of hart's-tongue fern), which has evolved an ability to withstand severe wave action and water surges along the side of gullies. *Porphyra* is a thin, lettucy-looking red seaweed with tattered edges (considered a delicacy in parts of Wales). You could think sometimes that *Mastocarpus* is a nursery for young mussels, so abundantly do they cluster about its bushy branches.

Although there is little variation in species composition over time, there is some variation in relative abundance, and in the vertical range over which particular species occur. A notable example of temporal variation is the abundance and ubiquity of barnacles in late summer and early autumn, which is due to the establishment of new colonies. The most abundant barnacle on this exposed shore is the acorn barnacle *Chthamalus stellatus*.

The rocks underwater are covered as densely with vegetation as they are on land, but of a completely different kind. Divers in the waters of Clew Bay come back to the dryland world with accounts of 'plumose and dahlia anemones in more colours than I have ever witnessed … strewn in their own colour group colonies like a living wallpaper, sponges providing their off-orange contrast to the surrounding psychedelics'.[6] It can hardly be described as vegetation of course, because although they are just about as firmly anchored as the plant life that clothes these same rocks on land, sponges and anemones are animals!

What lies beyond

Clare Island is really only a detached fragment of the mainland; indeed, it was this aspect of the island that Praeger found so fascinating. In this respect Clare Island is quite unlike the islands that attracted the attention of the biologists interested in the way the isolation they afforded might influence colonisation and the course of subsequent evolution within their confines. Many of these are volcanic islands, new-born rock on ocean ridges and in volcanic arcs. Clare Island, in contrast, like most of Mayo and Connemara, is made up of ancient rocks of many ages, annealed and welded together in the course of a long and tortuous geological history. Like those inspiring volcanic islands, it too is young, but not because its isolation is due to its having newly risen out of the sea. Rather it is young because a rise of sea level has isolated it but a moment ago in geological time, abandoning it to the merciless waves, which slowly and steadily deepen and widen its fretted edges on all sides and in the course of long future ages will take it beneath the waves entirely.

The rock of which the island is made continues westwards beneath the sea to the edge of the continental shelf, where it plunges into the abyss. To north and south it continues towards and becomes the mainland, on the east side mantled beneath Clew Bay's school of drumlins and all the products of their decay.

At the time of the Original Survey we could see almost nothing of this hidden sub-sea landscape. All that was available was what the *Helga* dredged from the bottom, which allowed the naturalists of the Original Survey to sketch the submarine landscape in the broadest of strokes. Twenty-eight hauls of the dredge in waters ranging from ten to forty-two fathoms deep, taken all around the island, gave us our first picture of what the life of the deep ocean is like.[7] Particularly characteristic of this area was the abundance of the brittle star *Ophiactis balli*, which lives under rocks and ventures forth after dark (see page 121). The rocky bottom west of the island was inhabited by massive colonies of the sponge *Cliona celata* (Figure 6.3b). North and north-east the bottom is also strewn with loose boulders; south of the island there are many patches of gravel and loose stones. In the lee of the island to the east and north-east there is a large area of clean sand on the bottom—trawling ground—as far as the outer edge of the drowned drumlins, where there is a great reef of boulders.[8]

Fig. 6.3a Brittle stars *Ophiactis balli* (Peter Southwood (CC BY-SA 3.0)).

Fig. 6.3b Sponge *Cliona celata* (Matthieu Sontag (CC BY-SA 3.0)).

Today we have geophysical tools and techniques for exploring the sea floor the naturalists of the Original Survey could not have dreamed of, and these enable us to image the topography of the ocean floor with stunning accuracy. However, these tools have not yet been used to explore the waters around Clare Island, as they have for Achill to the north, and Inishbofin and Inishark to the south, and for Clew Bay itself, where the rock is still cloaked with shifting sediment derived from the breakdown of glacial till and the sand and gravel, mud and silt hacked and rasped from the rocks of the coast by the erosive power of the sea. In some respects, these new techniques allow us to see the bedrock more clearly than on land, because it is less densely covered in vegetation, and every

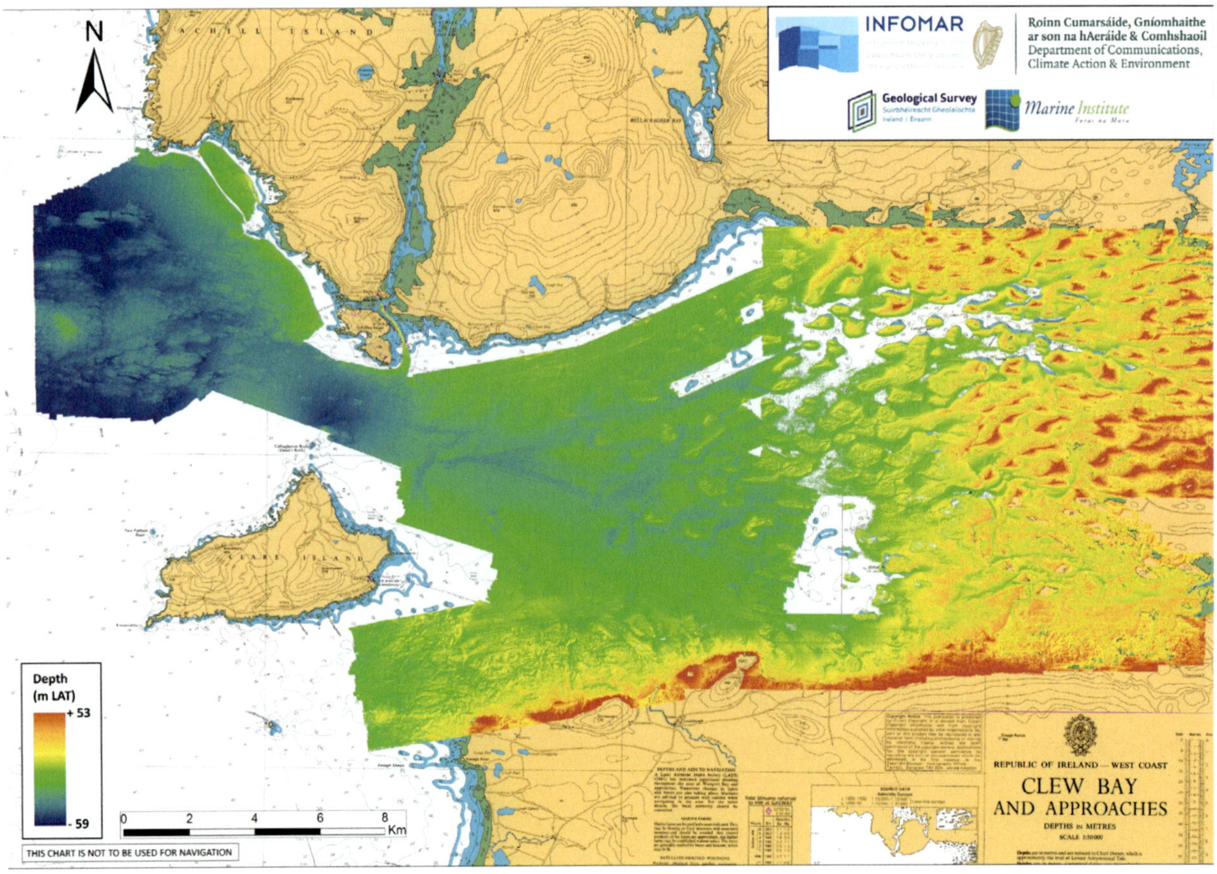

Fig. 6.4 This chart displays seabed bathymetry (collected using marine acoustic surveys) merged with land topography collected with airborne LIDAR. The rainbow colour scale covers heights ranging from 59 metres below to 53 metres above Lowest Astronomical Tide (LAT). INFOMAR is a joint venture between Geological Survey Ireland and the Marine Institute, funded by the Department of Communications, Climate Action and Environment. The bathymetric coverage shown here is a combination of data collected by INFOMAR, University College Cork and ICRAG (the Irish Centre for Research in Applied Geosciences). Background imagery is Chart No. 2667 provided by the United Kingdom Hydrographic Office.

crack is in plain view. Painting the colours of the geological timescale on this naked surface must however wait until we can take them in hand, and subject them to our geoforensic questioning.

The INFOMAR programme (INtegrated Mapping FOr the Sustainable Development of Ireland's MArine Resources) is a joint venture between the GSI and the Marine Institute. It is the successor of the Irish National Seabed Survey (INSS), and its task is to map the seabed down to the 200m contour approximately. Data for Clare Island from INFOMAR programme is not expected until some time around 2020.

Seaweeds

As trees are to a forest, so the larger seaweeds are to the flora of the ocean. There are two major types of these seaweed forests: those of the shore area between the tides (the *littoral*) and those below that, from which the sea never retreats (the *sublittoral*). And just as in the forest on land trees provide shelter and accommodation for a great diversity of plants of lesser stature and the bewildering menagerie of animal life that depends on them, so it is in these forests of the sea. The seaweeds that dominate the forests of the shore are the wracks or fucoids; those that dominate the sublittoral are the kelps, which extend from the sublittoral fringe into the subtidal to depths of about 32m. But whereas in forests on land the routine of life from one day to the next beats to the pulse of the rotating earth, here it beats to the pull of the moon controlling the ebb and flow of the tide.

The number of seaweed species in the seas around Clare Island outnumbers land plants, but because they lack the colourful flowers our eyes respond to, this diversity is less easy to appreciate on first acquaintance. Moreover, the forests in which they grow are largely hidden from view, and we can seldom see them properly in their watery element: what the air is to a forest of trees, the sea is to the forests of kelp and wrack, which are ancient beyond any land forest. They were here when nothing lived above the edge of the sea. It was out from under their shelter that the earliest land plants made their first timid excursions 430 million years ago, when the rivers that left us the sandstones and mudrock formations of Knockmore flowed from bare mountains across arid plains.

The forest between the tides

The wracks look very like each other when all you see of them is a floppy blanket on the rocks, but a closer look will show that there are several species, each of which dominates a particular band of the intertidal zone, to whose characteristics it has precisely adapted itself.

And though it looks a chaotic mess when the sea abandons it, the fronds lift themselves up with the returning tide, and the individual plants space themselves out so that there is plenty of room for all the creatures that live here to crawl and swim

Fig. 6.5a Serrated wrack (*Fucus serratus*) (Michael Guiry).

Fig. 6.5b Knotted wrack (*Ascophyllum nodosum*) (Michael Guiry).

Fig. 6.5c Carrageen Moss (*Chondrus crispus*) (Michael Guiry).

as they go about the business of their lives. How different these tangles of wrack look when the rising tide lifts them up, and that limp and slippery tangle is transformed into a swaying submarine forest, with all the room in the world for the astonishing variety of life that lives here, all of it lying low as it listens for the whisper of the returning tide. Only then is it possible to distinguish the plants of the understory in this aquatic woodland. The role played on land by herbs and shrubs is taken by a great variety of lesser seaweeds. The role of lichens is filled by encrusting algae, which cover the rocks with vivid patches of green, red and pink. The ground floor is often a lawn of Irish or Carrageen Moss (*Chondrus crispus*), which grows in a dense and tangled turf below the zone of the rockweeds and mussels, the labyrinthine spaces between its branchy fronds and stems forming a city in which live creatures of bewildering variety: brittle stars and sea urchins, baby starfish, tube-dwelling amphipods, nudibranchs and moss animals. There are also abundant epiphytes and epizoites: frilly red clumps of *Vertebrata* and *Polysiphonia* on the trunks, and an entire miniworld of creatures that live on the leaf blades.

Fig. 6.5d *Vertebrata lanosa* (Michael Guiry).

Fig. 6.5e *Polysiphonia stricta* (Michael Guiry).

The forest below the tides

Below the low-water mark of the spring tides the forests are very different. The trees here are the kelps, in some oceans as imposing as any rain forest on land, with stout stalks, and stems that carry the long, leafy blades tens of metres towards the light. They are anchored against the tug of the surging tide by a tangled holdfast whose gnarled fingers clench the rock from which only the greatest storms can dislodge them, and in the nooks and crannies of which a cornucopia of creatures find shelter. Where these forests are well established fishes move with little hindrance between the dark, silent stems. In the seas around Clare Island immense numbers of creatures of all sorts have been recorded in the miniature cave systems formed by these holdfasts: myriads of copepods and isopods, but most of all the segmented bristle-worms; more than 80 species of polychaetes along with various other kinds of worms, amphipods and other small crustaceans, halicarid mites and nematodes.[9]

Most of the creatures that live in these seaweed jungles are filter feeders, harvesting the immeasurable bounty of the plankton. A few of the great multitude of creatures that feed in this way are large and easy to see (such as clams, barnacles and mussels) but the great majority are much smaller, and many are tiny. If so many filterers are at work, how close to astronomical must be the constellations of tiny creatures on which they feed?

Praeger was extremely fortunate in securing the services of Arthur Cotton of Kew for study of the marine algae.[10] Cotton made several long visits, spending about half his time on the island and the remainder on the mainland or dredging.[11] Nowhere else in the world had the seaweed flora been described in such detail at that time. It remains the most detailed account for Irish shores, and is still a mine of valuable information. Cotton's outstanding fascicle in the Original Survey's proceedings is much more than a check list of species; it constitutes a pioneering contribution to the study of phycosociology (the study of algal communities). The list of marine blue-green algae produced by him in 1912 remains the most thorough survey of the group from any location in these islands.

In Cotton's pioneering account of these communities a 'Rocky Shore Formation' is subdivided into (A) an Exposed Coast Series and (B) a

Fig. 6.5f *Alaria esculenta* (Michael Guiry).

Fig. 6.5g *Laminaria digitata* (Michael Guiry).

Fig. 6.5h *Saccharina latissima* (Michael Guiry).

Sheltered Coast Series, each of which is in turn subdivided into (a) a Littoral Region and (b) a Sublittoral Region, each made up of a suite of different 'Associations'. Each of these is then discussed in considerable detail.

One of the most prominent communities is the Fucaceae Association of Exposed Coasts. On most rocky shores, wracks occur in characteristic horizontal bands or zones (from *Pelvetia caniculata*, channeled wrack, at the upper end, to *Fucus serratus*, toothed wrack, at the lower). The 'standard' succession from the upper limit of high tide down to the lowest limit of low tide is as follows:

Pelvetia caniculata (Channeled wrack)
Fucus spiralis and *Fucus guiryi* (Spiral wrack)
Ascophyllum nodosum (Knotted wrack)
Fucus vesiculosus (Bladder wrack)
Fucus serratus (Serrated wrack)
Fucus ceranoides (Horned or Estuary wrack)
(admixture of fresh water necessary)

The degree of shelter versus exposure has an important influence on the pattern of zonation. Under the fully exposed situations so prevalent on Clare Island—which is more or less along the entire south shore—only three species occur: *Pelvetia*, *Fucus spiralis* and *Fucus vesiculosus*.[12] The orange-yellow fruiting bodies make the *Pelvetia* band easy to identify in summer and autumn. On more moderately exposed coasts (on the slopes at Kinnacorra near the Castle and at several places on the south shore for example) *Fucus serratus* appears in the association above a zone where *Himanthalia elongata* (thongweed) is prominent, forming a broad band extending down to the *Laminaria* belt. *Ascophyllum nodosum* requires moderate shelter, and is only seen in one location, at Ooghbeg on the southern shore.[13] The wrack association harbours a rich assemblage of lesser algae, as a forest on land its shrubs and herbs, and also a characteristic suite of epiphytes.[14]

Subtidal waters are dominated by the *Laminaria* association, which forms a splendidly developed girdle round the island. This association also exhibits a characteristic zonation (most dramatically seen on vertical shores) but because the association is always under water this is not so evident. *Alaria esculenta* (bladderlocks) and *Laminaria digitata* (oarweed) are uppermost and often visible at low tide, followed by *Saccorhiza polyschides* and *Saccharina latissima* (formerly *Laminaria saccharina*: sugar kelp), which can usually be seen in calm weather. This last grows lower down, but is hardly ever visible. Like the wracks, these greater seaweeds are covered with a great diversity of smaller algae and host a rich epifauna.[15]

Cotton compiled a list of 437 species and 36 varieties—just 12–15 short of the entire algal flora of Ireland at that time. This included 92 species and 15 varieties new to Ireland, 18 species and 4 varieties new to Britain, and 5 species new to science. Advances in taxonomy reduce this total to 222 species of true algae (i.e. excluding the 50 or so cyanophytes in the list), 66 new to Ireland, 13 to Britain. Only one of the three new species of marine algae described in the Original Survey have survived taxonomic revision. Cotton recognised the same mixture of northern and southern forms that characterise so many other groups.[16]

Comparable time and effort were devoted to littoral, sublittoral and benthic algae for the New Survey of Clare Island by Fabio Rindi, Michael Guiry and Christine Maggs of NUI Galway and Queen's University Belfast, who visited the island on numerous occasions between 1990 and 2001, collecting mainly on the more accessible north-east, east and southern shores where the algal assemblages are best developed. The New Survey recorded 208 taxa belonging to 204 species, compared to Cotton's total of 231 belonging to 222 species, giving a combined total of 295 taxa belonging to 285 species. Of the 35 species newly recorded, one is new to science[17] and another new to Britain and Ireland.[18] It should of course be remembered that the marine biologists of a century ago were without the benefit of SCUBA and had to rely on dredging![19]

In spite of this apparent richness, the algal flora of Clare Island would be considered comparatively depauperate and of low diversity, but for an island of its size a count of 285 species of true algae out of an Irish total of 580 (including cyanophytes) is impressive.[20]

The differences between the checklists of the two surveys are probably not significant in terms of environmental change. Most of the associations described by Cotton can still be recognised easily, and for a few species the details recorded by him allow a comparison between then and now. The Galway team were particularly struck by the ecological fidelity of knotted wrack (*Ascophyllum nodosum*) which is widely distributed on the shores of the mainland but in Cotton's day only occurred at Portnakilly on Clare Island ('Oobeg on the south shore'): exactly where it still is today.[21]

As part of their study for the New Survey, Fabio Rindi and Michael Guiry looked at patterns of variation in the distribution of bladder wrack and its epiphytes on the shores of Clare Island, focusing on three sites: Portlea, Kinnacorra and Portnakilly.[22] Bladder wrack hosts a distinctive assemblage of lesser epiphytic seaweeds, which changes over the course of the year and from place to place, but is much more varied and abundant in spring and summer than in autumn and winter. The species that contribute most biomass to this assemblage are *Elachista fucicola, Vertebrata lanosa, Porphyra umbicalis* (Laver), *Pyropia* (formerly *Porphyra*) *leucosticta, Spongonema tomentosum* and *Ulva compressa*. The commonest species is generally *Elachista fucicola*, which is especially abundant in the summer months, and grows mainly on the upper parts of the wrack.

Portnakilly differs from the other two sites mainly because *Porphyra umbicalis* and *Ulva compressa* are abundant there in spring and summer, but absent or rare at the other locations. The two species are generally more or less absent everywhere over the winter months—indeed, in mid-winter most plants are nearly devoid of epiphytes. On the other hand, *Polysiphonia lanosa* and *Pyropia leucosticta* are common at Portlea and Kinnacorra but absent from Portnakilly. The dominant epiphyte at Portlea is *Vertebrata lanosa*, which is rare at Kinnacorra, where the dominant species is *Elachista fucicola*. These differences in the species composition of the assemblages are thought to depend to a large extent on the degree of exposure at the different sites. The distribution of *Vertebrata lanosa* seems to be more site-dependent than other species (it does not occur at very exposed sites), and shows little seasonal variation.

A patchiness in the extent of epiphyte cover on bladder wrack (on a scale of metres to tens of metres) is typical of these communities. The spatial differences—plants with a heavy epiphyte cover can be found very close to others that have almost none—are thought to be due to variation in the availability and dispersal of propagules.

The marine fauna off Clare Island

> If we go deeper into the subject of the sea, and the creatures that dwell in it, a romance unfolds that leaves the mermaids and sea-serpents and rotting hulks forgotten and unwept.
>
> Praeger, *Beyond Soundings* (1930), 45

Some 25 of the 68 fascicles that comprise the Original Clare Island Survey dealt with groups of plants or animals that are largely or exclusively marine-based. The New Survey has been more ecological than taxonomic in its treatment of the waters around the island. Partly for this reason many of the groups of animals examined by Praeger and his team do not feature in the New Survey. The other reason was the sudden death in 2004 of Brendan Keegan of NUI Galway, who had outlined an extensive programme of research in sublittoral ecology when the New Survey was being planned in 1990, and for whom no successor was appointed.

In the pages that follow the animals are arranged roughly in order of increasing evolutionary 'complexity' in accord with standard taxonomic practice, while emphasising that a less 'advanced' form is no barrier to diversity, beauty or complexity.[23]

Phylum Granuloreticulosa
Class Foraminifera

Foraminiferans are tiny creatures that live in sand and silt at the bottom of the sea, or attached to rocks or algae.[24] Two families are important constituents of the marine plankton. They range in size from 10 microns to several centimetres. Each little zooid constructs a cluster of spherical chambers around itself—partially fused, with internal partitions—using tiny particles of lime abstracted from seawater, and bound neatly together with its own organic cement. When the occupant dies these are chalky white, but during life are often brightly coloured: red, orange, yellow or brown.

The shell is perforated with tiny holes (the *foramina* that give the group its name) through which protoplasmic threads called filopods emerge. These interweave to form a net which the creature uses to trap the organisms on which it feeds: algae, ciliates, radiolarians, nematodes and small crustacean larvae. They are also used to gather material for

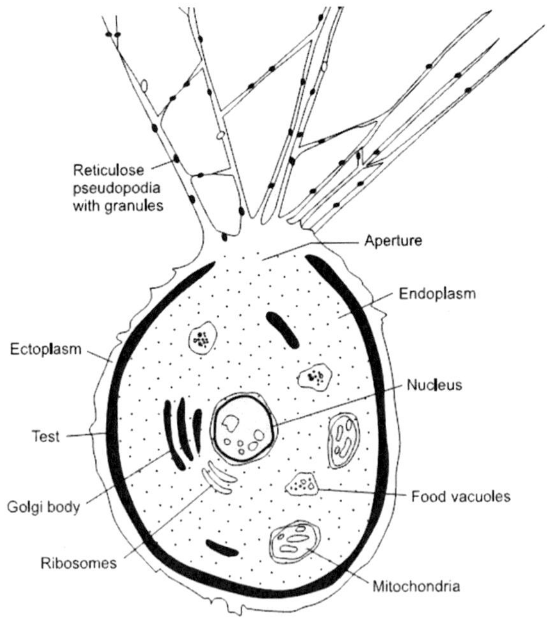

Fig. 6.6a Foraminiferal structure.

the shell, and to swim. They have a remarkably complex life history involving an alternation of haploid and diploid generations and a unique form of nuclear homosexuality. Some shallow water forms may harbour photosynthetic symbionts.

Forams (as they are affectionately known) are extremely important constituents of marine sands and clays, not only today but in the geological past. They first appeared in the Cambrian, but have been really important since the Triassic. No fewer than 40,000 species are known as fossils. The nummulitic limestone of the Egyptian pyramids is largely made up of forams. They are also useful as indicators of changing conditions in the world's oceans. The results from the Original Survey supported a growing awareness at the time that many southern species earlier regarded as almost subtropical were gradually extending their range into the North Atlantic and the North Sea by way of the west coast of Ireland.[25]

Edward Heron-Allen and Arthur Earland contributed the fasicle on the Foraminifera for the Original Survey. They collected samples on Clare Island in August of 1911, collected and examined material from the shores of the mainland and also dredged from the wider Clew Bay area. The beaches of the island and adjacent mainland proved to be ideal collecting ground, and although beaches are not extensive on this rocky coastline, 'in these few bays where shore sand accumulates, it stretches in broad, smooth sheets, often bounded at one or both ends by alga-covered rocks from which the tide recedes gently and slowly, leaving a rich strewing of shell debris and Foraminifera on the surface of the sand'.[26] Which means we can share in, and make our own, an early experience of D'Arcy Wentworth Thompson:

> In days gone by I used to see the beach of a little Connemara bay bestrewn with millions upon millions of foraminiferal shells, simple *Lagenae*, less simple *Nodosariae*, more complex *Rotiliae*: all drifted by wave and gentle current from their sea-cradle to their sandy grave: all lying bleached and dead: one more delicate than another, but all (or vast multitudes of them) perfect and unbroken.[27]

Material was collected from 37 stations ranging from Achill to Inishbofin, each of which was described and a summary of its foram fauna given. Subsequent examination of the material 'occupied us without intermission for some eighteen months'.[28] This dedicated work resulted in the meticulous catalogue that constitutes the largest volume of the Original Survey: 188 pages, accompanied by thirteen splendid plates and two foldout maps or charts (Figure 6.6b). Of the total of 299 forms recorded 14 were considered to be new species, 32 new to both Britain and Ireland, including even species that had hitherto only been known as fossils. One example was the discovery of *Bolivia gramen* on the beach at Clare Island:

> The discovery of this beautiful little form at Clare Island is one of the most noticeable features of our report ... originally described as recent from Cuba, and subsequently as a Miocene fossil from Vienna. So far as we know there are no subsequent records of the species in the recent condition.[29]

Almost all stations yielded long and interesting species lists. Fifty or sixty of these occurred at most stations, but nearly every station had something of special interest peculiar to itself. Two species of *Globigerina* were among the most abundant and universally distributed species. These are of particular interest in that they belong to a family of pelagic forams, an important constituent of the plankton of the open sea, often occurring in

(b)

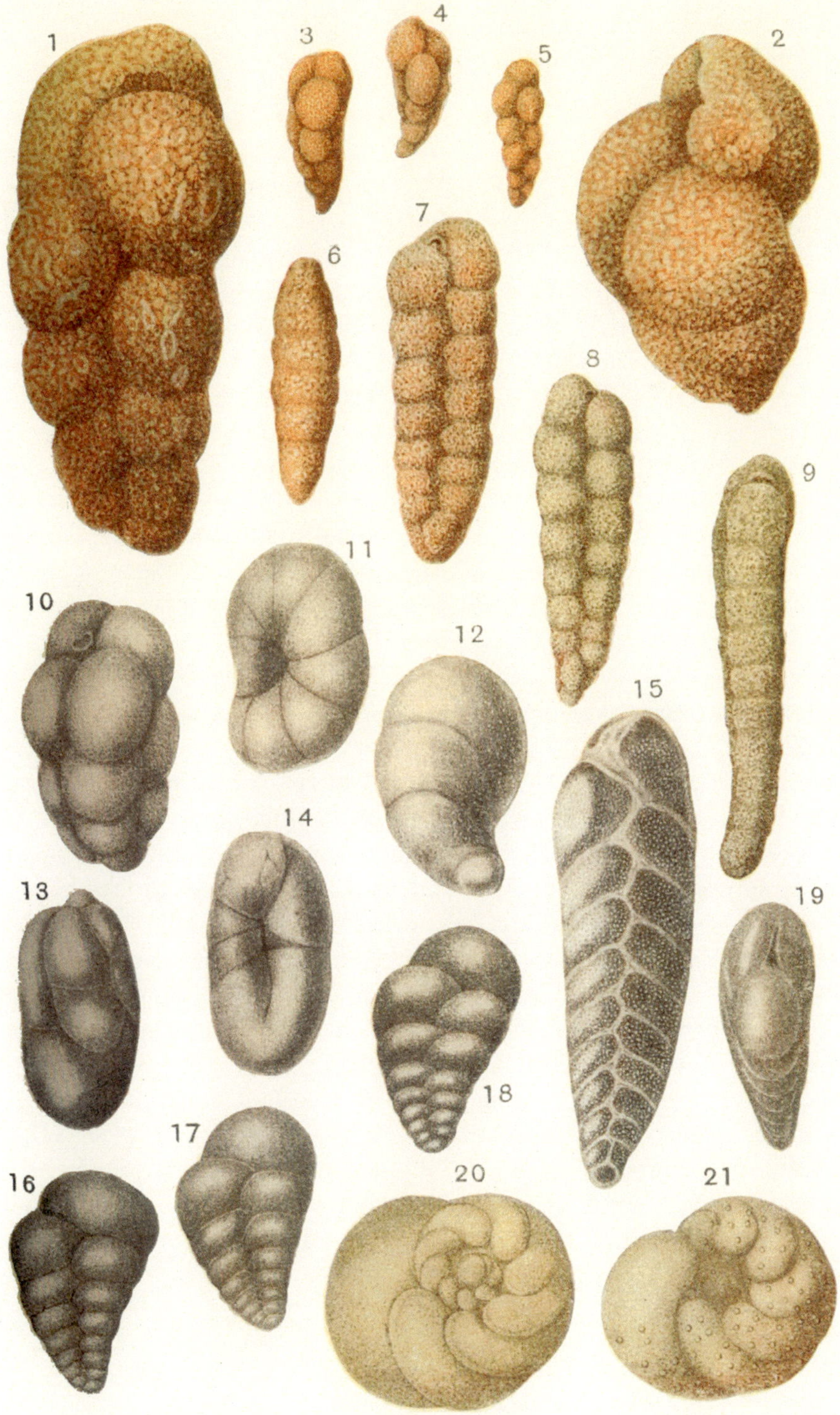

(c)

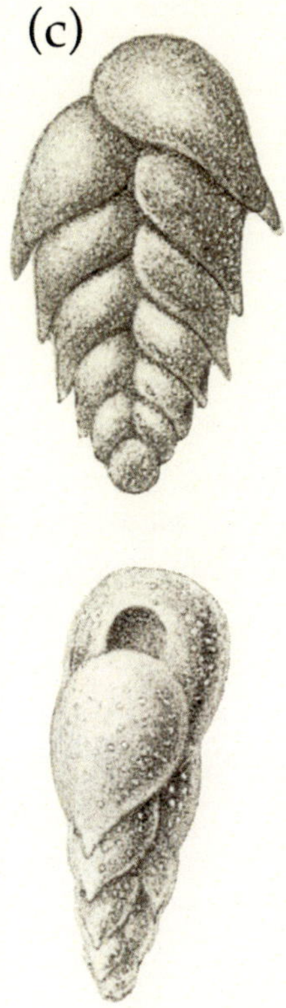

Fig. 6.6b Edward Heron-Allen and Arthur Earland's Plate IV (Clare Island Survey Part 64, 1913).

1–5.	*Verneuilina polystropha* Reuss sp., x120
6.	*Clavulina obscura* Chaster., x75
7–9.	*Gaudryina filiformis* Berthelin., x75
10.	*Verneuilina pygmaea* Egger sp., x120
11–12.	*Bulimina minutissima* Wright, x120
13–14.	*Bulimina subteres* Brady, x120
15.	*Bolivina aenariensis* Costa sp., x75
16–19.	*Bolivina inflata* sp. nov., x200
20–21.	*Pulvinulina punctulata* d'Orbigny sp., x120

Fig. 6.6c *Bolivia gramen*, x120

uncountable zillions. *Globigerina* ooze covers some fifty million square miles of the ocean floor today, and has been even more widespread in the past: forams are among the most important constituents of the Cretaceous chalk for instance.[30]

Two gatherings of sand were made on Clare Island: one from the beach, either as scrapings from the surface or 'light elutriated material, baled from wave-ripples as they broke upon the sand'.[31] A second gathering was made among the coralline algae in rock pools at the Strand. Most of the species here were forms that lived attached to the coralline algae. The beach on Clare Island proved to be of particular interest because it contains material derived from the rough bottom of the deeper water that extends out to the west, which is not easily dredged.[32] A striking find was a shell of *Discorbina mediterranensis* (one of the most abundant and characteristic species throughout the area) 'which was found to be full of young shells, already composed of two to three chamberlets', an important insight into one stage of the complex life cycle of these creatures, which at that time was only beginning to be understood.[33]

Phylum Porifera (sponges)

Sponges are the simplest multicellular animals, consisting of a network of linked individual cells embedded in a framework of silica or lime. Many are rather formless in appearance: just patches or encrustations on rock, but others take the form of fantastic fans, cups and tubes. A sponge may not appear to have a great deal in the way of anatomical complexity, but it does have an elegant arrangement for the circulation of water from which the individual cells extract oxygen and the miniscule creatures on which they subsist. The water is wafted into the body of the sponge by way of numerous openings through the co-ordinated beating of whip-like flagella on the individual cells, and expelled through a large hole (conspicuous in some species) called an osculum.

All are beautiful, diverse, and (it is generally believed) one group is ultimately ancestral to all other Eumetazoa ('higher' animals). Many sponges have symbiotic algae, which are partly responsible for the fantastic colours they often display. They are one of those groups which have been at home in the sea since before ever there was life on dry land.

The degree of exposure has a lot to do with their shape. In exposed situations, they tend to be flattened or rounded, but in the calmer waters of a rock pool the same species often have a more plant-like form with branching stems. This is one of the reasons they are difficult to name, and also because identification depends to a considerable degree on the shape and arrangement of their splinter-like spicules (which is a task for the microscope). For example, the *Haliclona simulans* that forms an encrusting network on rocks along the shore develops branching masses offshore. Their colour is also very variable. *Clathrina coriacea*, for example, is pure white and sulphur yellow in the sea cave near the lighthouse on Clare Island (see below), but when it grows in crevices between the Lithophyllum in rock pools it is a beautiful vermilion in colour. The number of species worldwide is thought to be somewhere between 5,000 and 10,000, of which only 150 occur in freshwater (Figure 6.7).

Jane Stephens collected marine sponges on Clare Island for the Original Survey in 1909–11, and also studied material from the wider Clew Bay area. Although many sponges live between tide

Fig. 6.7a *Clathrina coriacea* (Parent Géry (CC BY-SA 3.0)).

Fig. 6.7b *Myxilla incrustans* (Cwmhiraeth (CC BY-SA 3.0)).

marks on the island's shores, especially hiding away among the Lithophyllum of the rock pools (including several rare species), the shore fauna proved to be rather poorly developed in comparison with the more sheltered locations of the mainland. One of several things that struck her about the sponge fauna on Clare Island was the way sponges with siliceous skeletons are poorly represented compared with mainland locations where the bedrock is limestone. Calcareous sponges, on the other hand, are well represented on the acidic rocks of the island.

Conditions are more sheltered on the eastern side of the island, and Stephens was almost lyrical in her description of a particular cave near the lighthouse (Figure 6.8):

> The greatest contrast on Clare Island to these exposed situations was afforded by the "Seal Cave" to the north of the Light-house Cave on the north-eastern side of the island. In this cave the sponges grew in far greater luxuriance than at any other part of the island. The upper end of the cave was free from sea-weed, and there its walls were lined by a broad band of sponges, extending to about six feet beneath low-water mark. This band consisted chiefly of yellow *Halichondria panicea*, with very large patches of pure white *Clathrina coriacea*. In one place masses of *Esperiopsis fucorum*, of a rich orange-red colour, covered many square feet of rock. The pools at the upper end of the cave, which could only be reached at low water on a calm day, were crowded with sponges, sea-anemones, and hydroids, the bright colours of which contrasted with the pale purple background of incrusting seaweed which covered the surface of the rocks. In these pools *Clathrina coriacea*, pure white and sulphur yellow in colour, occurred in numerous patches, with many small specimens of *Sycon raphanus* and *Leucandra Johnstoni*, and a number of very fine specimens of *Leocandra Gossei*. With them grew large specimens of *Myxilla incrustans*, pale yellow in colour, *Esperiopsis fucorum* of a bright orange-red, and *Stylostichon plumosum* varying in colour from a livid red to pale orange-red. *Hymeniacidon caruncula*, *Ophlitaspongia seriata*, *Spongelia fragilis*, and other common species also occurred in these pools, while large specimens of *Pachymatisma johnstonia* grew under an underhanging ledge of rock.[34]

Stephens recorded 64 species, two of which were described as new to science (one of them, *Leucandra cliarensis*, named for the island); twenty were new species for Ireland, and seven of these had not up to that time been recorded for Britain either. *Lecosolenia* species occur as small, variously coloured patches (sometimes compared to minute bunches of bananas). *Leucosolenia variabilis* is abundant off Clare Island and between tides, often in very exposed situations. *Sycon* species resemble tiny vases, growing up to 5cm tall in clusters, and live under stones, in pools and hidden among the seaweed. Several species occur on the island. *Sycon ciliatum* (up to 3cm tall) is common between tide marks. It is creamy yellow in colour, grows in clumps and has a shaggy appearance (rub the surface gently and the spicules will come off on your fingers). A crown of longer, stiff spicules surrounds the osculum. *Sycon raphanus* (up to 2cm tall) looks like a petrified hedgehog, brownish in colour, usually growing on seaweed. It is not uncommon in rock pools under the shelter of Lithophyllum. *Sycon compressum* occurs between tide marks, often in exposed situations. *Leucoselenia* and *Sycon* are calcareous sponges.

Encrusting sponges form lumps and patches on rocks. *Halichondria panicea*, the breadcrumb sponge, is the most widespread species and grows nearer to high water mark than any other sponge. It grows in thick-piled carpets, and is dotted all over with what look like a myriad little volcanoes 2cm thick in patches approximately 20cm across, in varying colours up to bright green in the spectrum. Not too far below it on the shore you can expect to find species of *Hymeniascidion* (*sanguinea* and *caruncula*) and *Ophlitaspongia seriata*.

Oscarella lobularis looks like a coalescence of huddled brownish lobes, an osculum visible at the tip of each lobe (0.5cm thick in patches up to *c.*10cm across). It is common between tide marks, and especially abundant under the large boulders between Kinnacorra and the harbour, and among Lithophyllum in rock pools. *Pilochrota lactea* is fairly common in rock pools and remains invisible 'until masses of the Lithophyllum are broken up, when it is found growing in rather thin, white patches in the numerous crevices of the calcareous sea-weed'.[35]

Cliona celata is one of the boring sponges (i.e. boring as in excavating) that live in tiny holes excavated in soft rocks or shells. It has not been seen on the island, but is common at extreme low

Fig. 6.8 The seal cave.

tide on the islands in Clew Bay and around Westport Bay, where it bores into the blocks of limestone strewn on the shore. *Ophlitaspongia seriata* is a thick, encrusting sponge with regular spaced oscula, brick red in colour, and extremely common, even in exposed situations.

Phylum Cnidaria (Coelenterata)

Few animals are more familiarly associated with the life of the seashore than the sea anemones which turn the rock pools into gardens aflame with their brilliant hues. A sea anemone is in fact a colony made up of numerous individuals known as polyps, but they are organised in a much more complex way than the individuals in a sponge. One of their most astonishing and distinguishing features is the specialised structures called nematocysts, which harpoon prey, at the same time injecting a paralysing toxin. Other nematocysts act as knives, others again as threads that wrap around spines or other projecting surfaces.

Sea anemones are one of the three main classes in the phylum Cnidaria, in which there are some 10,000 species altogether. There are two stages in the life cycle: the attached flower-like *hydroid* stage exemplified by sea anemones, and the free-swimming *medusa* stage exemplified by jellyfishes. One or other stage comes to be dominant in the three different classes of the Cnidaria. In the Scyphozoa (jellyfish) the swimming umbrella of the medusa, which has batteries of nematocysts decorating its floating ribbons, is dominant; the polyp stage plays only a minor role in the life cycle. In the Hydrozoa the adult colony produces reproductive polyps, which give rise to free-swimming medusae. These float off into the sea and eventually produce eggs and sperm, which are shed into the water where fertilisation takes place. The resulting eggs hatch to give baby hydroids that settle down to renew the cycle. In the sea anemones (class Anthozoa) the attached muscle-tube of the hydroid stage with its floral corolla of upright tentacles is dominant; the medusa stage has been abandoned.

Jane Stephens studied the cnidarians of Clare Island for the Original Survey, recording a total of sixty species along the shorelines between the tides, and in offshore waters down to thirty fathoms for the entire area between Blacksod Bay and Ballynakill Harbour. She recorded around half of these from Clare Island itself, either on the shores or in deeper water immediately offshore.[36]

A textbook account of the Cnidaria will usually begin with the Hydrozoa, then look at the Scyphozoa, and finish with the Anthozoa. But it is more useful perhaps to start with the larger kinds you are more likely to see first: the sea anemones, where the individual animal is a single, large hydroid colony.

Class Anthozoa

Over a thousand different species of sea anemones have been described in the world's oceans. They grow in great profusion in the rock pools on the island, but the number of species is relatively small. The most characteristic include the beadlet anemone (*Actinia equina*), the snakelocks anemone (*Anemonia sulcata*), *Sagartia elegans*, *Urticina fellina*, *Cereus pedunculatus*, and the tiny, brilliantly coloured jewel anemone *Corynactis viridis* ('very abundant in the "Seal Cave," north of Portlea, where it covered one of the shallow pools at the end of the cave'.) (Figure 6.9).[37] *Corynactis* is a small anemone—usually grass-green in colour although they also come in other colours—that grows in crevices and the like, often a crowd of them together. Each has 100 or so translucent tentacles with dark tips, so that you can hardly see them as individuals when they are all huddled together. *Sagartia* is another small sea anemone.

Actinia is a little anemone, up to 3cm tall and comes in different colours: red, brown, orange or green. It lives on the mid to lower shore, and when the tide is out it shrinks to a rounded blob of stiff jelly (which is therefore all you usually see). When the tide returns it unfurls its 200 tentacles, which are arranged in six circles. It gets its name from a circle of 24 blue spots round the inside of the top margin. *Anemonia* lives lower down on the shore

Fig. 6.9a Snakelocks anemone (*Anemonia sulcata*) (Linél (CC BY-SA 3.0)).

Fig. 6.9b Jewel anemone (*Corynactis viridis*) (© Timur Kholodenko).

Fig. 6.9c Polyps of *Alcyonium digitatum* (Bernard Kaye).

Fig. 6.9d Devonshire cup coral (*Caryophillia smithii*) (Parent Géry).

and offshore to 23m or so. It is somewhat larger than *Actinia*, brown, grey or green in colour, up to 10cm tall with 170 longer, wavy tentacles, arranged in six circles.

Sagartia elegans comes in several distinct colour forms, treated by Jane Stephens as three distinct species (*S. venusta, S. miniata, S. nivea*). It has a column up to 6cm tall, with up to 200 tentacles and ejects white threads when disturbed. *Anemonia sulcata* and *Sagartia venusta* in particular often grow to a large size in the pools. *Cereus pedunculatus* is the daisy anemone. It is trumpet-shaped, generally orange to buff grey, with 750 short brown tentacles. Other anemones that are not uncommon include *Cylista undata* (along the eastern and southern shores) and *Cribrina verrucosa* in pools near the harbour and elsewhere.

Class Alcyonacea – Soft Corals
Alcyonium digitatum is widely distributed between tide-marks but not common (it is more frequent in deeper water). It forms white spiky cactus-like colonies up to 20cm high.

Class Ceriantharia
Cerianthus Lloydi has been taken offshore of the lighthouse, but only in small numbers (it is a mud-burrower).

Class Madreporaria – True Corals (hard calcareous skeleton)
Jane Stephens recorded two large specimens of *Caryophillia smithii* (Devonshire cup coral) at extreme low water in the Seal Cave. This is a tiny white or pink cup-coral that lives in a limy goblet 1–2cm high and the same across the top; it has 50 or so tentacles and is usually found in overhangs or deep cracks.

Class Scyphozoa
The stalked jellyfish (family Lucernariidae) are a very distinct family with several representatives on the east shore, but none of them is common. They are, in essence, upside-down jellyfish that attach themselves temporarily to a substrate. *Haliclystus auricula* is an elegant pink, tiny jellyfish that lives attached by a basal sucker to seaweeds on the lower shore. The eight radii of its inverted umbrella have clusters of tentacles at their tips, and wart-like anchor structures. *Lucernaria campanulata* is rather similar, usually cherry red though it sometimes wears pale green, and lacking the wart-like anchors. *Depastrum cyathiforme* has a horseshoe-shaped 'holdfast'.

Subclass Discomedusae[38]
Jellyfishes are of course pelagic creatures, but three species are washed ashore, often in considerable numbers: *Pelagia noctiluca (perla), Chrysaora isosceles* and *Aurelia aurita*.

Fig. 6.10 *Tubularia indivisa* (Bernard Picton (CC BY-SA 4.0)).

> On one occasion jelly-fish, belonging most probably to this species [*Aurelia aurita*] were seen in immense numbers off the north-west coast of Clare Island. They formed dense shoals several hundred feet in length, and were clearly seen from the cliffs, from a point about 600 feet above the water.[39]

Class Hydrozoa

Most hydrozoans are very much smaller than sea anemones—often indeed, they are diminutive—consisting of much smaller hydroid polyps that are usually linked together in branching colonies. *Hydra* itself is unusual in that each individual is a single polyp.[40] Many of the littoral hydroids have in some measure freed themselves from dependence on the sea because the medusoid generation is reduced (*Tubularia*) or absent (*Sertularia*), but the two generations persist in many of the hydroids of the sublittoral zone. *Laminaria* often has beautiful growths of hydroids.

The Hydrozoa are divided into two main orders, in one of which the horny skin that covers the body (known as the perisarc) extends up to form a protective cup (theca) around the polyp. These are the *thecate* Hydrozoa. In the other *athecate* order this theca is absent. The shoreline hydrozoan fauna is rather poor, although a small number of species are abundant.

Athecata

Coryne grows as little rose-coloured thickets that rise up from what look like creeping 'roots' (typically a few cms high, but it can be up to 15cm). It is found mainly in deep pools on the eastern and southern shores, where Jane Stephens recorded two species (*C. pusilla* and *C. vaginata*).

Tubularia has two circlets of slender tentacles, like filiform petals. It occurs in thousands where the water is calm and undisturbed. It forms dense colonies on rocks on the lower shore, again developing from creeping 'roots', but here the erect stems (up to *c.*18cm tall) are unbranched—though often plaited—tipped with red-pink polyps, with medusoids that look like little bunches of grapes. Stephens recorded *Tubularia indivisa* 'spreading in large sheets over the walls of the "Seal Caves" to the north of Portlea'.[41]

Fig. 6.11 By-the-wind-sailor (*Velella spirans*) (Mark A. Wilson (CC0 1.0)).

Thecata

Sertularia is the most frequently encountered thecate hydrozoan. It forms large branching white or pink colonies of delicate fern-like form, again arising from what look like creeping 'roots'. *Sertularia pumila* (common sea-oak) is a widespread species that grows in moss-like colonies up to 45cm high; it is common between tide marks everywhere, and is especially abundant on *Fucus*. Another species commonly encountered on the eastern shore is *Aglaophenia pluma*, which has tiny flat ferny colonies up to 5cm tall, often found growing on stones, shells and seaweeds.

A third order of the Hydrozoa, the CHONDROPHORA, look something like jellyfish because they have taken to the open sea and adopted a similar pelagic way of life. Nevertheless, they are hydrozoans, the polyps specialised in various ways to provide floats for the colony, for feeding, defence and reproduction, or to act as part of a linking structural framework. They have distinctive sail-like structure known as a siphonophore. By-the-wind-sailor (*Velella spirans*) is found washed up on the south shore of the island in some numbers from time to time (Figure 6.11).

Phylum Platyhelminthes

Class Turbellaria (flatworms)

Flatworms belong, along with flukes and tapeworms, to the phylum Platyhelminthes, which sounds very important, but in Greek means just that: flatworms. Flatworms in this *sensu stricto* belong to the class Turbellaria. The reason they are so flat is because they lack respiratory and circulatory systems, so they need to absorb oxygen through the skin, and have their internal organs to be close enough to the surface for such diffusion to be adequate.

For the Original Survey Roland Southern confined himself to the flatworms, explaining that 'the latter group is of much greater interest than the parasitic Platyhelmia [i.e. flukes and tapeworms] from the faunistic point of view, as the life-history of the latter is so closely involved with that of their hosts'.[42] As usual, he cast his net further afield than the shores of Clare Island to include the wider Clew Bay area and Blacksod Bay, searching the littoral area on the east side of the Mullet Peninsula more thoroughly than anywhere else, and he was mainly concerned with the marine species, which constitute the great majority.

There is a certain air of imperturbability and calm about flatworms. The species found on Clare Island tend to come in rather drab colours that help them blend into the background, but many species are like diminutive birds of paradise (due allowance being made for other differences). Other than that, they seem to eschew (for the most part) the extravagant ornamentation that characterises groups of animals that exploit the potential of their basic blueprint to develop adaptations that better fit them for the particular worlds they inhabit. Not so the flatworms. A flatworm is always unmistakably a flatworm, instantly recognisable as such, their anatomical conservatism presenting the turbellarian taxonomist with the occasional headache. Anyone can draw a flatworm; there are no visible appendages or bristles, no complicated mouthparts, although they are carnivores for the most part. The mouth is on the under side at the end of an eversible pharynx with which it smothers and sucks in its prey.

They seem to have patented their mode of locomotion, because no other creature gets about as they do. They do not walk, hop, jump or fly. Instead, they seem to float or glide, and they appear to have just the one gear. They achieve this rather ghostly progression by the co-ordinated beating of thousands of cilia on their lower surface. Indeed, it is the little whirlpools this sets up that give them their official name of 'Turbellaria' (turbellae = disturbances in Latin).

In their quiet way they hold their own, making their more-than-modest contribution to the tally of living diversity in their watery domain. There are 4,500 species, but many are not terribly well known and they get little attention in field guides. Marine forms tend to be more leaf-like, and can swim with graceful undulating waves. For example, one of our commonest shore-living species is *Leptolplana tremellaris* (1.5–2.5cm, brownish), which hides under stones or in other dark corners of the shore as it awaits the return of the tide. But it is so well camouflaged that you will hardly notice it until it moves, and then it is very hard to detach, so closely is it appressed to its substrate. It can progress serenely across a surface in the inimitable flatworm fashion—even upside down on the surface film—but it can also swim through the water with all the elegance and serenity (and rather in the same way) as a skate.

Southern's list enumerated fifty species, only five of which were freshwater species; 29 were new Irish records at the time and five new to Britain and Ireland. Only fifteen or so of these species are specifically listed for the island itself, nine or so being littoral species recorded from weeds and rock pools on the south-eastern shore, and the five freshwater species from among vegetation in the streams and pools on the island. Among them are *Procerodes ulvae*, 'found under stones, almost up to the high-water mark, and often occurs in immense numbers';[43] and *Convoluta convoluta*, a tiny flatworm that feeds on juvenile settling mussels and harpacticoid copepods (interestingly, its green colour is due to the presence inside of symbiotic diatoms).

Phylum Nemertina (ribbon worms)

Being a 'worm' is obviously a very successful way of living, when you consider the number of groups of animals that have adopted it: all variously worm-like on the outside, but often radically different on the inside. Here, then, is another. Nemertines are worms *par excellence* (if our standard is the almost bristle-free earthworm): they are long, thin and slippery but not segmented, varying in length from a few centimetres upwards, but they have incredible powers of extension and contraction. Five metres is not uncommon in *Lineus longissimus* (the record is 54m). They are carnivorous and very voracious—they will eat almost anything. One of their most distinctive features is a long, thin proboscis that is folded up in a pouch above the mouth. It shoots this out with great force and rapidity to lasso prey, which it then sucks into a mouth that can be distended to deal with larger captives.

In contrast to his somewhat casual approach to the roundworms, Southern paid particular attention to the ribbon worms. He recorded 31 species, two of them new to science, one to 'the Britannic fauna' (by which he meant Britain and Ireland) and a further seven new to Ireland. However, he does not appear to have spent much time around Clare Island itself; most of his littoral collecting was done on the east side of the Mullet Peninsula in Blacksod Bay, and he carried out extensive dredging in Clew Bay and Blacksod Bay. However, we can be sure that most of the species he recorded will be found on the shores and in the offshore waters of the island.

Some 1,150 species of nemertines are known; these are mainly a marine group, although there are some terrestrial and freshwater species.[44] They are often diversely and brilliantly coloured, the most popular hues being white, yellow, green, deep purple and various shades of red and pink. Many are to be found under stones and in the holdfasts of seaweeds. Some of those found under rocks have often tied themselves into such knots you wonder they can disentangle themselves. By far the most eye-catching of these is the dark purple bootlace worm or sea-snake (*Lineus longissimus*) mentioned above. *Lineus ruber* (the red-line worm) is found coiled under stones that rest on sand or gravel; in the upper part of the shore it is usually bright red, but near low water mark green or purple forms are found.[45] It feeds on bristle-worms, which it swallows whole. *Lineus bilineatus* (7–15cm but can be much longer; pinkish-brown with pale cream stripes down the middle) spends the hours when the tide is out under stones or buried in sand or mud.

Among other species likely to be encountered are *Cerebratulus fuscus* (up to 10cm, usually greyish-brown, but very variable, which is very common in clean and muddy sand and gravel; its flat rigid body, with transparent borders, is very characteristic. *Emplectoneema Neesi* (10–15cm, brown above, with lengthwise streaks, off-white below) is one of the most abundant and widely spread species under stones, in rock crevices and in *Laminaria* holdfasts. *Nemertopsis flavida* (3–4cm, flattened, pale cream with pink down the middle) is another very common species in *Laminaria* roots and on the shore. *Oerstedia dorsalis* (*c*.2cm, brown or yellowish, darker stripe down the middle of the back) is very abundant in weeds along the shore.

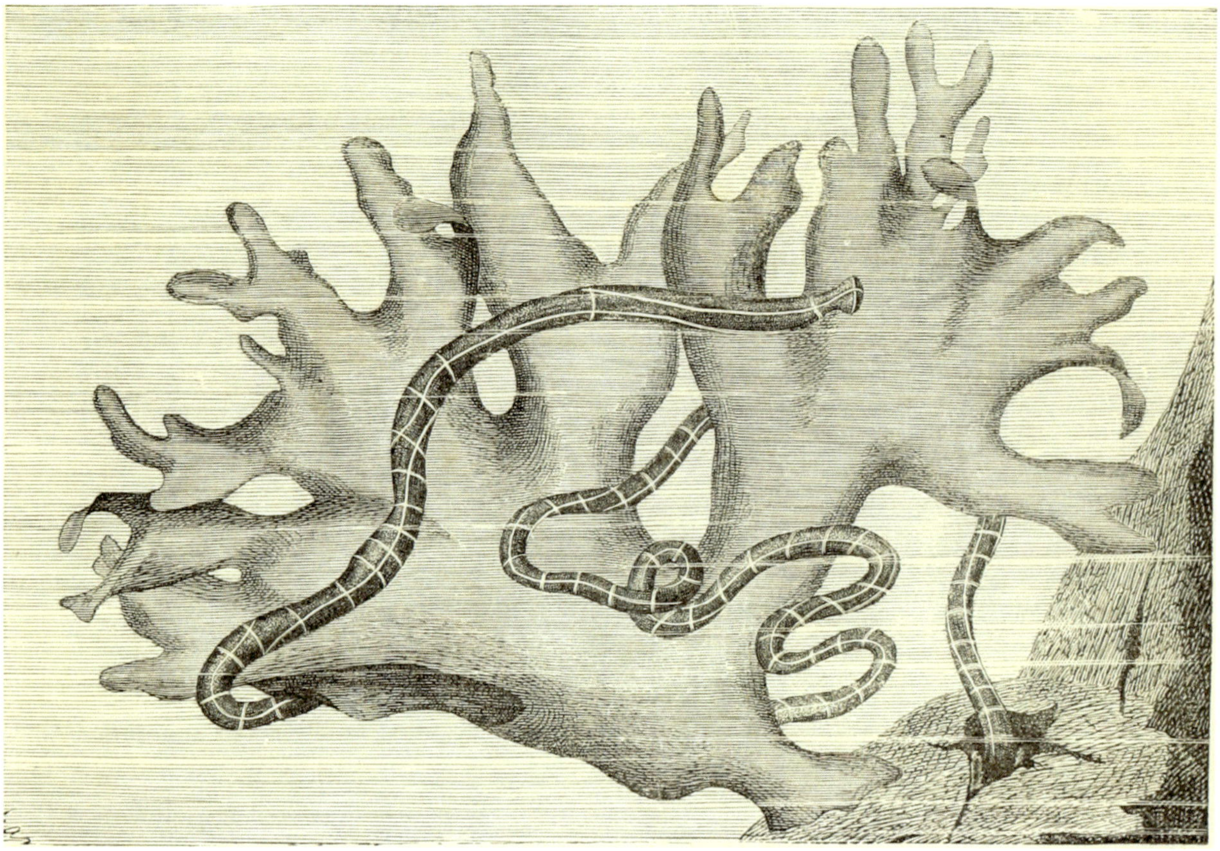

Fig. 6.12 *Tubulanus annulatus* (The Royal Natural History by Richard Lydekker, 1896).

Cephalothrix rufifrons (*c.*10cm, a very stretchy, pale cream-coloured worm) is very common among weeds, under stones and among corallines in the *Lithothamnion* pools, as well as in *Laminaria* holdfasts. *Amphiporus lactifloreus* (3–8cm, white or pink, proboscis as long as body) is also a lower shore species, found under stones and in shingle, and among *Laminaria*. *Prostoma flavidum* (1–2cm, delicate pink or grey) is found in seaweeds, under stones, and often in pools. The most beautiful of these worms perhaps is *Tubulanus annulatus* (15–40cm, red with a white stripe down the middle and white bands), which lives in a tube made of mucus under stones on the lower shore and on sandy flats (Figure 6.12).

Drepanophorus spectabilis (up to 8cm, flattened, red-yellow with five conspicuous longitudinal ivory stripes) is another common species under stones, in rock crevices and among shells. Some species (such as *Micrura purpurea*) often take up residence in empty bivalve shells.

Phylum Nematoda (roundworms)

Roundworms are tiny, and for the most part lack the external ornamentation of other wormy creatures, which makes them particularly difficult to identify. But they are everywhere, and, although invisible to us, they play an enormously important role in the natural economy of the world. Some 28,000 species have been described, and this is thought to be less than 1% of the total number. In a cubic metre of topsoil there are a million roundworms. Many species are free-living inhabitants of soil and water, where they are enormously important in recycling the elements necessary for plant and animal growth, but more than half of the species known are parasites, many with the most extraordinarily complex life cycles. Because they live such hidden lives, there are no field guides to nematodes, although you walk over them in their millions everywhere you go. As much as 90% of all life forms on the ocean floor is comprised of nematodes.

Rowland Southern contributed several studies to the Clare Island Survey, but he collected roundworms only incidentally—they were not 'systematically collected for their own sake, but obtained as a by-product of the search for other groups'. They turned up along with everything else he was collecting, so he did not even have to go looking for them; 'land and fresh-water Nematodes whilst searching for Oligochaeta in

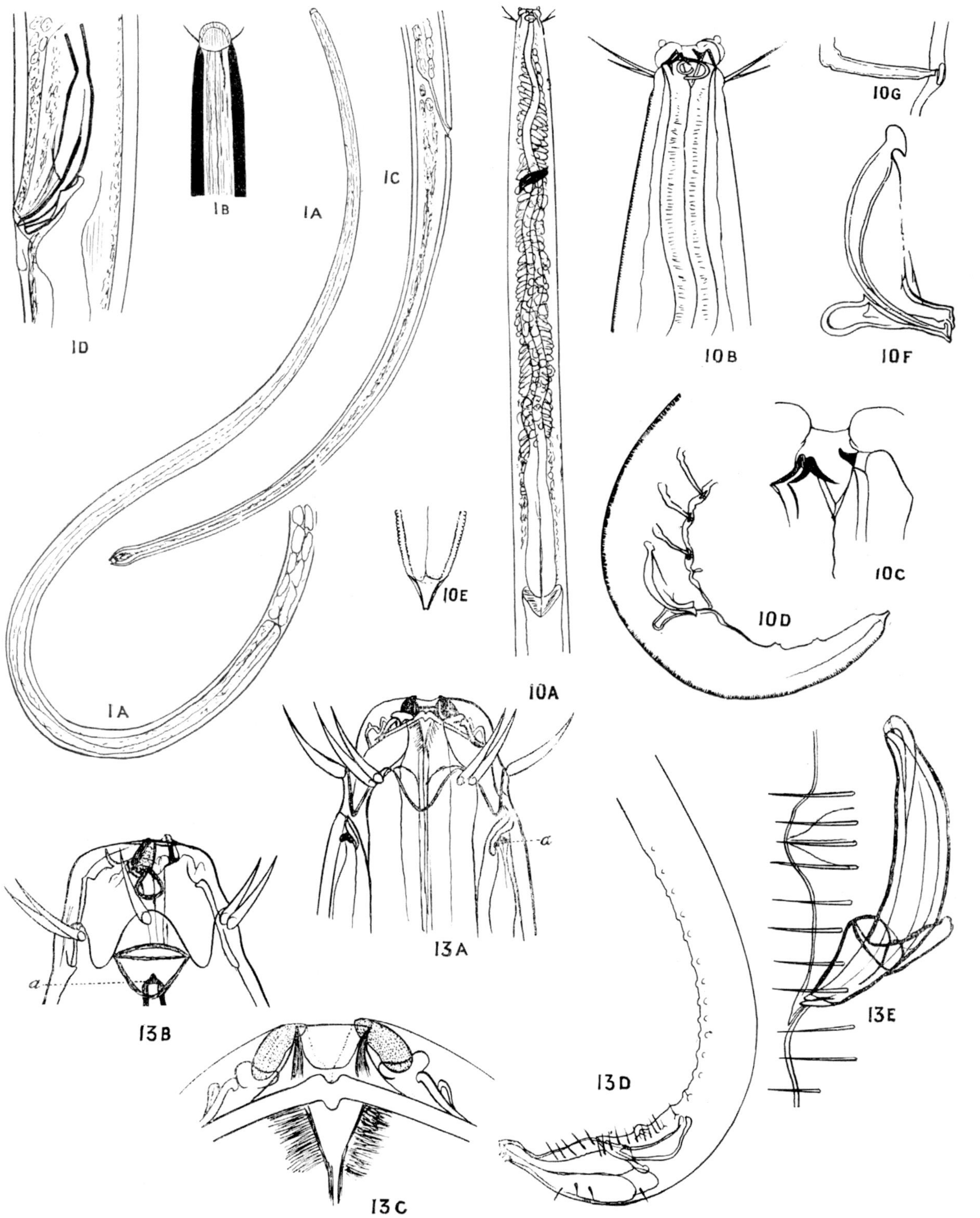

Fig. 6.13a Rowland Southern's nematodes (*Clare Island Survey* Part 54) (1914).

1. *Nuada leptosoma* gen. et sp. nov.
1A. Anterior end of male, ×150.
1B. Head of female, ×860.
1C. Tail of female, ×330.
1D. Region of the male pore, ×860.

10. *Dagda bipapillata* gen. et sp. nov.
10A. Anterior end, ×230.
10B. Head, ×860.
10C. Buccal cavity and teeth, ×1870.
10D. Tail of male, ×330.
10E. Tip of tail, ×860.
10F. Spicules, ×860.
10G. One of the supplementary organs, ×860.

13. *Fiacra longisetosa* gen. et sp. nov.
13A. Head, dorsal view, ×530. a = chitinous aperture of gland-duct leading into lateral sense-organ.
13B. Head, lateral view, ×530. a = as before.
13C. Dorsal view of buccal cavity and teeth, ×1330.
13D. Posterior end of male, ×86.
13E. Spicules, ×330.

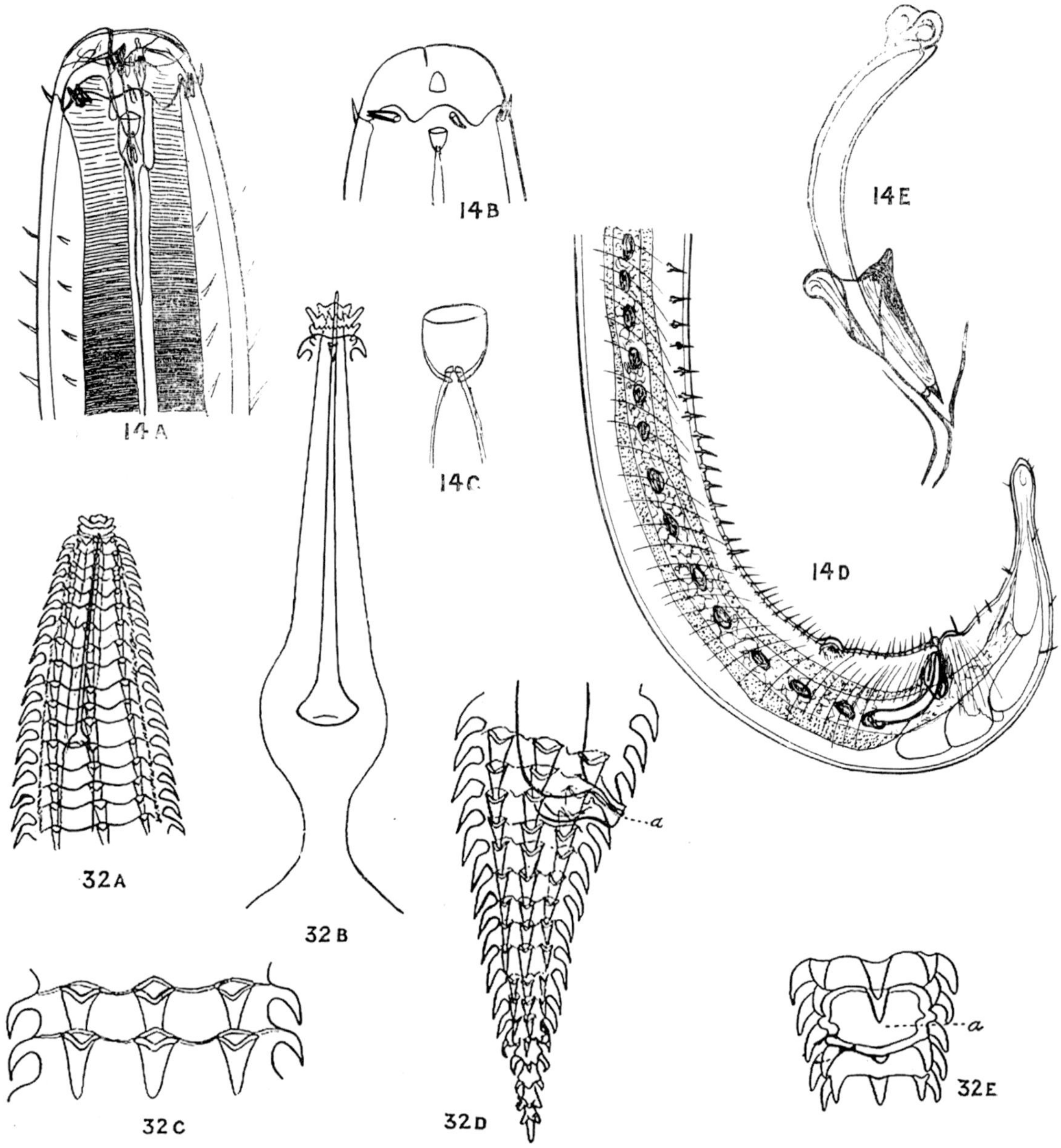

Fig. 6.13b Rowland Southern's nematodes (*Clare Island Survey* Part 54) (1914).

14. *Fiacra brevisetosa* sp. nov.
14A. Head, in optical section, lateral view, ×330.
14B. Head, showing surface characters, lateral view, ×330.
14C. Lateral sense organ, greatly enlarged.
14D. Posterior end of male, ×86.
14E. Spicules, ×330.

32. *Ogma Murrayi* gen. et sp. nov.
32A. Anterior end, lateral view, ×330.
32B. Head and oesophagus, with spear, ×660.
32C. Two segments from the mid-body, lateral view, ×660.
32D. Posterior end, lateral view, ×530. a = anus.
32E. Ventral view of anal and adjacent segments, ×530. a = anus.

soil, moss, water-weeds &c; the marine Nematodes and Kinorhyncha picked out of the *débris* at the bottom of bottles containing Polychaetes'.[46]

Because of this, the account of the nematodes in the Original Survey is little more than a preliminary foray into this dense thicket of biodiversity, and there has not been a great deal of progress since. Southern concentrated on 'new and strange forms', and ignored the all-important parasitic species. Most of the time he spent on roundworms was devoted to examination of a single haul of the dredge on the *Helga* in a bank of shelly sand and gravel off Crump Island in Clew Bay. This proved to be 'an apparently inexhaustible source of new and interesting species', mostly of microfaunal groups such as copepods, halacarid mites, polychaetes and nematodes.[47] Because this particular habitat had not been investigated properly at the time, the great majority of the nematodes proved to be new species. By way of contrast, there was little new in collections made on the shore or in shallow water.

Southern enumerated sixty species of marine nematodes along with eighteen land and freshwater species. He also listed five kinorynchs and two chaetognaths collected in the same lucky dip fashion as the nematodes (these strange worms are discussed below). The list included one new family, eight new genera, 28 new species and 26 others that had not previously been recorded from either Britain or Ireland; all except three species were new Irish records. Most of his account is taken up with discussion of the finer points of taxonomic detail, especially for the new species. The bewildering novelty meant that he had to come up with names for new genera and species, and he took the opportunity of naming four of his new genera after heroes in Celtic mythology: *Fiacra, Nuada, Dagda* and *Ogma*.[48] (Figure 6.13b).

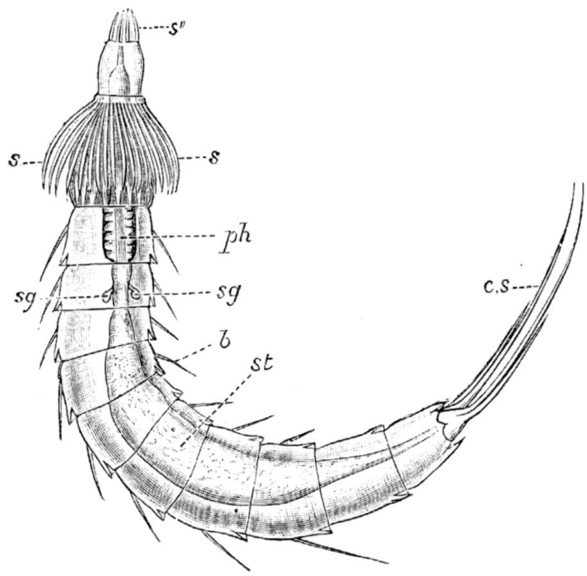

Fig. 6.14a *Echinoderes* (x210)
b, bristle; cs, caudal spine; ph, pharynx; s and s1, the spines on the two segments of the proboscis; sg, salivary glands; st, stomach.
The Cambridge Natural History, Volume II (1896), 236

Phylum Kinoryncha (mud-dragons)

In passing we must mention two other groups of worms that belong to the marine fauna around the island, although you are unlikely to meet them in the ordinary run of things. Kinorynchs are as far as you will get from our notion of what to expect in a mere worm. They live in mud or sand at the bottom of the sea, all 180 of them, the largest no bigger than 1mm in length. For all that, they have managed to acquire the name of mud-dragons. The most extraordinary thing about them is a retractable head that is covered with spines, which they use to excavate their passage through the sediment that constitutes their entire world. The five species of kinorynchs caught by Southern during the Original Survey were the first ever recorded from Britain or Ireland, and one of them (*Pycnophyes Zelinkaei*) was a new species (Figure 6.14b).

Phylum Entoprocta

Entoprocts are superficially rather similar to bryozoans and were once lumped in together with them ('proper' bryozoans being distinguished as ectoprocts). Entoprocts have a simple excretory system and the anus is encircled by a ring of tentacles (among other differences). They are now generally considered to be a separate phylum, with no more than 150 species known worldwide. The Original Survey recorded only three species in the waters around Clare Island, but one of them was *Loxosoma phascolosomatum*, a good example of how specific these little animals can sometimes be about where they will live. It is usually found at the tip of the tail of the peanut worm *Phascolosoma*, and is remarkable for being one of the few entoprocts that are not colonial.

Phylum Chelicerata (Chelicerates)
Class Pycnogonids (sea-spiders)

If you are really lucky you may spot a sea spider: a tiny creature that has the four pairs of legs of its (very distant) spider relatives, but seems to have nothing else, so attenuated is its body. For all its tiny size, it is carnivorous, feeding on sea anemones and other coelenterates (Figure 6.15).

Pycnogonids constitute a distinct class in phylum Chelicerata (the other two are arachnids and horseshoe crabs). There are about a thousand species, all of them marine, and as creatures go these are as weird as they come. They look like long-legged spiders that practice aquatic tai chi, so slowly and deliberately do they move. The rest of the body is so small there is room for nothing else; branches of the stomach extend into the legs, as do the reproductive organs: this is where the ovaries and testes lie, and they open through pores on the legs. The male gathers up the fertilised eggs, glues

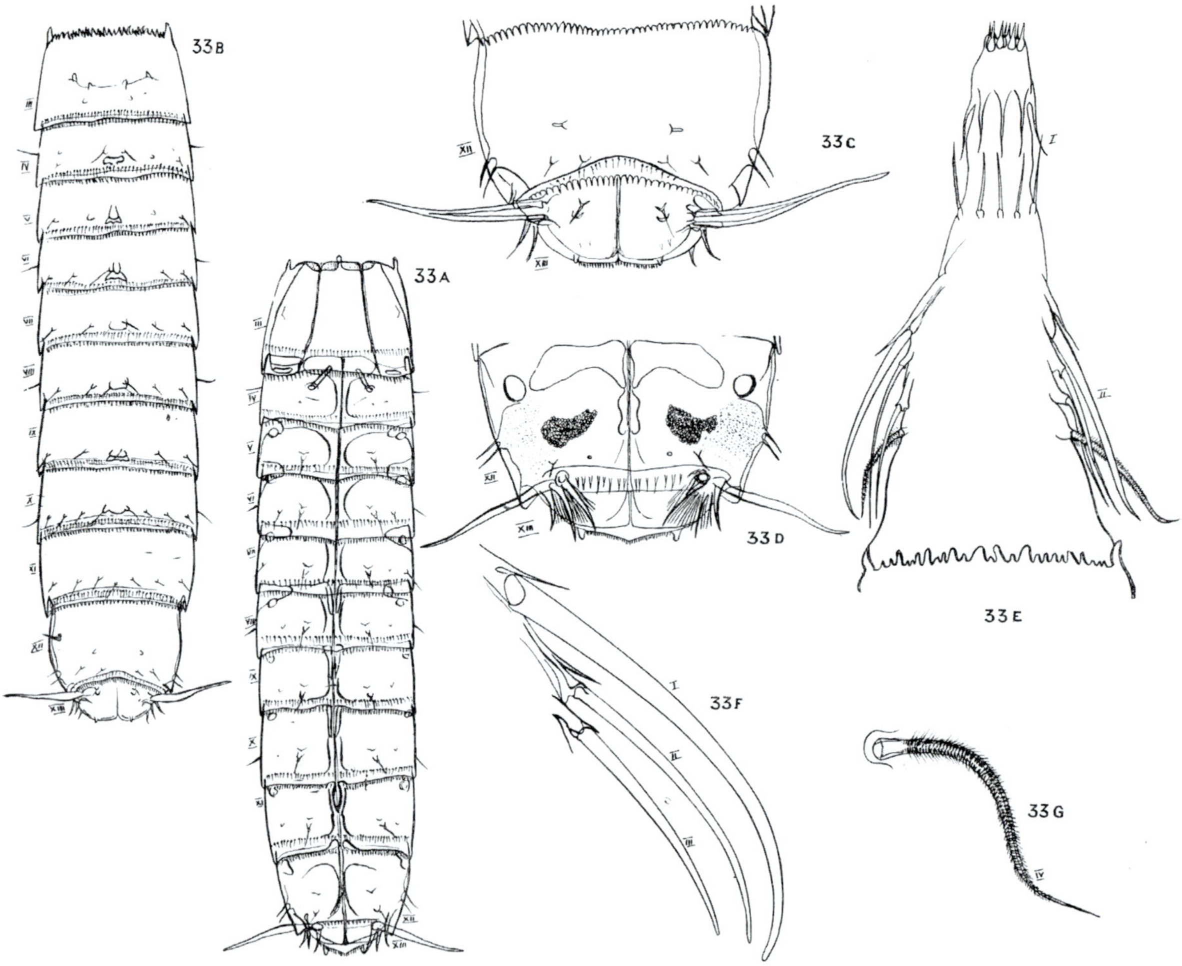

Fig. 6.14b *Pycnophyes Zelinkaei*: Southern's new species of kinorynch.

33A. Ventral view of entire male specimen (x150).
33B. Dorsal view of entire male specimen (x150).
33A. Ventral view of entire male specimen (x150).
33C. Posterior end of male, dorsal view (x330).
33D. Posterior end of male, ventral view (x330).
33E. Expanded proboscis (x230).
33F. Spines from the three anterior rows on the second segment of the proboscis (x530).
33G. Feathered spine from the posterior row of the second segment of the proboscis (x530).

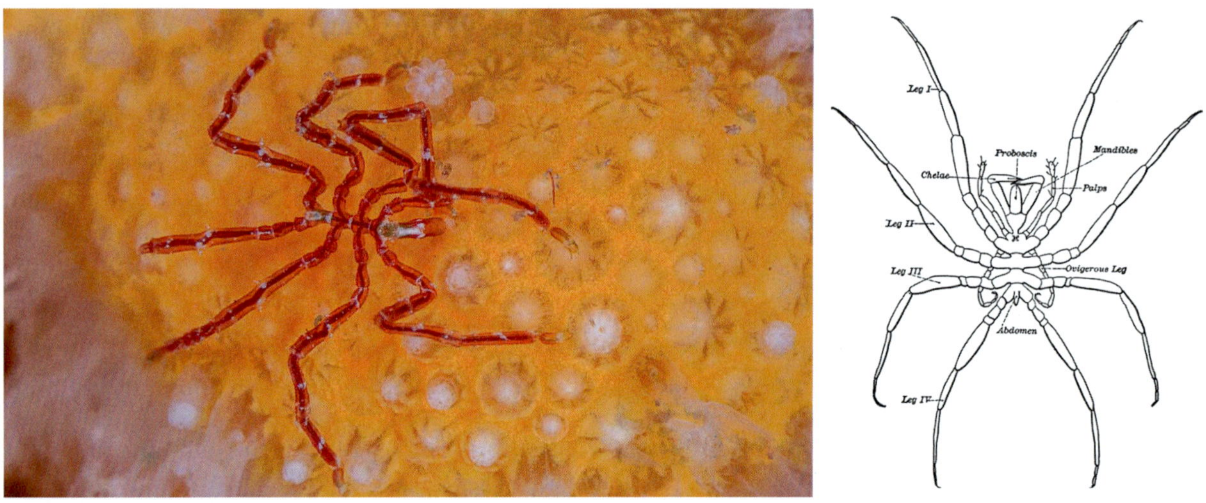

Fig. 6.15 Sea spider (*Obisium maritimum*) (photo © Timur Kholodenko, illustration by T.H. Morgan).

them into a ball which he then carries around between a pair of special ovigerous 'legs' behind the head (just in front of the four pairs of walking legs).

George Carpenter (who studied the group for the Original Survey) gathered records for eight species of sea spiders in the waters of the wider Clare Island area. Four of these were taken close to the island, but he believed all eight to be pretty generally distributed. If they are carefully looked for (and they are well worth the effort) several of these will be found on the lower shore. *Pycnogonum littorale* is a dull-coloured species that lives under stones, in debris and with sea anemones. It has a swollen-looking body (2cm long) and balloon-like legs, and lacks palps or chelicerae. *Nymphon gracile* has very long, slender legs two and a half times the length of the body (which is 0.5mm long). *Endeis laevis* is a rather more colourful species that lives among seaweed and hydroids, on which its larvae are parasitic. It has long, slender, spiny legs, three to four times as long as the trunk (*c*.5mm).

Phylum Crustacea
Class Malacostraca (decapods, amphipods, isopods)

Decapods

Decapod crustacea were described for the Original Survey by George Farran, who recorded 52 species from the shores and waters of Clew Bay and other waters around Clare Island and further afield.[49] However, most of these live in deeper waters offshore; you are likely to find only half a dozen or so in the intertidal zone of the shore. The most familiar of these deep-water decapods are the lobster (*Homarus gammarus*) and edible crab (*Cancer pagurus*). At the time of the Original Survey (and of course before and ever since) these were much sought after as human food.[50]

Crabs are perhaps the most familiar of shore animals, with those unmistakable mighty claws crafted in the laboratory of time to crush the shells of the molluscs that are so prolific on the sea floor. You will have to look hard to find crabs by day, because usually they are hiding from predatory gulls in crevices in the rocks and in the slippery

Fig. 6.16a *Cancer pagurus* (Matthieu Sontag (CC BY-SA 3.0)).

Fig. 6.16b *Carcinus maenas*
(Unterwasserfotografie von padidigital-U-P (CC BY-SA 3.0)).

Fig. 6.16c *Necora puber* (Bernard Kaye).

Fig. 6.16d *Leander serratus* (Massimiliano Marcelli).

Fig. 6.16e *Meganyctiphanes norvegica* (Uwe Kils (CC BY-SA 3.0)).

Fig. 6.17 *Nototropis swammerdami* (Hans Hillewaert (CC BY-SA 4.0)).

depths of the seaweed. But venture into the intertidal zone by night and you will find them in swarms.

The shore crab (*Carcinus maenas*) is the most common of the largest species (up to 10cm) in the intertidal zone. It is usually dark green in colour with red joints. The fiddler crab (*Portunus puber*) is also fairly common under stones. It has a fierce disposition; when disturbed it assumes a defensive stance that will make you think twice before picking it up! It is a swimmer, and uses its flattened and rounded back legs to act as oars or paddles. Another common intertidal species is the broad-clawed porcelain crab (*Porcellana platycheles*), which is found under stones on the lower shore. It is a tiny crab, with disproportionately large pincers; it looks as though it only has three pairs of walking legs instead of four (the small back legs are tucked under its round body) (Figure 6.16).

The common prawn (*Leander serratus*) is often found in rock pools, hiding in the seaweed and hard to spot because it is transparent (they only turn pink when you cook them). *Leander squilla* is also common in the pools, but only half as big (up to 5cm). On the other hand, the common shrimp (*Crangon vulgaris*) is a sand dweller, its speckled brown body ideal for camouflage in this environment. When the tide is out you will have to dig for it (in sandy spots around the harbour).

Isopods

Marine isopods and amphipods were studied for the original survey by Walter Medley Tattersall. Several marine isopods occur on the lower and middle shore, under stones and in seaweed, as well as in pools and wet sand, including several species of *Idotea* and *Jaera*. The most abundant species is generally *Idotea baltica*; it is very variable, which makes it difficult to distinguish from the several other species that live here. *Jaera marina* and *Jaera nordmanni* are common 'under stones on the shore between tide-marks, especially where a stream of fresh water enters the sea'.[51]

Amphipods

Walter Medley Tattersall recorded 32 species of shallow-water marine amphipods from the shores and waters around Clare Island. Few shore animals are more familiar than sandhoppers, the little creatures that hop and skip about madly if you enquire into a patch of rotting seaweed. The common species is *Orchestia gammarella*.[52] Most of the marine species live in shallow water offshore, on the bottom or in vegetation, but at night many take to the open waters, and some can be netted in prodigious numbers at this time. The order includes the familiar freshwater shrimp (*Gammarus*) and many species look rather like them, the most common on the middle shore being species of *Marinogammarus* (mainly *M. marinus*). There are also several species of *Gammarus*; *Gammarus locusta* is very abundant under stones on the middle and lower shore. *Amphithoé* looks somewhat similar, and is rather special because of its ability to change colour depending on what sort of seaweed it is living in.[53] It usually wears a cream-coloured jacket ornamented with brownish-red splotches. It weaves a cosy nest from plant fragments and silk, and from this it surveys its surroundings, straining such food as the currents bring their way. Each lives a solitary life, but in the breeding season the males seek out the (far more abundant) females. The young are cared for in a brood-pouch formed by appendages on the mother's abdomen, and she will sometimes leave her nest to fan currents of water through the pouch. She continues to care for young in this way until they are capable of spinning their own nests, at which juncture she is impatient to be rid of them. They, for their part, are reluctant to leave, and will hang around for ages, making every effort to get back to the security of home.[54]

Nototropis swammerdami, 'one of the most abundant and characteristic Amphipoda in the district' lives in clean sand (Figure 6.17).[55] *Jassa falcata* is often encountered on the middle and lower shore, its name (*falcata*) a reference to its knife-like nippers (*falx* means talon in Latin).

Other marine amphipods look very different. *Corophium bonelli* lives in u-shaped burrows in mud and muddy sand; it is only a centimetre or so long, but has disproportionately huge antennae. It can be enormously abundant, and is a favourite food of wading birds. The Capreliidae are like animated bits of leggy string, for all the world like miniature stick insects; and they walk with what looks even more like painful deliberation than the latter. Their prehensile front legs are like pen-knives. A number of *Caprella* species live in seaweed on the lower shore. Some amphipods have a more specialised lifestyle. *Hyperia galba* lives under the umbrellas of jellyfish, *Orchomenella nana* in the shells of dead crabs.

Class Cirripedia (barnacles)

The life stories and evolutionary histories of the plants and animals of the shore are often so elaborate and complex you could scarcely credit their possibility if they did not actually exist! But of no creature is this truer than barnacles, which live in dense—and apparently lifeless—concentrations on the rocky shores of the island. Each is like a little volcano whose slopes are six limy plates, the crater of which is sealed with a sliding door of four more plates. When the tide is in they come to life; the doors open, and a feathery plume emerges, delicately filtering the tide for the tiny organisms on which they feed (you can watch this for yourself if you put a rock with barnacles on it in a basin of salt water and let it stand for a while).

It is hard to believe that barnicles are on the same branch of evolution's family tree as crabs, lobsters and woodlice, but when barnacles first hatch their larvae look very like those of other marine crustaceans: tiny things with one eye and few limbs. These *nauplius* larvae are tossed about in the sea for a few days, during which time they moult several times, gradually changing into quite a different sort of creature, which has a pair of new eyes and six pair of limbs. It is these *cypris* larvae that settle down for life, anchoring themselves to rock or other substrates with their antennae, then walling themselves inside their chalky armour. Their legs undergo a process of hairy transformation into the net they use to filter food from the sea. Remarkably, fertilisation takes place internally; barnacles are hermaphrodite, so the eggs are fertilised by sperm introduced by the elongated penis of an adjacent animal that cannot be more than 3–4cm away. They take four months to develop, at the end of which the minute nauplius larvae hatch out. When this happens in early spring these are often the commonest animals in the inshore plankton.

Three species of acorn barnacles occur on Clare Island. There are large populations of both *Chthamalus stellatus* and *Chthamalus montagui*

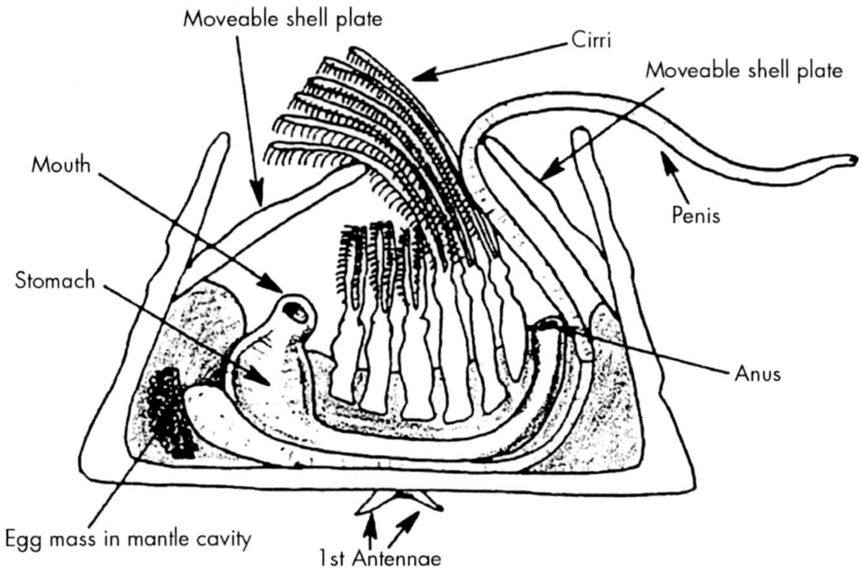

Fig. 6.18 The structure of an acorn barnacle (University of Southern California).

Fig. 6.19 Common barnacles: (a) *Chthamalus stellatus* (Michael Maggs (CC BY-SA 3.0)); (b) *Chthamalus montagui* (Auguste Le Roux (CC BY-SA 3.0)).

(Figure 6.19). The largest populations of *Chthamalus stellatus* develop on the most exposed shores, whereas *Chthamalus montagui* prefers more sheltered situations (as at Portnakilly), as does the third species, *Semibalanus (Balanus) balanoides*.[56] *Semibalanus* is a species of northern affinity, *Chthamalus* of southern affinity.

Marine entomostracans (Branchiopoda, Ostracoda, Copepoda)[57]

Very little is known of the marine entomostracans around Clare Island, which were studied for the Original Survey by G.P. Farran. He recorded 4 cladocera, 65 ostracods and 159 copepods, 30 of which were more or less planktonic forms, the remainder being littoral or living on the bottom.[58]

*****

W.M. Tattersall also provided accounts of several other groups of small and much more obscure arthropods for the Original Survey. Obscure and unfamiliar they may be, but some are of great ecological and economic importance. Various species of Euphausiacea (krill), for example, abound in the open seas around Clare Island, often in incredible numbers. *Meganyctiphanes norvegica* is the main food of mackerel, salmon, herring and trout in springtime; in the spring of 1900 the stomach contents of a whale caught off Iniskea consisted entirely of this diminutive creature.[59] Cumacea – tiny, scorpion-like shrimp-relatives most of which burrow in sand or mud – are also often superabundant and an important food of many fishes.

Some of the strangest creatures of the shore have to be looked for very carefully. Few are stranger than *Nebalia bipes*, a tiny shrimp-like creature which, when you catch your first startled glimpse of it after it has been transferred to a jam jar filled with water, might look like an attempt to adapt the pangolin body-plan to marine conditions on a diminutive scale! Currently it is classified as a leptostracan crustacean (the first species in this order to be described).[60] It is about 1cm long, orange or greenish-yellow in colour with startled-looking red eyes on stalks. It lives among decaying seaweed and the like, often in prodigious numbers 'and seems to prefer the neighbourhood of black, evil-smelling mud'. Swarms of them can sometimes invade lobster pots.[61] Mysidacea (opossum shrimps) are another important group of free-swimming shrimp-like creatures that find themselves confined in tide pools on occasion. They have large black eyes on stalks and are easily recognised by their habit of hanging vertically in the water (e.g. *Macromysis flexuosa* and *Sirella* spp.). In the view of several of the naturalists on the original survey 'a striking feature of the marine fauna is the blending of northern forms with southern species from the Mediterranean, the latter element somewhat preponderating'.[62] 'The shallow-water marine fauna of the west of Ireland is a blending of boreal and Mediterranean species, in very nearly equal proportions.'[63] This applies not only to the decapod crustaceans, but also to much less familiar groups such as the Cumacea, Euphausiacea, Mysidacea, Isopoda and Amphipoda.

Fig. 6.20a A selection of bristleworms: from *A Monograph of the British Marine Annelids* (1910).

Fig. 6.20b *Pisione remota* (© Hans Hillewaert (CC BY-SA 4.0)).

Fig. 6.20c *Eulalia viridis* (© naturamediterraneo.com).

Fig. 6.20d *Nereis pelagica* (Alexander Semenov (CC BY 2.0)).

Phylum Annelida (segmented worms)
Class Polychaeta (bristleworms)

Many of those who worked on the marine fauna for the Original Survey extended their geographical remit to include the waters beyond Clare Island. Rowland Southern collected polychaetes for his report from inshore waters stretching all the way from Blacksod Bay to Ballynakill Harbour, including a 'Clew Bay' area which he defined as the open waters east of a line joining Achill Head to Inishark. The polychaete fauna of these waters was the richest of any comparably small area in the world at the time, totalling 249 species in the whole area studied (Figure 6.20). This included sixteen species new to science (including one which he assigned to a newly minted genus: *Praegeria remota*—now *Pisione remota*), 36 new records for the entire marine area around Britain and Ireland and an additional 77 new Irish records.[64] In the wider 'Clew Bay' area there were 110 species, and 36 species around Clare Island itself.[65] Worldwide 9,000 species have been described so far.

Polychaetes are among the most fabulously ornamented creatures of the sea, for all the world earthworms in fancy dress—many adorned with colourful frills and fans, combs and flounces of mesmeric variety. And that variety of form is paralleled by a great diversity of lifestyles. Traditionally they were divided into free-living carnivorous or 'errant' forms, and 'sedentary' forms that live in burrows, tubes and galleries they construct for themselves out of lime (serpulids), sand-grains or seaweed (terebellids) and a parchment-like secretion of their own (sabellids), often under rocks; some are commensal or parasitic. They occur at all depths of the sea more or less abundantly, but most are inshore species, living between the tide marks or in shallow water down to 20 fathoms.

The Clare Island errant polychaetes include several species of *Syllis*, small millipede-looking worms that live among seaweed, stones and sand on the middle and lower shores (which is where most of the littoral polychaetes will be found). In our two species of *Lepidonotus* the body is covered with an armour consisting of 12 pairs of overlapping scales, which gives them a somewhat woodlouse-like appearance (and on account of which they are known as scaleworms). *Haemothoé* and *Lagisca* have 15 pairs of scales. The family Phyllodocidae are represented by *Phyllodoce* and *Eulalia*, which have a pair of paddle-like structures on each segment, enabling them to swim as well as walk (and which has earned for them the name 'paddle worms'). *Eulalia viridis* is a beautiful shiny green colour (Figure 6.20c).

The nereids (family Nereidae) or ragworms are the dragons of the polychaete world. There might not appear to be anything marvellous about them when they are freshly dug from the mud in which they wait out the drought of the fallen tide, but when the sea returns and they emerge to swim to new pastures there is about them all the magic of any dragon of fable.[66] They include the magnificent *Nereis viridis*, which is a glorious green colour and can be up to a metre long (not to be found on Clare Island shores, alas). Nereids have black, hooked jaws that mark them for the predators they are. They can often be found hiding under stones or in seaweed, especially among tufts of *Corallina* in rock pools. Crabs and fish are fond of them, in spite of the tusk-like jaws with which they seize their prey. Many species apart from the green dragon of the deep are brightly coloured. *Nereis pelagica*, which lives among the seaweed and stones of rocky shores (typically 8–10cm long, with 80–100 segments), has a bronze-green body with a red line down the back (its dorsal blood vessel). Like many other nereids, its body undergoes dramatic change in preparation for the breeding season, when it adopts a pelagic existence (its 'heteronereis' form). *Nereis pelagica* sheds its eggs and sperms into the open sea.

Perinereis cultrifera is a fearsomely magnificent worm 10–25cm long, glinting green-bronze in colour. When its proboscis is everted (enbling you to see the pair of scimitar-like jaws), you can see what look like a pair of eyebrows just behind, which are in fact two rows of tiny teeth. *Platynereis dumerilii* is like a small yellow millipede; it has exceptionally long feelers which it uses to explore the waters beyond the threshold of its cavern among kelp holdfasts. Unlike *Nereis pelagica*, it incubates its eggs within mucous tubes which it makes in the mud. *Nephthys ciliata* has abandoned colour for a long, elegant body aglow with pearly radiance. It has a characterisic long tail thread and a flattened body that enables it to swim with distinctive lateral waves passing down its body.

Glycera alba belongs to a family of small, thin worms (Glyceridae) generally found in pockets of sediment on the lower shore. The body tapers at both ends, and has up to 150 segments, but each has 2–3 circular bands around it, which gives it a very close-ringed appearance. They often coil up when touched. *Nerine cirratulus* is often associated with lugworms (*Arenicola*). It has a blue-green body and long, thin feelers.

By far the most familiar of the sedentary polychaetes are the lugworms (*Arenicola*): fat, greenish or yellow-black worms whose cylindrical casts lie in little heaps on the wet sand when the tide recedes. Just as with their earthworm cousins (which have polychaete ancestors) these tubes of sand have passed through the body in the course of feeding. We have two common species. *Arenicola marina* is the more familiar; it has a distinctive rat-like tail that lacks the little bunches of red gills that adorn the rest of the body. *Arenicola ecaudata* has no tail, and prefers more gravelly places. Lugworms are a favourite food of many fishes and are regularly sought after by fishermen on this account.

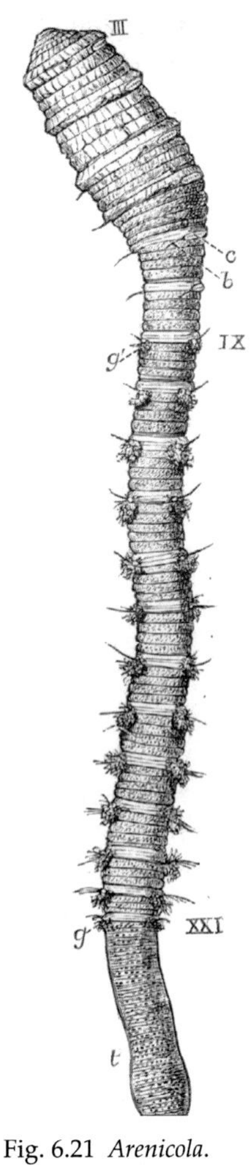

Fig. 6.21 *Arenicola.*

Fig. 6.22a *Spirorbis* (Paul Adams).

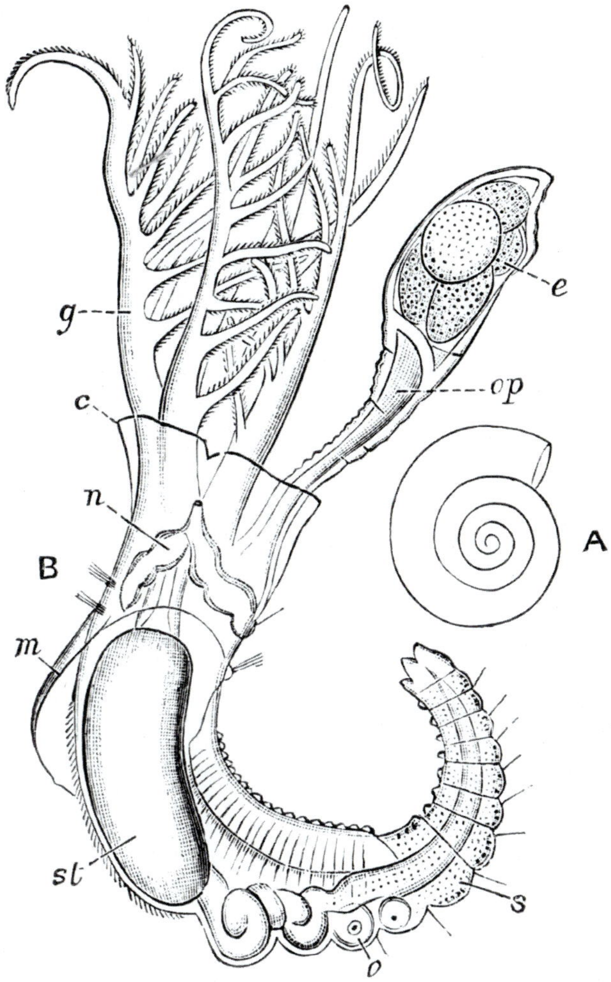

Fig. 6.22b The *Spirorbis* animal.

A: shell; actual size about 5mm across. B: the animal itself, x50; c: peristomial collar; e: eggs in the brood pouch at the end of the operculum; g: gills; m: thoracic membrane; n: the single pair of thoracic nephridia; o: ova in the anterior abdominal segments; op: operculum; s: spermatozoa in the hinder abdominal segments; st: stomach (Edouard Claparède, *Recherches anatomiques sur les annélides observés dans les Hébrides*, 1861).

Immensely fragile though they are, *Polydora* are rock-boring worms, able to excavate their burrows in soft rock, usually above cracks (or old shells). These burrows are u-shaped, with a pair of upstanding sandy turrets rising 1cm or so above each opening.

Audoninia cirratulus (family Cirratulidae: redthreads) is an extraordinary-looking creature (although it needs to be coaxed out of its burrow and admired in water for you to appreciate its beauty). It is very long (*c.*8–20cm) with as many as 300 segments, colourful in a jacket that ranges from reddish-brown to orange-yellow or bronzy green. Every six or seven segments there are two bunches of eight curly thread-like red feelers, and most segments also have red gills, which are also long, thread-like and curly. *Cirratulus cirratus*, which is often found in the mud under stones, is like an underdeveloped *Audoninia*, about half as long and less thready-looking. *Dodecaceria concharum* is a common little borer that forms galleries in oyster shells.

Flabelligera affinis lives in a translucent case of mucus under stones or in seaweed thickets. Through this you can view its pale green body (with the red line of the stomach). *Potamilla torelli* is a large worm, sometimes over 10cm long, but it hides in a long tube, most of which is hidden in cracks or between bedding planes in rock. Its most eye-catching feature is its bunches of elegant, feathery tentacles. *Fabricia sabella* is a tiny yellow or pink worm that frequently wanders away from the horny, transparent tube it constructs among algae or ascidians. *Scolelepis vulgaris*[67] is a small thread-like worm that lives in mud or sand, recognisable by what look like coiled horns on the head. *Capitella capitata*, which looks like a blood-red earthworm, makes a mucus tube in mud or muddy sand.

One of the most striking polychaetes is *Spirorbis*, a tiny plumed worm that constructs a chalky spiral in which to live on rock or seaweed, generally in the company of many companions

Fig. 6.23 *Spirobranchus* (*Potamoceros*) *triqueter* (Cwmhiraeth (CC BY-SA 3.0)).

(Figure 6.22). When the tide is in it pushes out its head and unfurls the delicate plumes of its tentacles to filter its food from the sea. As the sea recedes it folds up its plumes and withdraws into its spiral tunnel, sealing the entrance with a special little stopper. *Potamoceros triqueter* makes an irregular tube that is triangular in cross section, and increases in size as the worm-mason inside grows (Figure 6.23). Though small and seldom seen, it is startlingly beautiful: 'very handsome, the thorax being deep blue, the abdomen red in the female, whitish in the ripe male. The branchiae ['tentacles'] are barred and spotted with blue, orange, and white; the operculum is calcareous, and furnished with a couple of horn-like processes'.[68]

One of the most remarkable things about these tiny creatures is the precision with which their reproductive cycle is synchronised with the phases of the moon. Every two weeks, at the moon's quarter phases, it takes a new batch of fertilised eggs into its brood chamber and nurtures them there, each one in a little cellophane-like balloon, until they have developed into larvae and are ready for independent life. At the same time, it sheds the batch that was fertilised a fortnight before and is now due for release. In this way, the parent worm ensures that the little ones will find themselves in calm waters as they seek out somewhere to settle down, without having to face the buffeting of greater tides.

Classes Oligochaeta (earthworms amd bristle-worms) and Hirudinea (leeches)

Most oligochaetes and leeches are terrestrial or freshwater animals, and will be found on pages 153–4. Rowland Southern recorded nine species from the shores of Clare Island during the Original Survey; these were usually living under stones, in weeds from rock pools or in muddy gravel, including a species of *Tubifex*. Tubificid worms live head down in tubes made in the mud, waving their tails ceaselessly to extract oxygen from the water; their bodies contain haemoglobin, which greatly increases their efficiency at this. Southern got excited over what he considered a new species of enchytraeid from weeds in rock pools on the island (*Enchytraeus cliarensis*: now *Marionina cliarensis*), and *very* excited over 'a remarkable Oligochaete' which was recovered from a bottom of sand and shells in 24 fathoms in Clew Bay, part of 'a rich and peculiar micro-fauna, consisting of an association of species, many of them new to science or very rare'. This he named *Grania maricola*.[69]

Phylum Sipuncula (sipunculids: peanut worms)

Sipunculids are odd-looking worms of uncertain taxonomic affinity, rather strange to our notion of what an 'animal' should look like. They can fold up their plump, cylindrical bodies like a telescope, and at the front end they have a distinctive structure called an introvert which they can extend when feeding and retract within the mouth at other times. Most are deposit feeders, much as earthworms and burrowing polychaetes are, and must play a comparable role in turning over the sandy sea bed.

Only a few hundred species have been described. Several species of *Physcosoma* occur in the waters around Clare Island, the most characteristic littoral species being *Physcosoma* (*Phascalosoma*) *granulatum*, which is very common on rocky shores, living under stones and in crevices of the rocks. Another common species is *Phascolosoma Johnstoni*, which lives under stones on the shore and in *Laminaria* roots.[70]

Phylum Mollusca (molluscs)

Nathaniel Colgan spent three weeks studying molluscs for the Original Survey in the summers of 1909 and 1910.[71] He recorded 102 species (of which 77 were living) from the shores of Clare Island and Clew Bay. Between 1993 and 1996 Julia Nunn of the

Fig. 6.24a *Placida dendritica* (Sylke Rohrlach (CC BY-SA 2.0)).

Fig. 6.24b *Elysia viridis* (Parent Géry(CC BY-SA 3.0)).

Ulster Museum looked at marine molluscs for the New Survey. She recorded 63 species from the shores of Clare Island (19 of them new records for the island) and 104 from the shores of Clew Bay (23 of them new).[72] Overall, the New Survey confirmed the finds of the Original Survey, which recorded 54 or 55 species on Clare Island, but although 43 of these were re-found, 12 were not. The most interesting new discovery was the rare and magnificent red sea snail *Aeolidella sanguinea*.[73] Among the most widespread and beautiful species are two green sea slugs—*Placida dendritica* and the tiny *Elysia viridis*—abundant because of the luxuriant growth of their food plant, *Codium tomentosum*, which inhabits deep pools in the mid-shore (Figure 6.24).[74] *Placida dendritica* is abundant at low tide, always on *Codium*, 'and so closely assimilating in colour with the varied tones of green assumed by the seaweed as almost to defy detection. The presence of the animal was usually betrayed by the creamy spawn-clusters deposited chiefly round the base of the main stems'.[75]

Sixty-three species is a lot, but it would be higher if Clare Island was less exposed and had greater habitat variety along the shore: a number of common species that might be expected are absent because conditions are so demanding in the stormy winters experienced on the island. At a good location along the eastern or southern shores you can find over 30 species, but no more than a dozen or so can be expected on the extremely exposed shores of the west and north, where hardy species such as *Brachystomia eulimoides* are dominant, usually in the company of mussels.

The shores of the mainland have a richer fauna, including 70 species that are not seen on the island. The total number of species for the wider Clew Bay area today stands at 137 species, but (as the Original Survey also noted) any given species is likely to be more abundant on the shores of the mainland than on the island.[76] On the other hand, a small number of species that occur on Clare Island have not been seen on the mainland shores of Clew Bay in recent times.

Characteristic Island Species

Mussels

The common mussel (*Mytilus edulis*) is among the most widespread, abundant and easily recognised shore molluscs, found mainly in the middle shore and below, often in dense colonies, which feed by filtering small organisms from the tide with their much enlarged and highly specialised gills (Figure 6.25). Mussels are much sought after as food, not only by ourselves, but by gulls and otters, crabs, whelks and starfish; among fish the ballan wrasse (*Labrus mixtus*) is particularly partial to mussels. Predation by ballen wrasse is thought to be responsible for high mortality in mussels.[77] Buckets of them were at one time scraped off the rocks and

Fig. 6.25 Common mussel (*Mytilus edulis*) (Andreas Trepte (CC BY-SA 2.5)).

used as fertiliser. The horse mussel (*Modiolus modiolus*) also occurs, but almost always as small specimens in rock pools, hiding away among the *Corallina*.

These dense congregations of tightly anchored mussels, all huddled together shell to shell, might look like permanent fixtures. However, careful observation shows that there is change over time: individuals lost to the community for one reason or another are quickly replaced by new arrivals (which of course all look the same to human eyes, so we do not notice).[78]

Limpets

The common limpet (*Patella vulgata*) can be found everywhere in the inter-tidal zone. The beautiful blue-veined limpet (*Patella pellucida*) is much more restricted in its occurrence; it often occurs in kelp, which is its food plant. It is only half as big (20mm along the base), with a transparent shell, yellowish-brown with kingfisher-blue stripes (Figure 6.26).

You might think the shelly wigwams of the common limpet are fixed for all time, but when the tide is in and night falls, off they go on their grazing expeditions, returning to the same spot when they have finished. Although they appear to do very little, limpets are among the most important controls on the growth of seaweed. They can live for more than ten years, over which time they can wear an almost circular depression if the rock is soft. On irregular rock surfaces they are able to adjust the rate of growth around the edge of the shell so that they fit more tightly into their circular foundations. A watertight grip is necessary to keep the limpet moist when the tide is out, but—proverbial though this grip is—oystercatchers and gulls have learned to break it.

Fig. 6.26 Blue-veined limpet (*Patella pellucida*) (Andy Cowley (CC BY-SA 1.0)).

The limpet removal studies

Between October 1992 and the end of 1994 an experiment was carried out in a series of tide pools at Portnakilly to see what the effect of removing the dominant limpet grazers would be.[79] First of all four control pools were chosen; five others were cleared of all their limpets (two species: the common limpet *Patella vulgata* and the rough or China limpet *P. ulyssiponensis*), and a further six were blowtorched to remove everything. Three of these were restocked with *Patella vulgata* and the remaining three with *P. ulyssiponensis*. When things had settled down after these catastrophic events there was a dramatic and clearly visible increase in the cover of microalgae and diatoms in the manipulated pools, followed by a bloom of the green alga *Enteromorpha*. At first this was higher in the restocked pools, but after a while it became most prominent in the pools that had been cleared of their limpets. There was moreover a different level of cover between the two sets of restocked pools, presumably due to some difference in habitat preference or behaviour between the two limpet species, with *P. ulyssiponensis* apparently the more voracious grazer.

The brown alga *Scytosiphon lomentaria* made its appearance in all the disturbed pools. The *Ralfsia* crusts we met on page 81 are in fact an alternate phase in the life cycle of *Scytosiphon*, but interestingly, whereas *Scytosiphon* turns up in habitats when grazing pressure and competition is reduced, *Ralfsia* depends on herbivores to prevent it from being smothered by other algae.

Another study tried to shed light on the different habitat preferences between the two species of limpet that live on the shore. *P. ulyssiponensis* is normally found on more exposed shores, at lower tidal levels than the common limpet. The common limpet is more tolerant of desiccation than *P. ulyssiponensis*; higher up the shore you only find it in tide pools lined with crustose algae otherwise dominated by *Corallina*, and if they occur together here, *P. ulyssiponensis* tends to be dominant.

This time a series of mid-shore tide pools were cleared of all limpets in each of three years. Juvenile limpets were found to colonise these pools in great numbers (coming in with the tide), but the only ones to stay there were *P. ulyssiponensis*. This is because common limpets just use the pools as a nursery ground; when they grow up they migrate out of the pools to find a place they feel more at home as adults. In a parallel series of experiments a series of pools were cleared of all limpets and then restocked with marked individuals of just one species. In these restocked pools transplanted *P. ulyssiponensis* remained, but the common limpets tended to wander out of the pools to feed.

One reason for the preference of *P. ulyssiponensis* for these pools is the prolific development here of the crustose coralline algae (CCA) that constitute a high proportion of their diet. The presence of these algae has been shown to encourage enhanced larval settlement in several groups, including corals, bryozoans, polychaete worms as well as limpets. The removal of CCA from pools does not seem to make much difference to the subsequent migration into them of either adult or juvenile common limpets, but it does make a difference to *P. ulyssiponensis*, recruitment of which—both adults and juveniles—is reduced.[80] Adults of *P. ulyssiponensis* also migrate into CCA pools at higher rates than into those without.

Periwinkles

Six periwinkles live on Clare Island, all of them herbivores, grazing on minute vegetation on the surface of rock and seaweed with a tightly-wound ribbon of minute chitinous teeth. This ribbon—the *radula*—unrolls gradually to expose new teeth as the old ones are worn away. The cumulative effect of their grazing is as great as other major forces of erosion, wearing down the surface by as much as a millimetre every two years.

The small periwinkle (*Melarhaphe neritoides*) is a characteristic species of the splash zone, where it lives in shaded crevices in the rock (often in the company of the rough periwinkle): 'abundant in many stations on the E. and S. of the island on shaded rock-faces or in rock-clefts, associated with *Littorina rudis* [= *saxatilis*] and often several feet above high-water mark; also among *Lichina* with the minute bivalve molluscs *Lasaea rubra* and *Turtonia minuta*'.[81] It is almost never completely underwater, and is of all shore animals the most adapted to life on land. A study carried out for the New Survey showed that although eggs and sperm are produced throughout the year (the sexes are separate), the peak in maturity and spawning coincides with the winter storms, so that the spring tides will carry its egg capsules out to sea, where the larvae need to be for their early development.[82] Spawning is thought to occur rhythmically every fortnight so as to coincide with the spring tides: the only time the animals are submerged and can shed their eggs into the sea.

Fig. 6.27a Common periwinkle (*Littorina littorea*) (S. Rae (CC BY 2.0)).

Fig. 6.27b Rough periwinkle (*Littorina saxatilis*) (S. Rae (CC BY 2.0)).

The abundant (and edible) common periwinkle (*Littorina littorea*) is generally distributed on the eastern and southern shores of the island (Figure 6.27). It is found across a wider extent of the intertidal area than any other shore animal, though it prefers the lower end, and in spite of being gathered in enormous numbers for food remains abundant everywhere. Like the small periwinkle it releases its eggs into the sea, at night and on the flood tide, but in triplet-containing capsules. The rough periwinkle (*Littorina saxatilis*) is found further up the shore, often where the sea only comes every fortnight with the spring tides. In between times its only contact with the sea is the constant spray; it is commonly found in crevices with the small periwinkle at Ooghcorragaun.[83] It is viviparous, so that the young are spared the hazards of an early planktonic life, but sacrifice the possibility of wider distribution as the price of this.

In periwinkles that spend longer periods out of the water, the gills become reduced. The species lowest on the shore—*L. littorea*, *L. obtusata* and *L. fabalis*—have the normal gill within the cavity behind the head through which water circulates. But in *L. saxatilis*, and still more in *Melaraphe neritoides*, the gill is reduced while the cavity becomes more vascular and both in structure and function more like the 'lung' into which the gill chamber is converted in the true land snails.

The flat periwinkle (*Littorina obtusata*) lives among the larger fucoid seaweeds (mainly *Fucus vesiculosus* and *Ascophyllum nodosum*), on which it lays its eggs on gelatinous masses. The characteristic empty yellow shells are abundant on the tideline. *Littorina compressa* is a species found only on south-facing shores of Clare Island, but is abundant where it does occur. Another abundant and widely distributed species is *Littorina fabalis*, which lives mainly on *Fucus serratus*. *Littorina arcana* has only recently (1978) been recognised as a new species, so it is obviously 'new' to Clare Island, where it is probably common but overlooked.

The banded chink shell (*Lacuna vincta*) inhabits the lower shore; it is a little periwinkle, with a shiny yellow shell that has reddish bands. A number of species belonging to the large family Rissoidae are common; these are tiny snails with tall-spired shells. *Rissoa parva* is abundant on weeds at the lower end of the shore; another characteristic species is *Rissoa interrupta*.

Whelks

The common dog whelk (*Nucella lapillus*) is easily recognised by its thick, sculptured pointy spire-shell (Figure 6.28). It is abundant in the barnacle zone and in mussel beds; it is the main predator of barnacles, in whose shells it bores a hole with a long, powerful proboscis: and of mussels, whose valves it simply forces apart. The whelks are themselves preyed upon by herring gulls, which seem to prefer the dark-coloured form characteristic of the more exposed north and west coasts, and in whose diet they are second only to mussels on the island.[84] The egg capsules of dog whelk are in evidence most of the year. Young dog whelks occur lower down than the adults, feeding at first apparently on the tubeworm *Spirorbis borealis* before moving further up the shore when they reach a centimetre or so in size. The thick-lipped dog whelk (*Hinia incrassata*) is common along the low-water mark.

Fig. 6.28 The common dog whelk (*Nucella lapillus*) (Martin Talbot (CC BY 2.0)).

Topshells

In spite of appearances, topshells are more closely related to limpets than they are to periwinkles, and share their rasping grazing strategy. Topshells are distinguished from winkles by having a circular operculum and mother of pearl (nacreous) inner layer, whereas winkles have an oval operculum and dark inner layer.[85]

Several topshells are common on Clare Island, and are another good example of zonation among related species. The common or painted topshell (*Calliostoma zizyphinum*) is generally distributed on the lower shore and below. The grey topshell (*Gibbula cineraria*) grazes the middle and lower shores, whereas the flat or purple topshell (*Gibbula umbilicalis*) prefers to stay around the middle, but its range extends to the top of the lower shore.

Nathaniel Colgan was puzzled by the absence of the thick topshell or toothed winkle (*Osilinus* [now *Phorcus*] *lineatus*), which was a common mid-shore species elsewhere along the west coast at the time of the Original Survey, and for which he searched carefully (Figure 6.29). This was a temporary absence however, because it occurs today at several localities. It seems likely that the severe winter of 1895 brought about its disappearance from the island.[86]

Phylum Bryozoa (moss animals)

There are few groups of multicellular creatures in which what you can see with the naked eye prepares you less for the wonder and complexity revealed by the microscope than the bryozoans, well over 4,000 species of which have been described. They are an ancient group, abundant as

Fig. 6.29 (a) The common or painted topshell (*Calliostoma zizyphinum*)
(b) The grey topshell (*Gibbula cineraria*)
(c) The thick topshell or toothed winkle (*Osilinus* (now *Phorcus*) *lineatus*)
(H. Zell (CC BY-SA 3.0)).

Fig. 6.30a *Flustrellidra hispida* (Ar rouz (CC BY-SA 4.0)).

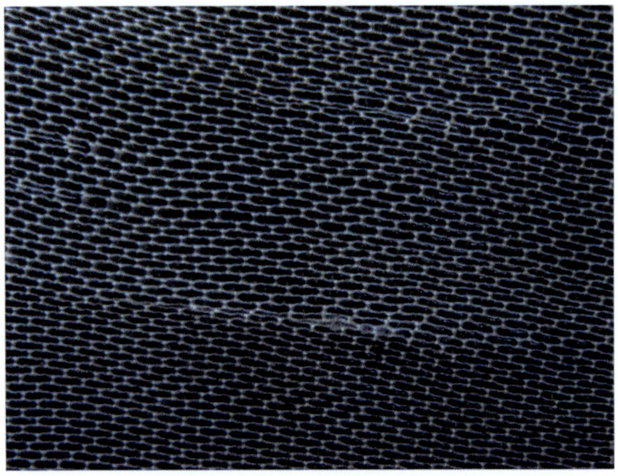

Fig. 6.30b Sea lace (*Membranipora membranacea*) (Bernard Kaye).

fossils, particularly in the Carboniferous Limestone. One of the species you are most likely to encounter on the shore looks like nothing more than a dry piece of a brown paper-like substance thrown by the tide up on the beach, but of this very species the eminent Victorian botanist Joseph Hooker wrote: 'For curiosity and beauty, I have not, among all the plants and vegetables I have yet observed, seen any one comparable to this seaweed.' [87]

Many common species take the form of a honeycomb-like film (or mat) on the surface of rock or seaweed, in which each cell of the honeycomb houses an individual animal (zooid). Other bryozoans are like diminutive shrubs hanging from the blades of kelp or wrack, or on the surface of a rock, the individual zooids arranged in serried fashion along the branches (the individual compartments are called zooecia). Some freshwater species take the form of gelatinous masses around stems and twigs that have fallen into the water. Each of these little animals is highly complex, if built on a body plan very different from more familiar creatures. The zooids have only one opening, so mouth and anus have to be close together.

Frustrellidra [*Flustrella*] *hispida* is exceedingly common on *Fucus* between tide-marks (Figure 6.30). If you hold in your hand a piece of wrack on which the soft brown 1–2mm-thick encrustation of the creature is growing, there is no sign of life, because all the zooids have withdrawn into their cells and closed the window. But place it in seawater and wait for a while, and you will see a delicate bluish halo developing at the surface, as thick as the encrusting film itself. If you look very carefully you will now see that a circle of about 30 long, delicate tentacles forming a graceful funnel has appeared at each window. However, at the slightest vibration—the casting of a shadow even—they retreat, all together, in a flash, before emerging cautiously once more when they think it is safe. When the tentacles are in action the colony has a fuzzy, moss-like appearance, whence the common name for these animals.

Because of their general habit—and in particular the crown of tentacles—bryozoans look rather like the hydroids we have met on page 97, but they are considerably more advanced than they are, at more or less the same level of complexity as the rotifers (see pages 201–3). The circlet of tentacles arises from a highly distinctive circular or horseshoe-shaped ridge called a lophophore, which is controlled by a band of muscle to protrude or retract the tentacles.

The zooids are all identical, but the colony can produce specialised zooids for particular tasks, such as hatcheries for fertilized eggs for instance, or even as legs, which allow species such as *Plumatella* or *Scrupocellaria* to move about (very slowly!), but more commonly for defensive purposes. These defensive zooids are called avicularia from their resemblance to tiny bird-beaks (*avis* is a bird in Latin). The colony grows by simply budding off new individuals, but they also produce sperm and eggs in testes and ovaries. Cross-fertilisation takes place, but quite how they do it is still a mystery. The fertilised eggs hatch to produce free-swimming larvae of the trochophore type found in many groups of invertebrates. Like sponges, bryozoans produce special overwintering buds called statoblasts.

Albert Russell Nichols recorded 75 species from the wider area of the seas around Clare Island for the Original Survey (including 3 entoprocts: see page 103), and many of these are common on the shore (most bryozoans live in shallow water).[88] The species to be found on Clare Island shores include *Membranipora membranacea* (sea lace), which occurs everywhere on the fronds of *Laminaria* (less commonly on *Fucus*), its lace-like encrustation often forming large white patches.[89] Each cell in the honeycomb houses an individual moss-animal; there may be millions on a single plant. Colonies of *Scrupocellaria reptans* are common between tide-marks, especially on sandy shores. As the name suggests, it is a mobile species. *Alcyonidium hirsutum* is generally distributed, growing on algae between tide-marks, where it often occurs along with *Membranipora* [*Flustra*]. It forms gelatinous sponge-like masses with irregularly lobed and fingered margins.[90] *Crisia cornuta* and *Crisia eburnea* are both common on *Laminaria* roots, as well as other seaweeds and shells between the tide-lines, forming delicate branched white tufts no more than a few centimetres high.

Cribrilina cryptooecium forms white or silvery encrustations on stones between tide-marks. *Tubulipora phalangea* is an encrusting bryozoan that often grows on shells such as oyster. The colony is divided into distinctive radiating lobes, with the zooids arranged in transverse rows in each lobe. Colonies of *Cauloramphus spinifer* are crust-like patches on stones between tide-marks. The zooecia have an operculum and spines, with tall, slender stalked avicularia among the spines.

Callopora lineata and *Callopora dumerilii* are among the most characteristic encrusting bryozoans of rocky shores, where they form irregular sheets on shells, stones, seaweeds and the undersides of stones. They are very often found in *Laminaria* holdfasts. Among other species that are particularly associated with *Laminaria* holdfasts are *Escharoides coccinea*, an extremely common encrusting bryozoan that forms extensive subcircular patches, red-orange in colour; it is also found growing on stones between the tides. Another is *Microporella ciliata*, which again also forms white encrusting colonies on stones, shells and other seaweeds on the shore.

Among the many other species that form encrusting patches on rocks, shells and so on, are *Membraniporella nitida*, which occurs as irregular sheets that spread over the surface of stones between tide-marks. *Hippothoa divaricata* is often found on shells between tide-marks, forming small runner-like colonies with ovicells. *Haplopoma impressum* is another common encrusting species on stones between tide-marks, the colonies typically occurring as small patches. *Escharella immersa* forms extensive white to grey thin sheets on hard substrates along the shore. *Fenestrulina malusii* forms bright white patches—never more than 1cm across—on a wide variety of substrates. *Cellepora pumicosa* is generally distributed on stones, seaweeds, shells and so on, along the tide-line and into deeper waters.

Phylum Chaetognatha (arrow-worms)

Arrow-worms look like torpedoes or darts, and are completely see-through. They have two pairs of transparent lateral fins and a tail fin that give them remarkable stability and arrow-like mobility over the short distance needed to impale their prey and inject it with toxins.

They are pelagic species for the most part, and have lived in the oceans since the Cambrian. There are only a hundred or so species today, but they sometimes occur in absolutely prodigious numbers. They are important in the diet of many

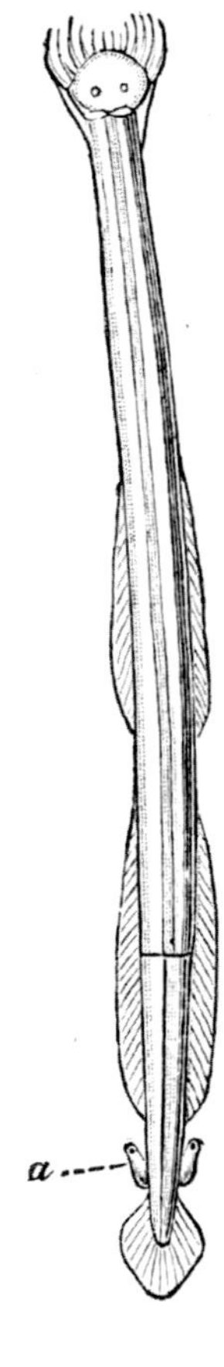

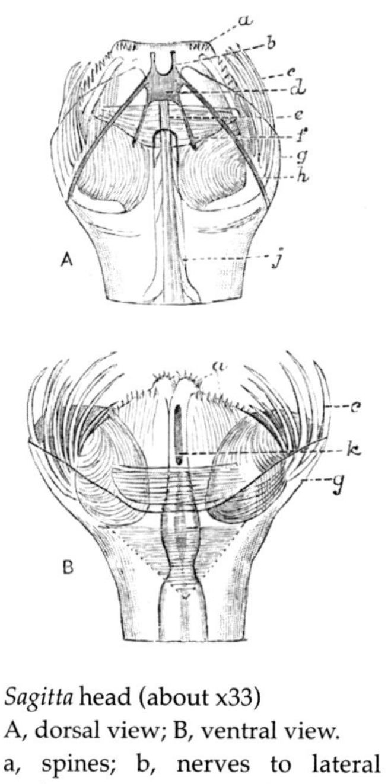

Sagitta head (about x33)
A, dorsal view; B, ventral view.
a, spines; b, nerves to lateral cephalic ganglia; c, hooks; d, cephalic ganglion; e. olfactory nerve; f, optic nerve; g, hood; h, commissure to ventral ganglion; j, olfactory organ; k, mouth.
The Cambridge Natural History, Volume II (1896), 188.

Fig. 6.31a *Sagitta bipunctata.*

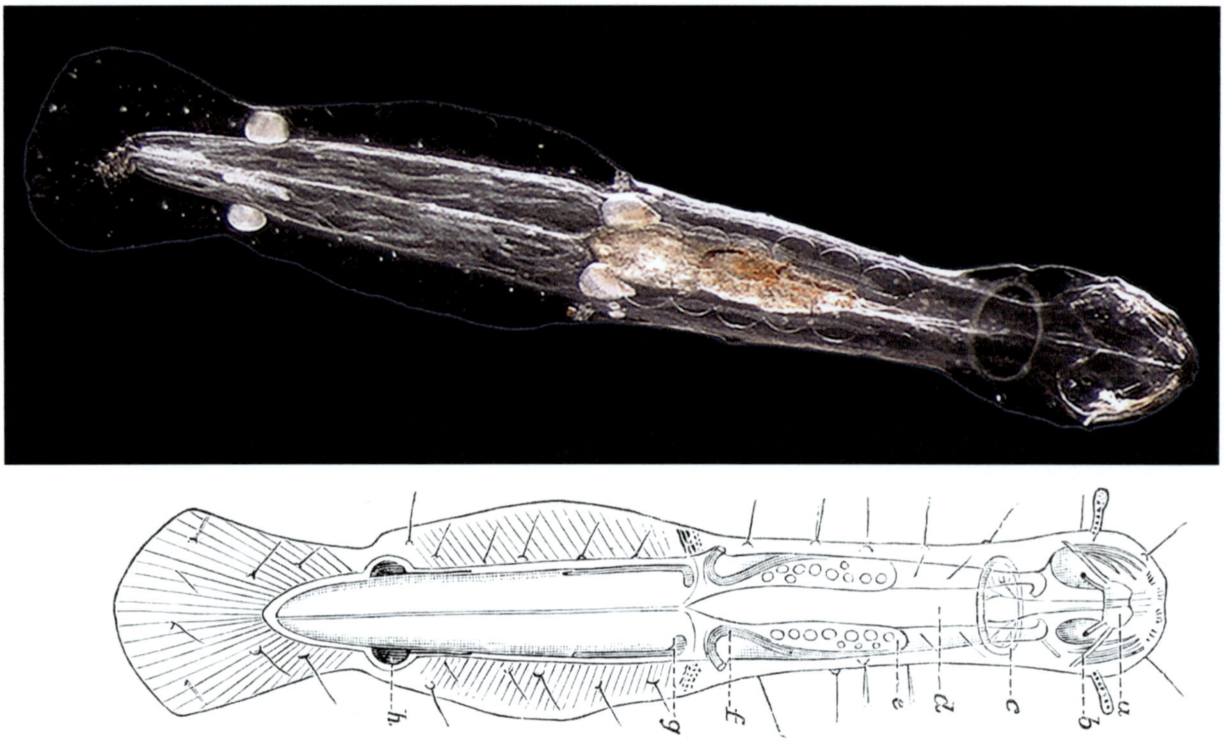

Fig. 6.31b *Spadella cephaloptera* (photo: Zatelmar (CC BY-SA 3.0)).
Dorsal view: a, cephalic ganglion; b, commissure to ventral ganglion; e, olfactory organ; d, alimentary canal; e, ovary; f, oviduct; g, testis; h, seminal vesicule.
The Cambridge Natural History, Volume II (1896), 189.

fishes (notably herring), but on the other hand they are themselves voracious predators of anything smaller than them (including other arrow-worms).

Spadella cephaloptera is the only littoral species; it lurks unseen among water weeds (in much the same way as The Predator in the popular science fiction films!). The other species recorded by Southern (1914b) was *Sagitta bipunctata* (Figure 6.31).[91]

Phylum Echinodermata

The echinoderms comprise starfish and brittle stars, sea-urchins and sea cucumbers, as well as the now largely extinct crinoids; there are about 7,000 living species in total. They were studied for the Original Survey by Albert Russell Nichols, who recorded some 32 species from Clare Island and the waters around it. He also collected in Blacksod Bay to the north and Inishbofin to the south, which added a further nine, fourteen being common to all three areas. Most of these species are sublittoral, and only a few are met with on the shore.[92]

The cushion star (*Asterina gibbosa*), which is fairly common onshore, is a tiny greenish starfish with short, stubby arms, tinged with yellow and red, about 6cm across. The other common starfish is *Asterias rubens*, which is reddish-brown to pale red or yellow in colour, and has a characteristic red beauty spot at the tip of each ray (Figure 6.32). It can grow up to 50cm, but the specimens found on the shore are generally small. A further ten or so species occur in the shallow waters offshore.

As adults, starfishes are benthic animals, but early in life they spend time as swimming larvae in the plankton before attaching themselves firmly for a brief sedentary phase. The larvae are like something from another world, delicate glass-like creatures with inexplicable horns and protuberances. During their sedentary phase, their tissues become reorganised in much the same way as those of an insect inside its pupal case. Starfishes are voracious predators of mussels and oysters; the starfish wraps itself round the shell and slowly forces it open before everting its own stomach and eating the flesh of the mollusc after it has been softened with digestive juices.[93]

Brittle stars have a pronounced central disk and slender spindly arms that they use to propel themselves in octopus-fashion; they resemble so many centipedes attached at regular distances around a little sea urchin. They are impossible to

Fig. 6.32a Cushion star (*Asterina gibbosa*) (Parent Géry).

Fig. 6.32b Common starfish (*Asterias rubens*) (Hans Hillewaert (CC BY-SA 4.0)).

handle, because unless they are plunged immediately into water they just fall to pieces. They are very easy to recognise as a group, but species determination can be very difficult (one of the easier key features is the number of spines on one side of one arm joint).

Brittle stars are scavengers, although they also capture worms and other small creatures with their flexible arms. In some places they form great assemblies on the sea floor. In the deeper waters off Clare Island large numbers of the brittle star *Ophiactis balli* can often be found under the rocks (Figure 6.3a). *Ophiothrix fragilis* is a tiny red brittle star, common among stones, seaweed and shells on the lower shore.

There are two species of sea urchin around Clare Island. The common sea urchin *Echinus esculentus* has red and white spines with lilac tips,

Fig. 6.32c Purple urchins (*Paracentrotus lividus*) among corraline algae in a rock pool.

Fig. 6.33a Colony of sea squirts (Bernard Kaye).

Fig. 6.33b The star ascidian (*Botryllus schlosseri*) (Parent Géry).

and feeds on algae and small encrusting animals. The most eye-catching of all the animals of the tide pools is the purple urchin *Paracentrotus lividus* (Figure 6.32c). It is often gregarious, living side by side in villages of dozens or even hundreds, each in a cosy hollow it has chiselled out for itself in the rock. It is usually purple, but olive-green and reddish-brown ones also occur. Roland Southern described the colonies of *Paracentrotus* as 'one of the most beautiful and characeristic features of littoral life on moderately exposed parts of the coast of the west of Ireland, where the water is very transparent and least affected by the adjacent land'.[94] *Asterina* and *Paracentrotus* are especially characteristic of exposed situations.

The other important group of echinoderms are the sea cucumbers (Holothuroidea), which indeed do look something like what you would imagine a cross between a spiky gherkin and a worm of some sort might look. But echinoderms they are, a group that abandoned the radial symmetry of their cousins for a more unidirectional approach to life. Unlike other echinoderms they have a front and a back end, with a mouth at the front surrounded by a circlet of retractable tentacles. Mostly they live offshore, so we seldom see them, but now and then they are encountered between and under stones on the lower shore. Nine species have been recorded in the waters off Clare Island; these are rather a conservative bunch in terms of surface ornamentation, but many of those that live in warmer seas are extravagantly and colourfully flamboyant. *Cucumaria lactea* is perhaps the most widely distributed (but is no more than 2.5cm long). *Cucumaria saxicola,* which is three times as long, lives in narrow cracks or the holes made by rock borers.

Phylum Urochordata

Urochordata comprise two classes: the tunicates or sea-squirts (Ascidiacea) and the salps or chain tunicates (Thaliacea)—the wonderful planktonic *Salpa* species, several of which are known to be abundant way out to sea. We should note in passing the related phylum Hemichordata, which contains the acorn worms, several species of which occur in the wider marine area: *Dolichoglossus ruber* and a species of *Balanoglossus*.

If you were asked to select from among the bewilderingly wide range of lesser creatures available as candidates those that are closest to our branch of the evolutionary tree, sea squirts would be unlikely to feature on the short list: not just because of the undignified sound of their common name, but because (it would appear) they look nothing like us. However, it is to the structure of the larva that biologists are looking when they tell us that these unassuming creatures are indeed our closest non-vertebrate relatives. The free-swimming larvae are as active as tadpoles and startlingly like them in structure. However, when they find a place to settle down, this structure is reorganised in radical fashion into that of the adult.

Sea squirts are enclosed in a thick, jacket-like skin (from which they take their more scientific name of tunicates). The body has two openings, an *inhalent* mouth through which water is wafted in, and another *exhalent* opening through which this is expelled after passing through an elaborate sieve mechanism to remove food particles. If the animal is disturbed the body wall contracts suddenly, and water is forced out in two jets through these openings: hence the common name of 'sea squirt'.

Many are solitary creatures, others colonial and enclosed in a communal 'tunic'. These colonial forms are often abundant on the stems of brown seaweeds, where they occur as clasping masses of firm jelly patterned with what appear to be star-like flowers: blue, crimson, orange or white in colour. Each petal is an individual animal, with its mouth at the tip of the petal and a common exhalent opening at the centre.

Tunicates and hemichordates were just about the only groups for which Praeger and his colleagues were unable to find a specialist to contribute to the Clare Island Survey, so it fell to George Philip Farran and Rowland Southern to bring together what information was available on these groups, the result being a check-list for the tunicate and hemichordate fauna of the wider marine area from Achill Island down to Ballynakill. From reports of earlier collections, they compiled a list of 35 species, but uncertainty about nomenclature means the actual number is less than this.[95]

Almost nothing in the Original Survey report for these animals relates to the island itself. However, many ascidians live attached to stones and seaweed in the littoral zone, and can be seen on occasion between the tide lines; others are frequently washed ashore from deeper waters. Through diligent searching, you are sure to discover such common solitary species as *Ascidiella aspera*, which is widespread in shallow water, and easily recognisable by its soft, cylindrical see-through body (up to 13cm tall). The commonest of the colonial forms is the star ascidian *Botryllus schlosseri* with its six-petalled white flowers (Figure 6.33).

Phylum Craniata
Subphylum Vertebrata

Fishes

George Farran prepared the report on marine fishes for the Original Survey, basing his check list for the deeper waters on the fisheries records of the cruiser *Helga*, and for the shoreline on specimens collected by members of the various expeditions to the island.[96] The fish fauna of the trawling ground north and north-east of Clare Island was better known than the area of rocky bottom on other sides, which is unsuitable for trawling. The resulting list of some 93 species he considered to be 'manifestly an incomplete one'.[97]

A number of shoreline species are commonly stranded in pools at low tide (and in some cases prefer to live here), among them the father-lasher (*Cottus bubalis*), a brightly-coloured and very spiny fish 12–18cm long (but sometimes up to 45cm). Although it looks conspicuous and showy as an illustration in a book, the irregular outline and mottled colour pattern are highly effective camouflage, making it almost invisible among the rocks and weeds. It feeds mainly on such things as crustaceans, prawns and shrimps, upon which it pounces from the shadows. A yellow or greenish pattern mottled with black makes the blenny or shanny (*Blennius pholis*), a small fish some 5–10cm long, almost invisible among the weeds and rocks along the shoreline, where it is common and easily recognised by its long, spiny fins. Its broad teeth are specialised for biting acorn barnacles and limpets off the rocks. When the eggs are laid the male mounts ferocious guard over them.

Fig. 6.34 Butterfish (*Centronotus gunnellus*) (Bernard Kaye).

Few fishes are more distinctive than the gunnel (butterfish, Peter-nine-eyes) (*Centronotus gunnellus*), which is common among rocks and stones on shore. It is very slippery and somewhat flattened laterally, with nine dark spots along each side of its eel-like body (it is typically 5–15cm long, but it can reach a length of up to 28cm). During the breeding season (December to March) both parents look after the developing eggs, curling their sinuous bodies around them as a sort of nest.

The small worm pipe-fish (*Nerophis lumbriciformis*) (10–15cm, but up to 23cm) is common on shore. It has an eel-like body which tapers smoothly from head to tail; it has no tail-fin, just a short dorsal fin set noticeably forward of half-way along the body. Pipe-fish are very particular about what they eat, their delicate mouth designed for picking off copepods and other small planktonic creatures. Here too parental care is the rule, but it is the male that carries the eggs around with him (glued in rows to his body) until they hatch. The

Fig. 6.34b Sunfish (*Mola mola*) (Nol Aders (CC BY-SA 3.0)).

Fig. 6.34c Basking shark (*Cetorhinus maximus*) (Greg Skomal / NOAA Fisheries Service).

larger worm pipe-fish (*Entelurus aequoreus*) is also found on the shore.

George Farran recorded only two specimens of the Cornish sucker (*Lepadogaster gouanii*) on the shores of Clare Island, but it is worth keeping an eye open for this remarkable fish. It is some 10cm long, and takes its name from the extraordinary sucker on its underside, which has evolved from a pair of pelvic fins that enable it to cling steadfast to rocks. It is a fish of southern disposition, so it breeds in summer. The mother lays her single layer of yellow eggs on rock or weed, and both parents look after these as they develop. Notice how many shore fishes take care of their eggs, in contrast to pelagic species.

Sand eels (*Ammodytes*) come inshore in spring and summer, and at low tide may be found buried in sand. There are two species, which generally live together in shallow water: the greater sand eel (*Ammodytes lanceolatus*) which is about 20cm long, and the lesser sand eel (*Ammodytes tobianus*) which is half this. Sand eels move like flashes of silver; they swim in shoals and are much preyed upon by seabirds and fish—and porpoises at times. Their defence is the speed with which they bury themselves in sand, which their shovel-like lower jaw and snake-like body enable them to do with remarkable agility. A great shoal of them may be hiding invisibly underfoot as you walk across the beach (see page 126).

Perhaps the most exotic fishes seen in the waters around the island from time to time are the sunfish (*Mola mola*) and basking sharks (*Cetorhinus maximus*), frequently seen off the island in autumn. Sunfish are unusual in that they feed largely on jellyfish. They often float at the surface, where their conspicuous dorsal fin makes them easy to spot; in fact they can even be taken in a net if approached with care! (Figure 6.34b).

Among species of commercial importance common in the waters around Clare Island at the time of the Original Survey were mackerel (*Scomber scombrus*), herring (*Clupea harengus*), whiting (*Gadus merlangus*), cod (*Gadus morhua*), pollack (*Gadus pollachius*), ling (*Molva molva*), turbot (*Rhombus maximus*), brill (*Scophthalmus rhombus*), plaice (*Pleuronectes platessa*), haddock (*Gadus aeglefinus*), grey mullet (*Mugil chelo*), horse mackerel (*Caranx trachurus*), John Dory (*Zeus faber*), dab (*Pleuronectes limanda*), sprat (*Clupea sprattus*), several species of gurnard (*Trigla* spp.), thornback ray (*Raia clava*), spotted ray (*Raia maculata*), blonde skate (*Raja brachyura*), spotted dogfish (*Scyllium canicula*) and conger eel (*Conger vulgaris*). Less common fishes included hake (*Merluccius merluccius*), halibut (*Hippoglossus hippoglossus*), lemon sole (*Microstomus kitt*), black sole (*Solea solea*) and common skate (*Raja batis*).

The absence of salmon and sea trout from the list is somewhat puzzling. Stories are told on the island of the periodic but unpredictable arrival of great shoals of sea trout pushed up against the shore—probably by predators such as dolphins: news of which in McCabe's pub precipitated a stampede to the beach and the launch of the currachs, from which great quantities of fish were taken in herring nets.[98]

Fig. 6.35 Ballan wrasse (*Labrus bergylta*) (Jonathan Couch (1877) *History of the Fishes of the British Isles*).

Harvesting the sea today

Rowland Southern remarked on the way the islanders of his day seemed to be less interested in the produce of the sea than they might have been, although the relative unimportance of the fishery at that time was at least partly due to the want of suitable boats and equipment and of an adequate harbour. According to Samuel Lewis (*c.* 1840) 'About 340 persons, who are also farmers, are occasionally employed in the fishery.' Facilities were much improved under the Congested Districts Board and by the time of the Clare Island Survey most families were able to add to their farming income by taking part in the autumn herring and mackerel fishing, by lobster fishing (in which about three-quarters of the islanders engaged), and by burning kelp, of which about thirty tons were exported annually—much less than in earlier years, when kelp was more profitable.[99] At the time of the Original Survey, lobsters (*Homarus gammarus*) were 'fished for regularly by the islanders,' and edible crabs (*Cancer panurgus*) 'taken in lobster pots' south and east of the island.[100] Hand-line fishing (trawling the lines behind a currach or yawl) for scad (horse mackerel), mackerel, ballan wrasse and pollock (or rockfish) was carried on in summer and autumn.[101] Lobster fishing has increased prodigiously in recent decades, in no small measure because GPS makes the exact location of good fishing marks so easy.[102]

There were two mackerel fishing seasons. The spawning 'spring fish' were taken in trains of drift nets some kilometres to seaward of the island in April and May; smaller 'harvest' fish were caught closer to shore between June and year's end, either in drift nets or on hand lines.[103] Today mackerel are eaten by everybody on the island during the fishing season (they are said to be particularly easy to catch near the sea-farm cages). Herring were often netted in large numbers at the end of September. John Dory (*Zeus faber*) were frequently taken by line fishing.

Ballen Wrasse (Conner, Bream) (*Labrus bergylta*) were 'caught on lines in great numbers by the Clare Island men and split and dried for food' (Figure 6.35). [104] It was much fished for in the kelp forests off rocks at low water mark, the lines provided with an ingenious crab-shell guard to prevent the hooks fouling in the weed.[105] The fish were normally salted; boiled in milk with an onion or two, they were much appreciated as a good 'kitchen' or relish with potatoes. Of all the fish in the waters off Clare Island why was ballen wrasse favoured? Because it is a substantial fish (often half a metre long) that lives in the kelp forests, where it could be fished with a line from rocks at low tide once the trick of not getting caught in the tangleweed was mastered. They feed on shelled molluscs and crustaceans; their prominent lips are adapted for picking up such shelly creatures, and their powerful teeth for crushing shell and carapace.

With the advent of the salmon farm, wrasse have taken on a new importance because they are 'cleaner fish', which peck harmful sea lice from the skin of salmon. The international company Marine Harvest is supplied with the fish by islanders, who take them from the wild. The capture method differs from traditional line-fishing, using instead a specially designed pot in which several wrasse are imprisoned in comfort, since the fish must not be harmed. In view of concern that this practice may seriously reduce populations of wrasse, research is under way in several salmon-producing countries to develop hatcheries where wrasse can be bred specially for salmon farms.[106]

Sand eels (*Ammodytes*) were much prized as food. They come inshore in spring and summer, and at low tide may be found buried in sand; they were 'sometimes caught in large numbers, of all sizes, on Clare Island strand'.[107] Conger eels (*Conger vulgaris*) often live under stones on muddier shores and can be semi-stranded at low tide, at which times they were fished for by the islanders.[108] The piked dogfish (*Acanthias vulgaris*), which was sometimes very abundant and caused considerable damage both to mackerel nets and the enmeshed fish,[109] was also sometimes eaten, and its oil was formerly used with rush-lights and lamps.

Today only lobster and crab are fished to any extent, and smaller numbers of crayfish; a few people still gather periwinkles. There is a ready market for shellfish at Roonagh, which is periodically visited by wholesaler buyers.[110] Salmon fishing was once important, in spring and summer especially, but salmon are hardly ever taken in the seas around Clare Island today. Over-fishing over the last 30 years has resulted in a fall off in the size and number not only of lobsters and crayfish, but of commercially significant species generally. The advent of French and Spanish long-lining boats has led to the virtual disappearance of several species.[111]

Sport fishing continues to be popular in the deeper waters off the island, where porbeagle, blue shark and bluefin tuna are sometimes taken, and pollock, octopus and John Dory widely fished for. The skate fishery in Clew Bay is slowly recovering with the introduction of mandatory catch and release.

The harvest of the sea continues to be an important, indeed vital, element in the rural economy of Clare Island, but today's enterprise is more a form of intensive farming than a continuation of traditional fishing. Clare Island Organic Salmon Farm, which is owned by Marine Harvest, was established in 1995, and the fish it rears have the reputation of being among the finest farmed salmon in the world. Initially there was much concern about potential negative impacts on the marine environment, which at times led to significant tensions within the community. Twenty years on most of the initial concerns have been allayed, and indeed it is not an exaggeration to state that without the farm Clare Island would be in serious decline.[112]

The life cycle of farmed fish is 18 months. The new smolts are introduced around Easter time and reared in cages south-east of the harbour until they are big enough for transfer to the waters off Portlea; great care is taken to prevent transfer of disease between generations. The operation is conducted along strict organic standards. The salmon are fed twice a day with a pelleted fish-meal and fish-oil feed made from the by-products of fish processed for human consumption, including herring and mackerel, as well as shrimp and plankton. Great care is taken to minimise waste; the little that escapes through the nets attracts a variety of different wild fish species, and constant scouring by the tides means that the bottom under the nets is completely clean. The most expensive ingredient in the meal is a special yeast named *Phaffia rhodozyma*, the fermentation activity of which produces the astaxanthin responsible for the pink colour of the flesh; necessary vitamins and minerals are also added. Typically, there would be fifteen cages, each with 75,000 fish as smolts.

The nets and other equipment have to be robust to cope with the exposed conditions, but this also means that the cages are constantly scoured by strong tides, against which the fish are constantly swimming, so that they have the strong tails and firm muscle tone of wild fish. Due consideration is given to the space requirements of naturally shoaling fish. The stocking density is two fish to a tonne of water, so that they have the same freedom of movement as in the open sea. Indeed, it is claimed that the 23,200km or so they cover in their caged lives is close to that of salmon in the wild.

Although data to assess the impact of the salmon farm on marine flora and fauna in any detail are absent, such indications as we have suggest no significant negative impact that can be attributed to its activities. Indeed, there is anecdotal evidence of a positive impact, due to the small amounts of feed that escape from the cages.

Fig. 6.36 Grey seals. From Millais (1904–1906), *The Mammals of Britain and Ireland.*

Marine vertebrates other than fishes

The only carnivorous mammals are grey and harbour seals, along with the otter—creatures of both land and sea—but more properly considered here since their food comes from the sea, and it is in the sea the grace and agility with which evolution has equipped them is manifest (Figure 6.36). The otter has been studied in detail on Clare Island, where it feeds on a varied diet of fishes, crustaceans and molluscs offshore, the composition of species varying with seasonal availability of its prey.[113] The main fish eaten are rockling, blenny and wrasse, while the predominant invertebrate species are mussels, limpets and isopods. Anecdotal evidence suggests that seals have increased significantly both in size and number in recent years in the general area; the south shore of Caher Island can sometimes appear to be covered with seals hauled out on the rocks. In salmon-fishing days they were the bane of the Clare Island fisherman, who sometimes had to watch helplessly as they took fish from the nets, leaving just the head behind.[114] A survey carried out in 2005 estimated the population of grey seals to be between 150 and 500.[115]

Offshore, the species of dolphin most regularly seen is the bottlenose dolphin. There have also been occasional sightings of killer and fin whales.[116] Whales, which have altogether cut their ties with the land their terrestrial Tertiary ancestors walked, are a rare occurrence in the waters around Clare Island.

The splendid leatherback sea turtle (*Dermochelys coriacea*), one of the ocean's most remarkable animals, visits the deeper waters around Clare Island from time to time, especially between July and October.[117] The sub-species that visits our seas ranges all the way across the Atlantic and as far south-east as Australia, returning to the north-east coast of South America and the Caribbean to lay its eggs, only migrating into the deeper waters of the oceans in pursuit of the jellyfish on which they feed.[118] In spite of its great size—over 2m long—it is not easily seen, but there have been numerous sightings in recent years of this great creature, whose form was shaped in the Triassic period of earth history, in the days of the first dinosaurs.[119]

> Being interested to see what kinds of fish were in the deeper water off Clare Island some years ago I experimented setting plaice nets on the bottom in quite deep water: say 50m plus, maybe 8 to 10 miles west of Beetle Head. We would set the net in the morning when we went fishing – making sure that we had a flowing tide most of the day as I am absolutely convinced that fish only feed significantly on flowing tides – and then we would pull in the net in the evening on the way home. On one such trip, on a beautiful summer's day, flat calm, on the way back we couldn't find the net and suspected that it had drifted. About two weeks later a friend of mine from Clare Island spotted what he thought was my plaice net on the pier in Inishturk, and obviously being careful with his query he got the story. We had set the net, and perhaps an hour or two later one of the Inishturk boats out fishing for lobster saw our buoys and of course went to investigate. However as they came near to the net the net continued to swim away. They couldn't figure out what was going on until they eventually caught up with the net and found that a huge leatherback turtle had snagged itself in the rope from the net: the rope was caught between its head and its flipper and the leathery carapace. Fortunately they were able to free the poor unsuspecting turtle. Note that he only caught in the rope and not in the actual net itself, which was way below him. Of course they pulled the net in and brought it with them, thinking it to be drifting. I never did find out what nice fish they got in the net.[120]

The plankton

Although we seldom get to experience it directly, its enormous importance requires a note at least on the plankton: the uncountable myriads of mainly tiny creatures that live at the mercy of ocean currents, and upon which the rest of life in seas and lakes depends. Not all of the creatures that constitute the plankton are microscopic; it includes such things as jellyfish. What they all have in common is that they are at the mercy of the current (although many have diurnal *vertical* migration cycles), unlike the free-swimming *nekton* and bottom-dwelling *benthos* (*planktos* is the Greek word for drifter or wanderer).

There are two fundamental groups, the *zooplankton* and *phytoplankton*: although there are lots of tiny creatures that defy any simple distinction between plant and animal and so straddle the boundary between the two. The phytoplankton are predominantly microscopic algae, and are the primary producers: orders of magnitude greater in terms of global productivity than all the forests, grasslands and peatlands on land. Upon them everything else depends, from the small animals that feed directly on them to larger ones at one or more removes up the food chain, all the way to the carnivorous fishes, birds and mammals at the top.

Many groups of plants and animals are permanent residents of the plankton: their entire life is spent here. These are the *holoplankton*, which includes most algae, copepods, tunicate salps and so on. Many other groups however only spend their early days as planktonic organisms, undergoing remarkable metamorphoses over the course of their life's journey, before settling down to a less buffeted life as bottom-living or shoreline citizens. These are the *meroplankton* (the larvae of many marine worms and snails, crustaceans, starfish, cnidarians, sea urchins and most fishes).

Almost nothing is known of the marine phytoplankton, nor of the Protozoa and other nanoplankton, creatures too small (less than 20 microns) to be retained in standard nets: essential though these are as the food of the early larval stages of most of the littoral fauna with free-swimming larvae, as well as holoplanktonic animals. Uncountable as the numbers of the smaller invertebrates may be at times, their numbers are dwarfed by the unimaginably numerous zillions of these tiny one-celled creatures.

George Farran prepared a summary of the records of marine plankton gathered during the course of the Original Survey that were scattered through the various fascicles dealing with the different taxonomic groups.[121] He divided the wider Clare Island marine area into three regions, each of which has a different planktonic assemblage. The area of open sea around the islands is characterised by a relatively low species diversity compared with both the open ocean and the shallower inshore area of bays and harbours: but it is extremely important in terms of the productivity of certain copepod species, the most abundant of which is *Calanus helgolandicus*, which occurs in enormous numbers from May to August, followed in abundance terms by *Metridia lucens* (Figure 6.37). Other species that are especially characteristic and hardly ever absent are *Calanus finmarchicus*, *Pseudocalanus elongatus*, *Parcalanus parvus*, *Centropages typicus* and *Oithona similis*. *Calanus finmarchicus* is the largest of these (up to 2mm long); a characteristic part of its strategy is that it spends the six coldest months in hibernation, often sinking to depths of 500–2,500m to do so before rising again to the surface when conditions improve.[122] These most characteristic species are reinforced by a number of others that have drifted in from deeper oceanic waters to the west, as well as some from waters closer to shore. A number of species of Cladocera also play an important role.

The shoals of schizopods that are the main food of the mackerel that gather along the coast in winter and early spring probably have their origin further out in the western sea as well.[123] There are also several characteristic amphipods, including the little green-eyed *Hyperia galba*, which turns up in large numbers when flotillas of its host, the mauve stinger jellyfish *Pelagia noctiluca* approach the shore. *Pelagia* is widely distributed in all the warm and temperate oceans of the world, and it frequently occurs in prodigious swarms; in 2007, a swarm covering an area of 25km² wiped out a 100,000-fish salmon farm in Northern Ireland. *Pelagia* is intensely phosphorescent, giving off a ghostly vapour that clings to one's hands. It can also administer a severe and painful sting. The awesome-looking sea-gooseberry (*Pleurobrachia pileus*, a comb-jelly) occurs regularly, as do several other cnidarians that have drifted in from seaward. Others that have their home in the wider region are common in summer and autumn.[124]

The most abundant planktonic mollusc is *Limacina retroversa*, which occurs along with the larvae of various bottom-living gastropods and bivalves, the most conspicuous (and frequently encountered) being *Lamellaria*, a sea slug the adults of which prey on ascidians (in the cloaca of which it lays its eggs). *Limacina* is a predatory sea-snail belonging to a group known as the sea butterflies (a reference to their flight-like style of swimming), but these waters also harbour its deadly enemy, the pteropod *Clione*, which feeds entirely on *Limacina*. The larvae of decapods are also common, and in summer the larvae of echinoderms are numerous. Salps turn up periodically in immense shoals that have come in from the open ocean (there was a remarkable invasion in August 1903).

Fig. 6.37a *Calanus helgolandicus* (Tone Falkenhaug (CC BY 4.0)).

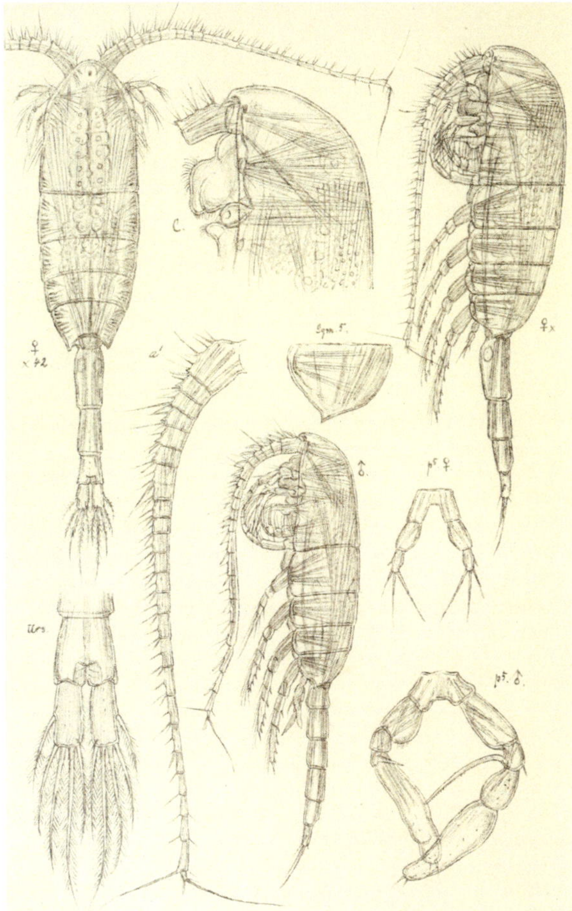

Fig. 6.37b *Metridia lucens* (G.O. Sars, *An account of the Crustacea of Norway*, c.1901).

An intermediate off-shore region between the mainland and the larger islands has a mixed fauna, made up of species derived mainly from the open sea, but with the addition of some typically shore forms. The great shoals of *Calanus helgolandicus* extend into these waters, as do shoals of *Pseudocalanus elongatus* and *Temora longicornis*. Conditions here seem to be optimal for the jellyfish *Aurelia aurita*. Other characteristic inhabitants include several mycids (*Gasterosaccus spinifer*, *Gasterosaccus Normani* and *Anchialus interrupta*) and the copepods *Corycaeus anglicus* and *Altheuta interrupta*.

Oceanic creatures from further out drift in on occasion, notably salps and the mauve stinger, and occasionally a few stray individuals of the remarkable siphonophore known as the by-the-wind-sailor (*Velella*), which looks like a smaller version of the spectacular Portuguese man-o-war.

One of the most noteworthy features of the region of the bays and harbours, where temperature and salinity are much more variable than in the other regions, is the prevalence of creatures that are meroplanktonic: the early stages of molluscs, decapods, barnacles and worms, some of which will become pelagic adults, while others will settle down to a more sedentary life. Along with these are reptant or sessile juveniles that will later become free-swimming forms that have an alternation of generations, one of which is planktonic (as with many cnidarians). There is also a host of species that stay quietly at the bottom during the day but come towards the surface at night, as many Harpactic copepods, Cumacea, amphipods and schizopods do.

Fig. 6.37c Sea gooseberry (*Pleurobrachia pileus*) (©Joe (jb1562) Flickr, CC BY-NC-ND 2.0).

The Euphausians that are so noticeable in the open sea are almost absent, but several species of Mysidacea—all of them nocturnal swimmers—remain abundant. Numerous amphipods also become active in the plankton at night. By day the Cumacea remain buried in sand, but when night falls they come to the surface in shallow water in large numbers. During summer, the larvae of decapods occur in large numbers, those of porcelain crabs perhaps being the most numerous. Along with them are molluscan veligers and the tailed larvae of ascidians.

In terms of numbers the copepods remain the most important organisms, but there are a number of significant cladoceran species. Several cnidarians are common, and although larval echinoderms occur they are not very numerous. The chaetognath *Spadella cephaloptera* is a permanent inhabitant, and the arrow-worm *Sagitta bipunctata* occurs at times.

Notes

1 Southern 1915, OS 67.

2 Myers 2002.

3 Southern 1915, OS 67, 49–77.

4 Southern 1915, OS 67, 51–52.

5 O'Riordan *et al.* 2002a.

6 A Touch of Class,' *SubSea* 27 (Spring 2005).

7 Southern 1915, OS 67, 98–100.

8 Southern 1914a, OS 47, 2.

9 Southern, 1915, OS 67, 2. On page 25 there is a list of hydroids and bryozoa that live on stems and fronds of *Laminaria* – in addition to those in the holdfasts.

10 Cotton 1912, OS 15.

11 *Ibid.*; two visits of a fortnight each in April and July of 1909, the same in June and October of 1910, and in 1911 a week in February, a fortnight in early May and another in late August. 'I have also availed myself freely of the help of specialists' (page 6).

12 *Fucus spiralis*, *F. guiryi* and bladderless forms of *F. vesiculosus* are not easy to tell apart even by the most expert.

13 Cotton 1912, OS 15, 23–26.

14 These are listed on pages 25–26 (Cotton 1912).

15 Cotton 1912, OS 15, 48–50.

16 Cotton 1912, OS 15, 161–64, 168–71.

17 Maggs *et al.* 1991.

18 Guiry *et al.* 1991.

19 Guiry 1997.

20 Rindi *et al.* 2002: 'Most subtidal assemblages are standard kelp forests, and intertidal assemblages, especially on the southern shores, show very few macroscopically observable species' (page 204).

21 Rindi *et al.* 2002, 203–04.

[22] Rindi and Guiry 2004.

[23] The taxonomic sequence adopted here follows Margulis and Schwartz 1998.

[24] In Latin *granulum* is a small grain, *reticulum* a network, and *foramen* a perforation.

[25] Heron-Allen and Earland 1913, OS 64, 4.

[26] Heron-Allen and Earland 1913, OS 64, 2.

[27] Wentworth Thompson 1942.

[28] Heron-Allen and Earland 1913, OS 64, 5; all the locations are shown on a foldout map at the beginning of the report.

[29] Heron-Allen and Earland 1913, OS 64, 69.

[30] Chalk is made up of foraminifera, coccoliths, sponge spicules and radiolaria etc. in a calcite mud.

[31] Heron-Allen and Earland 1913, OS 64, 9.

[32] The 100 fathom line lies some 70km west of Clare Island; the continental shelf is 320km west of the coast.

[33] Heron-Allen and Earland 1913, OS 64, 18.

[34] Stephens 1912a, OS 59, 3.

[35] Stephens 1912a, OS 59, 17.

[36] Stephens 1912c.

[37] Stephens 1913c, OS 58, 6; *Corynactis* is in the order Corallimorpha.

[38] Most jellyfish belong to this old (and now informal) order.

[39] Stephens 1913c, OS 58, 6.

[40] In Greek mythology Hydra is a many-headed monster. As soon as one of these was cut off two grew in its place unless the wound was stopped by fire. Its dispatch was one of the labours of Hercules. Medusa was one of the Gorgons, three monstrous sisters in Greek mythology. She was noted for the beauty of her locks, which the goddess Minerva changed to serpents because Medusa had violated the sanctity of her temple.

[41] Stephens 1913c, OS 58, 2, 4.

[42] Southern 1912, OS 56, 1.

[43] Southern 1912, OS 56, 14.

[44] Nemertes was one of the Nereides; see endnote 66.

[45] Southern 1915c, OS 55, 9; although the green forms are now regarded as a different species: *Lineus gesserensis*.

[46] Southern 1914b, OS 54. 1.

[47] *Ibid*.

[48] As a further example of the sort of education these scientists had: of the 26 references listed in the bibliography only 3 are in English.

[49] Farran 1914, OS 40.

[50] Lobsters were 'fished for regularly by the islanders', and crabs 'common, taken in lobster pots' (Farran 1914, OS 40, 4, 7).

[51] Tattersall 1912b, OS 43, 5.

[52] Tattersall also recorded the species from nests of the great black-backed gull and puffin on the Bills (Tattersall 1913, OS 42, 15).

[53] The early marine biologists knew their Greek. Amphithoé was one of the nereids (see endnote 66); which is admittedly taxonomically confusing, since Nereis is a worm).

[54] Recounted in Rachel Carson's *The Edge of the Sea*, 93-94.

[55] Tattersall 1913, OS 42, 11.

[56] O'Riordan *et al*. 2002b.

[57] Entomostraca is a somewhat old-fashioned name used to collectively describe the smaller crustaceans (as distinct from the Malacostraca).

[58] Farran 1913, OS 45, 1.

[59] Tattersall 1912a, OS 41, 3. Krill is the main prey of baleen whales, including the blue whale.

[60] They are the most primitive members of the class Malacostraca, with a fossil record that extends back to the Cambrian period.

[61] Tattersall 1912a, OS 41, 2.

[62] Tattersall 1912a, OS 41, 8.

[63] Tattersall 1912b, OS 43, 1.

[64] Southern's *Praegeria* has since (alas!) been relegated to synonymy with *Pisione*.

[65] Southern 1914a, OS 47.

[66] The nereids were sea nymphs in Greek mythology, daughters of Nereus and Doris. Many sea creatures are named for them (there were 50).

[67] *Scolelepis vulgaris* was originally named as a species of *Nerine* by Catharine Johnson (in 1827). In the plant kingdom *Nerine* is a genus of flowering plants in the Amaryllidaceae (daffodil family).

[68] Blaxland Benham 1896, 340.

[69] Southern 1913a, OS 48, 8.

[70] Southern 1915, OS 49, 2–3.

[71] Colgan 1911, OS 22.

[72] Colgan himself states that 51 species were found on the shores of the island between tide-marks.

[73] Nunn 2002. An annotated check of all the species she found is included on pages 168–83.

[74] 'Both of these occur in profusion between tide-marks on the island; and their prominence in its marine fauna, if one may be permitted to associate the idea of prominence with such elusive, protectively coloured species, is perhaps conditioned by the abundance on the island shores of their food-plant, the green seaweed, *Codium tomentosum*' (Colgan 1911, OS 22, 8).

[75] Colgan 1911, OS 22, 21.

[76] Colgan attributed the absence from the island of a considerable number of species found on the shores of the mainland to 'the present contrast in the physical conditions of the eastern and western divisions of Clew Bay' (OS 22, 7).

[77] Cussen *et al*. 2002.

[78] *Ibid*.

[79] Delany, Myers and McGrath 2002a.

[80] Delany, Myers and McGrath 2002b.

[81] Colgan 1911b, OS 22, 14.

[82] Cronin *et al*. 2002.

[83] Myers *et al*. 1990.

[84] The operculum is a plate used to close the aperture of the shell when the snail withdraws into it.

[85] Myers and McGrath 2002. *Osilinus* is near the northern limit of its geographical range in western Ireland. It has a short breeding season in mid-summer and is very sensitive to prolonged cold in winter.

[86] Cussen and Myers 2002.

[87] Quoted by Landsborough, *History of British Zoophytes* 1852, 346 (with reference to *Flustra foliacea*).

[88] Nichols 1912a, OS 53..

[89] For pictures and more information see http://britishbryozoans.myspecies.info/

[90] *Flustra papyracea* is recorded from Clew Bay, not so far from Clare Island itself.

[91] Southern 1914b, OS 54, 5, 72.

[92] Nichols 1912b, OS 57, 10.

[93] This can take a couple of hours, but they have all the time in the world. Once upon a time fishermen would cut starfish found preying on beds of mussels or oysters into pieces and throw them back in the sea: a singularly counterproductive approach to control since each piece can grow back into a full starfish as long as part of the central disc is present! As well as the suctorial tube feet they use to exert this sustained pressure on the two shells in order to prise them apart, the starfish surface is studded with batteries of elaborate tweezers with which they keep themselves free of debris and would-be squatters. These are called pedicellariae, and are partly responsible for the rough feel of starfish when handled.

[94] Southern 1915, OS 67, 57.

[95] Farran and Southern 1914, OS 21, 4.

[96] No attempt was made to study freshwater species.

[97] Farran 1912, OS 19, 1.

[98] Marcus Sweeney, pers. comm.

[99] Wilson 1911, OS 5, 44.

[100] Farren 1912, OS 40, 4, 7.

[101] MacCarthaigh (1999), 54–56.

[102] Marcus Sweeney 2016, pers. comm.

[103] Farran 1912, OS 19, 2; Lobster fishing only became important in the later nineteenth century.

[104] Farran 1912, OS 19, 4; 'Conner' was the local name.

[105] Described in Colgan 1911, OS 4, 23.

[106] Wrasse are thought to play a key role in controlling populations of sea urchins, which in the absence of such predators can seriously reduce the kelp forest that is such an important marine habitat.

[107] Farran 1912, OS 19, 4.

[108] Farran 1912, OS 19, 6.

[109] Farran 1912, OS 19, 6. The 'spurdog', once the bane of fishermen, seems to have disappeared from the waters around Clare Island today.

[110] Marcus Sweeney 2016, pers. comm. Down to a generation ago every schoolchild collected winkles: the sale of which paid for their schoolbooks.

[111] Spurdog, for instance (Marcus Sweeney, pers. comm.).

[112] Marcus Sweeney 2016, pers. comm.

[113] Curran-Cournane 2007.

[114] Marcus Sweeney 2016, pers. comm.

[115] Ó Cadhla *et al*. 2007.

[116] Beattie 2001, 12–13.

[117] Viney 2003, 179.

[118] Doyle 2007.

[119] There have been numerous records off the west coast, including Clew bay (Tom Doyle, pers. comm.).

[120] Marcus Sweeney 2016, pers. comm.

[121] Farran 1914, OS 66.

[122] Although for us they must remain little more than names, a great deal is known about the individual copepod species because of their importance in marine food chains.

[123] A now obsolete collective name for two groups of smaller Malacostraca: the opossum shrimps or mysids (Mysidacea) and the Euphausiacea; see page 108.

[124] E.g. *Philadium temporarium, Phialidium cymbaloideum, Obelia nigra.*

Opposite

Fig. 7.1 *Ramalina siliquosa* and other lichens on rocks of the Strake Banded Formation on the south coast.

Chapter 7
Life on land

The terrestrial flora

Introduction

The geological record of Clare Island tells of a time when there was no life on dry land. It was during the Silurian period of earth history that the algal ancestors of today's terrestrial plants made the first incursions onto dry land, having first made the transition from seawater to freshwater. Land plants must evolve the structures and metabolisms that are necessary to prevent desiccation, that will enable them to obtain their nutritional requirements from the soil, and they must develop reproductive strategies that are not dependent on the all-enveloping sea. Gradually plants evolved protective epidermal structures to prevent loss of water, vascular tissues to transport water and nutrients from the soil, assisted in this by symbiotic alliances with bacterial and fungal associates, resistant spores to prevent desiccation as they were wafted through the air. Fungi also proliferated in concert with this green invasion of the land, relying on their distinctive chemical versatility in deriving the materials and energy they need through the breakdown of organic substrates, confining their aerial excursions to the production of evanescent spore-producing structures, achieving in the process a much greater taxonomic and ecological diversity than they had done beneath the waves.

In the Silurian however this great epic of colonising the land was only beginning. The geologist inspired solely by the guiding Principle of Uniformity originally enunciated by James Hutton in the late 18th century—a guiding light in geological investigation to this day—looks with bewilderment at the sequence of sedimentary rocks that constitute the geological framework of the southern half of Clare Island: because there are no modern counterparts to act as guides in understanding their formation. They originated as fluviatile sediments certainly: but these rivers flowed across landscapes without soil, and their patterns of stream flow were very different from those that characterise today's world.

In the Bunnamohaun Formation however we find the rare remains of plants that resemble today's stoneworts (see page 30): specialised algae that lived in pools of freshwater on a slowly-greening earth. In this chapter we will review first of all the bryophytes that are the closest living descendants of early terrestrial ancestors, before moving on to look at the vascular plants that dominate today's flora, and the habitats they constitute: and then at the great variety of animals that make their homes in the vegetation and in freshwater.

Bryophytes

Although not a component of the flora that catches the attention of the casual visitor, bryophytes (mosses and liverworts) are among the most beautiful and fascinating of plants. Bryophytes are non-vascular plants, which means they lack the elaborate conducting tissues of ferns and flowering plants; this is what accounts for their diminutive status. They reproduce by means of spores or by producing asexual gemmae or tubers. Also, they lack the flowers that make it so much easier to identify the different species of more familiar plants.

In spite of the challenge their small size poses, mosses and liverworts are enormously diverse, often very beautiful and endlessly fascinating. Their more diminutive stature means that closer attention and the use of a hand lens—and in some cases a microscope—are required to study them. The British Bryological Society's splendid new field guide uses to best advantage the detailed descriptions and illustrations in Paton's *Liverworts* (1999) and Smith's *Mosses* (2004), and makes them accessible to anyone prepared to devote a little time and attention to their study.[1]

The bryophyte flora of Clare Island was first studied by Canon Henry William Lett in 1909 and 1910.[2] Little attention was paid to the mosses and liverworts of the island in the years that followed, although nearby Achill Island was visited by many bryologists on several occasions. During the course of the New Survey however, much more attention was given to the bryophytes of Clare Island, particularly by Donal Synnott, who has critically examined and re-evaluated the records of the earlier survey.[3]

The Original Survey recorded 74 species of liverwort and 139 mosses; eighteen of the liverworts were not refound in the New Survey, but fourteen of these are considered doubtful records. Of sixteen

Fig. 7.2 Bryophytes.

(a) Lustrous Bog-moss (*Sphagnum subnitens*) (Jukka Laine); (b) Swan's-neck Thyme-moss (*Mnium hornum*) (Hermann Schachner (CC0 1.0)); (c) Purple Spoonwort (*Pleurozia purpurea*) (Wikimedia Commons, Matt von Konrat); (d) Dilated Scalewort (*Frullania dilatata*) (Bernd Haynold (CC BY-SA 3.0)); (e) Blunt-leaved Bog-moss (*Sphagnum palustre*) (Jukka Laine).

mosses not refound in the later survey, ten were incorrectly identified and have had to be re-determined.[4] The number of species known from the island now stands at 101 liverworts, 185 mosses and one hornwort; of which 36 liverworts, 50 mosses and the one species of hornwort have been added since the first survey (Figure 7.2).

However, the difference in bryophyte diversity between the two surveys is not due to any significant change in the flora, but rather to the greater number of experienced field bryologists to have visited the Island in recent decades.[5] Of the 87 species added since the first survey only one—Heath Star-moss (*Campylopus introflexus*)—is sure to be a recent arrival. A species of wide distribution in the southern hemisphere and now widespread in Europe, it was first found in Ireland on Howth Head in Dublin in 1942. Achill Island, which is only 6km away from Clare Island, is much richer in bryophytes, but it has a greater range of habitats, ranging from coastal dunes and machair to high north-facing corries and block scree, as well as more extensive areas of the habitats found on both islands.[6]

Canon Lett considered Swan's-neck Thyme-moss (*Mnium hornum*), a nearly ubiquitous species that grows on acidic soil and rocks, tree bases and logs, to be the most frequent moss on Clare Island, 'from sea-level to the top of Knockmore'.[7] However, Tim Ryle's detailed sampling of the vegetation suggests that the commonest bryophytes on Clare Island are Springy Turf-moss (*Rhytidiadelphus squarrosus*), Heath Plait-moss (*Hypnum jutlandicum*), Lustrous Bog-moss (*Sphagnum subnitens*), Common Tamarisk-moss (*Thuidium tamariscinum*), Common Haircap (*Polytrichum commune*) and Tamarisk Scalewort (*Frullania tamarisci*).

Bryophytes have a conspicuous presence in the wood along the streamlet that flows into the sea at Portlea, which Canon Lett regarded as 'the survivors of the Mosses and Hepatics that abounded on the trees and ground of the primeval forest, the bleached stumps of which stud the ground near Maum and other places, and conjure up the idea of what a congenial habitat their presence formerly provided for our little plants ... the scanty survivors of plants that formerly occupied every nook and cranny that then were sheltered by the forest, but now lie exposed to every storm and hurricane'.[8]

Canon Lett was particularly impressed by 'the beautifully situated Lough Knockacurraun at an elevation of 600 feet above sea-level; on all sides but one it is surrounded by hills that reach 1000 feet higher. And here along the west and south margins of the lough the stones and rocks are covered with a profusion of mosses and hepatics, several of which are rare. These grow at the edge of the water, which in summer and winter alike is always at the same level and they are except for a few hours about midday out of sunshine, and at the same time are perfectly sheltered from any wind'.[9]

Although he did find a liverwort believed to be new to science, *Adelanthus dugortiensis*, on Achill and another, Lindenberg's scalewort (*Radula lindenbergiana*)—new to Ireland—on Corraun and on Clare Island, Lett had expected the bryophyte flora of the islands and the adjoining mainland to be richer than it was. He spent a total of three weeks on Clare Island, two in 1909 and one in 1910 but does not appear to have visited the higher ground, some of the specimens from which were collected by Praeger (Lett was 73 at this time). Despite changes of land use, many if not all of the bryophytes of Clare Island have survived. The report of a day spent by the British Bryological Society on the island in 1995 is rather more enthusiastic than was Canon Lett. It concluded: 'During the blustery crossing, with gannets above our heads, we were able to reflect on our good fortune on another favourable and productive day in such an excellent place.'[10]

The vascular vegetation of Clare Island

The flora of Clare Island was first surveyed by Praeger himself, who spent a week on the island with his wife in July 1903, finding the vegetation 'more varied and more interesting than was expected'.[11] He had finished work on *Irish Topographical Botany*, his *magnum opus* on the distribution of Irish plants, two years before this first visit, and his botanical interest was beginning to focus on plant communities and the new science of plant ecology rather than the distribution of individual species.[12] This dominated his thinking at the time of that first visit to Clare Island in 1903, and his account of island's vascular vegetation for the survey is an embryonic attempt at phytosociological description.

He spent a great deal of time studying the plants during the early years of the survey (between 1909–11), but 'much general work fell to me as organiser and secretary, otherwise the amount of time spent on the island would have permitted of a very detailed floristic study. As it was, little attention could be given to critical plants, and the time devoted to botany was spent mainly in ecological work, in mapping, and in studying the influence of man on the flora'.[13] He listed more than 360 species on his first visit, and added a further 30 in 1911 ('many of these being uncommon plants, and some of them of high interest') to give a total of 393. In addition to his work on the plant communities present on Clare Island, Praeger drew up an annotated list of 38 of the more important plant species.

The island has attracted numerous botanists over the intervening century, culminating in a detailed re-examination for the New Survey by Timothy Ryle and Matthew Jebb.[14] Praeger had divided the island vegetation into nine communities, but the study of plant communities (phytosociology) was at an early stage in his day. Ryle's survey utilised the approach and methods of modern phytosociology, which allows different communities to be distinguished on broader

Table 7.1
Comparison of the vegetation divisions included on Praeger's (1911) vegetation map with those utilised in the New Survey

Vegetation divisions (1911)		Main divisions and subdivisions (New Survey)		
	①	Beach complex		
Saltmarsh		Saltmarsh community		
		Plantain sward	*	*Armeria* community
Plantago sward			*	Pure sward
			*	Grass-rich sward
	①	Intermediate Coastal/Plantain		
	①	*Phragmites* reed swamp		
Natural Grassland		*Calluna* heath	*	Dry heath
			*	Wet heath
		Agrostis-festuca grassland	*	Coastal subtype
			*	Species poor
			*	Damaged
Calluna formation			*	Dry grassland
			*	*Calluna* rich
			*	*Ulex* community
			*	*Racomitrium* community
			*	Agricultural land (lazy bed dominated)
	①	Bracken community		
		Meadow vegetation	*	Wet grazing pasture
Farmland			*	Tall marsh vegetation
			*	Hay fields
			*	Crops
Calluna-Eriophorium	∞			
Calluna-Eriophorium-Juncus	∞			
	③	Bog vegetation	*	Blanket bog
			*	*Schoenus*-rich bog vegetation
			*	Miscellaneous deep peat communities
Arctic-Alpine Vegetation		Cliff community		
Woodland		Woodland communities		

① units were not included on the 1911 map.
∞ units were not recognised during the New Survey.
③ distribution of bog communities in the New Survey does not correspond with any community identified by Praeger (1911), because these were included in the *Calluna* formation.
* newly recognised community / subdivision in New Survey.

floristic criteria than the dominant species (the basis of Praeger's relatively simple classification). Ryle re-organised the vegetation into twelve 'divisions', many of which are subdivided into several community categories (Table 7.2). The vegetation however is a complex mosaic, and there is often overlap between communities. Ryle recorded a total of 384 species, which is 10% less than Praeger's figure, but if species newly recorded since Praeger's day are subtracted the reduction is 25%.[15] Jebb prepared a very comprehensive annotated checklist of the vascular flora to accompany Ryle's review of the communities for the New Survey.

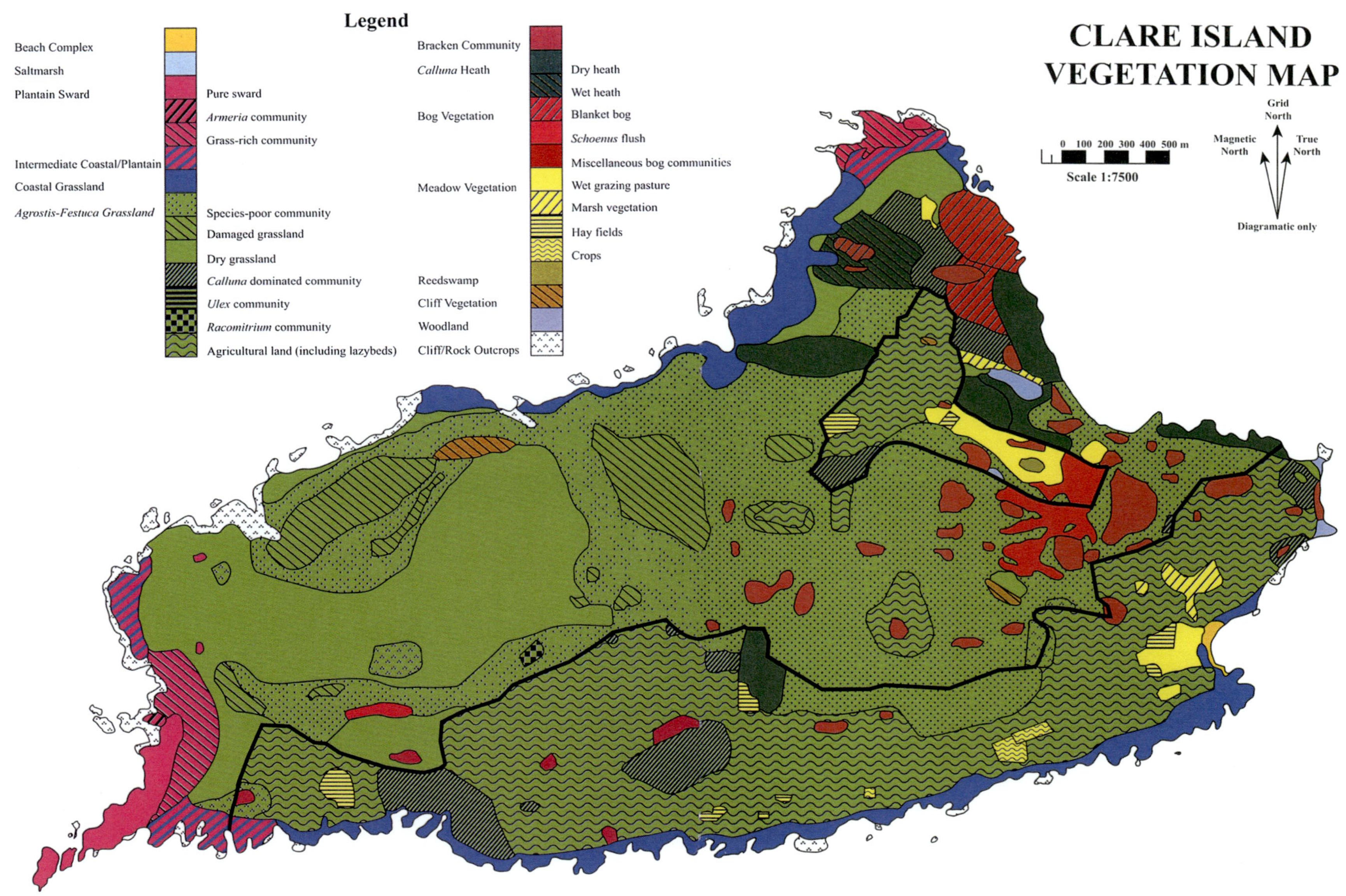

Fig. 7.3 Ryle's vegetation map from the *New Survey of Clare Island*.

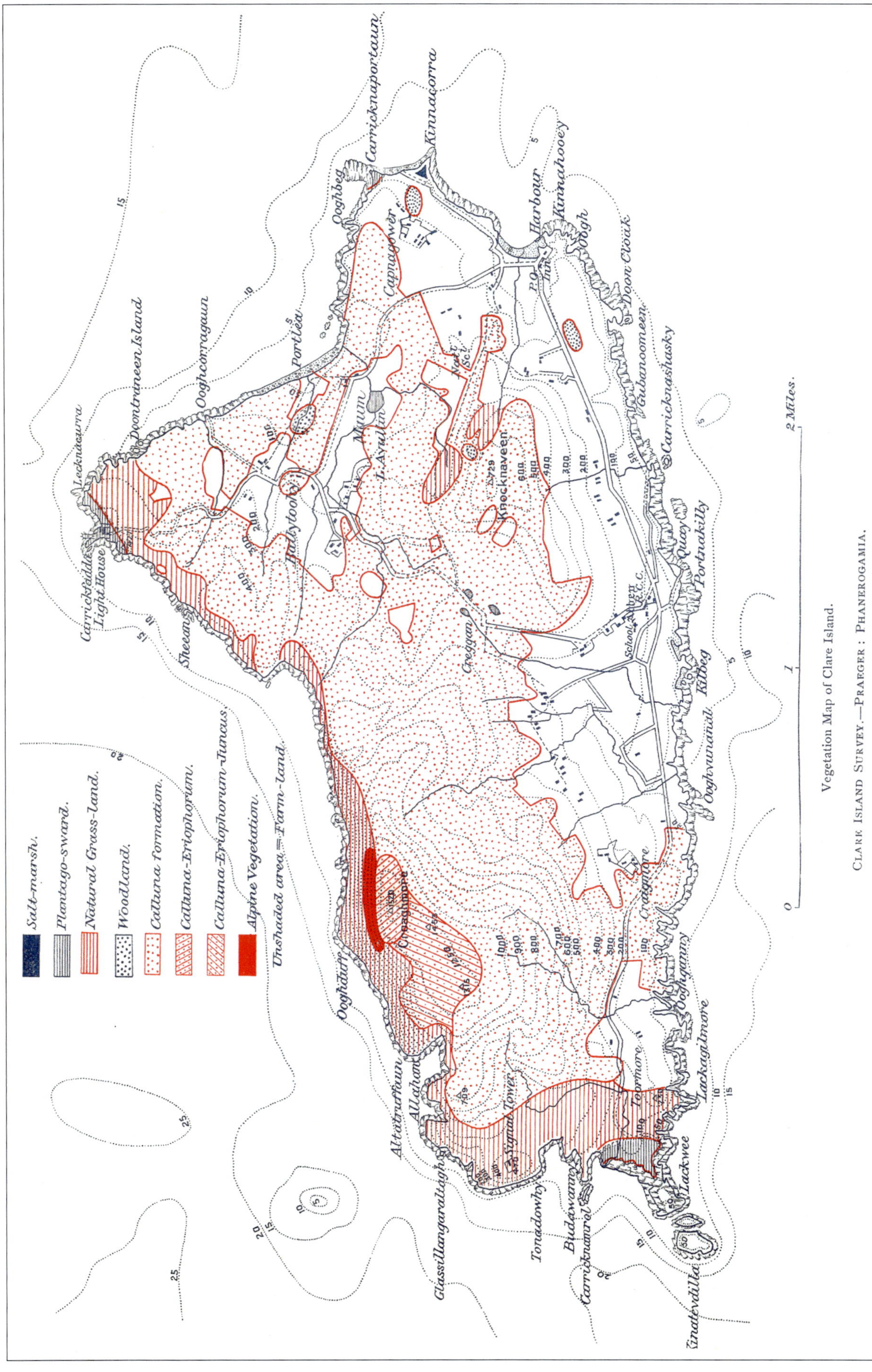

Fig. 7.4 Praeger's vegetation map from the *Original Survey*.

Fig. 7.5 Lassau wood.

A vegetation mosaic of acid grassland communities covers most of the island today, both on the commonage and former arable land (Figure 7.3). At the time of Praeger's surveys his 'Heather Community' dominated the drier areas of the commonage, accounting for nearly half of the total land area (Figure 7.4). This has almost disappeared as a result of overstocking, and been replaced by heavily grazed rough grassland. It now only survives in the few areas inaccessible to sheep. Only four patches of scrubby woodland remained on the island by the time of the Original Survey, covering less than 5 acres between them; the biggest was the two-acre wood near Portlea, where there was willow (mainly eared willow), hazel and birch, rowan, holly and a little oak, with a lot of bracken and bramble, and abundant honeysuckle (Figure 7.5).[16] Other than the wood at Lassau, which has been considerably extended by planting in recent years, woodland is almost absent today: 'We'd rinse the clothes by the river and spread them on the bushes. I don't think there's any bush now to spread clothes on; they're all withered away. The bushes are gone and the people are gone.' [17]

When people first made Clare Island their home in prehistoric times the island was extensively wooded. The wide extent of oak and pine forest beneath the bogs in the east and south of the island gives some indication of how widespread this was before the climatic downturn that saw the spread of bog across the island. The pollen record chronicles woodland history on the island for us, and shows its persistence into the early modern period (see Chapter 4). When the last woods fell to the axe it is impossible to say, but the almost total absence of reference to trees or woods in placenames suggests that during the period over which the island acquired its overlay of Irish toponyms woodland was no longer an outstanding feature of the landscape. The only tree to appear in a placename is rowan, and this solitary reference is to 'a high precipice on the sea-shore where it is said by the inhabitants a tree formerly grew'.[18] Other than this there are three or four references to areas of scrubby woodland.[19] The presence in the open of relict colonies of such woodland shade species as bluebell, wood anemone, wood sorrel and primrose may be seen as a distant echo of the island's former woodland cover, particularly where a covering of bracken shades them from the summer sun.[20]

The bogs of Clare Island

Ever since the disappearance of its woodland cover, turf has been the only source of household fuel on the island. Elsewhere we will see the extent to which the shallow moorland peats were cut away in the nineteenth century (page 271): by the time of Praeger's survey almost the entire surface of the moorland vegetation had been interfered with by turf-cutting. Numerous bogs remain in the hollows of the hummocky moraine, in topographical depressions in the commonage and on impervious glacial outwash on the eastern side of the island in Maum and Ballytoohy More: a complex mosaic of bog pools, cutaway hollows, drainage channels and cutaway ground.[21]

Peat formation is thought to have started some time around 4,000 years ago in numerous small hollows in the landscape of hummocky moraine.[22] The average peat depth on the island is less than 2m, but in some of the deeper hollows it can be as much as 7m. One of the largest areas of peat remaining is concentrated in and around Maum townland, where fossil pine and oak stumps, usually in definite layers, can be seen to overlie peat, showing that the bog had started to form before the forest reached its maximum extent (see page 65–7). Oak stumps occur frequently under bogs in the north-east of the island, especially between 200 and 400 feet above sea level. The pines are often between 1 and 3 feet across at ground level, which would mean a diameter of 1–2 feet at breast height. The oak is often 18 inches diameter at breast height, which would mean the trees were between one and two centuries old (See Chapter 4).

The grasslands

By far the commonest grassland class is the Nardetea. On most of the area occupied by Praeger's 'Grassland Community' a form of heathy grassland formally assigned to the Nardo-Caricetum binervis association is now found.[23] The most widespread species are sheep's-fescue (*Festuca ovina*), common bent (*Agrostis capillaris*), mat-grass (*Nardus stricta*) and sweet vernal-grass (*Anthoxanthum odoratum*), with viviparous fescue (*Festuca vivipara*) more frequent at higher altitudes, and Yorkshire fog (*Holcus lanatus*) locally abundant in drier places, often accompanied by early hair-grass (*Aira praecox*) and silver hair-grass (*Aira caryophyllea*). In more severely disturbed areas mat grass and heath rush (*Juncus squarrosus*) become dominant, sometimes accompanied by heath grass (*Danthonia decumbens*).[24]

Today a species-diverse sward similar to that found on the drier *Agrostis-Festuca* grasslands of the commonage covers most of the lazy beds which carried oats and potatoes in the nineteenth century; the extent of these gives us an accurate picture of the extent of arable land at its most intensively managed. Most are closely grazed, but the few cases that are preserved as meadow develop a taller sward that is often rich in species more characteristic of the order Arrhenateretalia. Among the most striking botanical features are meadows of dwarf spotted orchids in tightly grazed grassland, which flower in May in the company of equally diminutive lousewort, and which are most accessibly seen in various places beside or close to the Lighthouse Road.

Even in the short interval between his two reports Praeger noticed how the flora of the island was changing. The wholesale change in ownership and management that had taken place just before the Clare Island Survey was undoubtedly responsible for much change. In 1911 Praeger remarked that since his earlier visit in 1903 improved farming practice and increased numbers of sheep were having a clear impact on the upland vegetation. By 1892 the number of sheep on the island had grown to 3,000, along with 160 cattle, 74 horses and 16 asses. In 1905 there were 2,000 sheep. Sheep numbers remained at these levels more or less until the middle of the last century. In 1965 the number stood at 2,500, but then rose to 6,200 by 1975. This had disastrous consequences for the vegetation, because on Clare Island sheep are left on the commonage throughout the year. The number steadied at about 5,500 by the early 1990s. Today the number stands at around 4,000 (see pages 281, 284).

One small example of change noted by Praeger was the way pineappleweed (*Matricaria discoidea*) had extended its hold along the roads of the island between the time of his first visit in 1903—at which time it was only found 'for about half a mile along the roads from the Harbour'—and 1911, when it was 'nearly continuous from the Harbour for three miles along the chapel road, and is abundant in many places on the lighthouse road'.

Praeger was particularly interested in the various ways in which plants had arrived on

Fig. 7.6 Alpine saw-wort (*Saussurea alpina*): (Ivar Leidus, Wikimedia Commons, License CC BY-SA 4.0).

Fig. 7.7 Mountain sorrel (*Oxyria digyna*) (photo Kim Hansen, CC BY-SA 3.0); illustration (Lindman, 1922).

Clare Island in the first place, and in the role of human agency in this regard. He devoted more than thirty pages of his 112-page contribution to the Original Survey to the question, and later went on to publish a detailed account of the dispersal mechanisms of different species.[25] He paid great attention to the distinction between 'native' plants and others which owed their presence to human agency, either by being deliberately introduced or escaping into the wild. Of the 393 'wild' taxa he recorded on the island, he considered 338 to be 'true natives' and the rest to have been introduced as a result of human activities. He was particularly excited by the Arctic-Alpine vegetation on the cliffs of Knockmore: a tiny relic of the communities that existed more widely at the end of the Ice Age. He listed thirteen species in this community, but only some of these would be regarded today as true 'Arctic-Alpine' species. Six of the species on Praeger's list have not been seen in recent years; this could be because they have disappeared from the island, or because the cliffs are less accessible. In his day the scarp could be traversed 'by anyone with a good head by means of sheep-paths starting from its end at about 1000 and 1200 ft. respectively', but a century of erosion has made these paths much less accessible. We should also remember however that not everyone has Praeger's 'good head', and few botanists are as fit as he or as ready to take the risks he routinely took.

These Arctic species are among the most memorable of the island's botanical sights. Moss campion (*Silene acaulis*) grows in compact cushions on the cliffs and rock ledges of Knockmore and flowers in June and July. In Praeger's day it was 'extremely abundant, forming large cushions all over the cliff from summit almost to base'.[26] Purple saxifrage (*Saxifraga oppositifolia*), another magnificent cushion-like plant, produces its gorgeous blossoms earlier in the year (March-May). Other characteristic plants in this community of Arctic plants are mountain sorrel (*Oxyria digyna*), locally abundant on the cliffs, roseroot (*Sedum (Rhodiola) rosea*) and Alpine saw-wort (*Saussurea alpina*) (Figures 7.6–7.9).

The Plantago sward

The *Plantago* sward is an unusual and distinctively Irish maritime habitat, characteristic of many windswept western cliffs and islands, occurring on

Fig. 7.8 (a) Moss campion (*Silene acaulis*) (Isidre blanc (CC BY-SA 4.0)). (b) Purple saxifrage (*Saxifraga oppositifolia*) (Alastair Rae (CC BY-SA 2.0)).

Fig. 7.9 In Praeger's day *Sedum Rhodiola* grew in several places ('abundant all round the northern and western cliffs, from storm-level to 1,450 feet. Inland on rocks east of the signal tower at 800 feet, and sparingly on the cliff on the north slope of Knocknaveen'; OS 10, 29): but he noted that its distribution had been considerably altered by sheep (OS 10, 44).

Clare Island mainly just inland from the western and north-eastern cliffs (near the lighthouse) on shallow peaty soils. Sea plantain dominates the sward, covering up to 80% of the ground along with sheep's-fescue, together with as many as thirty or more other species, including buck's-horn and ribwort plantain, tormentil, eyebrights, heath bedstraw, English stonecrop, sheep's-bit, wild thyme, field wood-rush and early hair-grass. The association depends on tight grazing for its maintenance, and the dwarf character of the flowers. Where dangerous headlands have been fenced off an altogether different type of vegetation develops, in which heather is prominent (Figure 7.10).

The Irish heath community

One of the most interesting communities on the island is a small population of a plant association dominated by the very rare winter-flowering Irish heath (*Erica erigena*). This is confined to the steep till cliffs fronting the sea between Portlea and Cove. Here it is prolific and totally dominant, some of the plants being as much as a metre tall: but it is found nowhere else on the island. In Ireland *Erica erigena* is confined to parts of Galway and Mayo, and occurs elsewhere only around Bordeaux and in several places in the Iberian peninsula.[27] How it got here is something of a mystery: the most widely held view is that it was introduced in Early Christian times, and that it arrived on Clare Island

Fig. 7.10 Effects of the absence of grazing on headlands: Capnagower.

from Achill, where it is abundant. Be that as it may, it is intriguing that it is so strictly confined to this one location on the island: exactly where Praeger recorded it in 1911 (Figure 7.11).[28]

Many botanists have studied the flora of Clare Island since Praeger's day, and their records chronicle a process of continuing change. A quarter of the species recorded by Praeger had disappeared by 1986.[29] During the course of a four-day visit in 1984 Gerry Doyle and Peter Foss recorded 301 of Praeger's 413 taxa (393 'wild' plus 21 cultivated), including 56 of 86 species recorded by Praeger but not seen since 1930. But they also recorded 27 taxa that were *not* seen by Praeger, of which nine are garden escapes. Tim Ryle recorded 328 of Praeger's list among his total of 384; but although he added 11

Fig. 7.11 *Erica erigena* growing on the till cliffs in Capnagower.

species to those recorded by earlier botanists, he failed to find 79 taxa listed in earlier surveys. Forty-eight of the taxa recorded by Praeger have not been seen since 1980, but 71 have been added to his checklist.[30]

Some of the new arrivals are conspicuous and have an obvious ecological impact. The giant rhubarb (*Gunnera tinctoria*)—a native of South America—is the most dramatic example. Since its arrival in the 1960s it has colonised most of the water courses in the south of the island, requiring the establishment of an eradication programme, still ongoing (Figure 7.12).[31] Most new arrivals have had a more modest ecological impact.

Several other alien species have become an established part of the Clare Island flora in a less invasive way, but these foreign nationals should be considered an extension of the cultural rather than the natural landscape—they always occur around houses. All are hardy woody shrubs: *Fuchsia magellanica* from Patagonia, *Escallonia rubra* from South America, daisy bushes (*Olearia traversii* and New Zealand holly *Olearia macrodonta*), *Hebe speciosa* (*H. x franciscana*) and New Zealand flax (*Phormium tenax*).

Impressive as the total number of species on the island is, the absence of many species common

Fig. 7.12 Giant rhubarb (*Gunnera tinctoria*).

Fig. 7.13 Marsh St-John's wort (*Hypericum elodes*): one of the most characteristic and common flowers of the peaty ponds and ditches that are so widespread on Clare Island (Ghislain 118, Wikimedia Commons, License CC BY-SA 2.5).

Fig. 7.14 Sea spleenwort (*Asplenium marinum*) on cliffs at Glasoileán beside Beetle Head. Sea spleenwort is one of the few abundant ferns on the island: at its greenest and most luxuriant in rock crevices at the limit of the spray zone, but extending far inland to nearly 150m above sea level.

Fig. 7.15 The pink form of primrose (*Primula vulgaris*) on the bank of the Toormore stream in Bunnamohaun.

Fig. 7.16 Four Clare Island waxcaps.
(a). Bonnet waxcap (*Hygrocybe calyptriformis*); (b). Vermilion waxcap (*Hygrocybe miniata*) (Frank Gardiner (CC BY 3.0)); (c). Butter waxcap (*Hygrocybe ceracea*) (Stu's Images (CC BY-SA 4.0)); (d). Scarlet waxcap (*Hygrocybe coccinea*) (Damian Egan).

elsewhere is surprising; black medick (*Medicago lupulina*), wild arum (*Arum maculatum*) and meadow foxtail (*Alopecurus pratensis*) are striking examples. The flora of the island is however also distinguished by the presence of species that are rare elsewhere, most notably the Arctic-Alpine assemblage on Knockmore, and the thyme broomrape (*Orobanche alba*) that Praeger missed on his first visit, but later spotted growing sparingly on the cliff below the lighthouse, and of which Mr. Duffy, the head light-keeper, 'kindly went down on a rope and procured a specimen'.[32]

Notably, the pattern of distribution has changed for the majority of species that have remained on the island's checklist of plants since records began. Species that were once common are now less so, and *vice versa*. The Arctic-Alpine species that maintain a precarious hold on the cliffs of Knockmore were more widely distributed 10,000 years ago. In Praeger's day the stately royal fern (*Osmunda regalis*) was so widespread and vigorous that 'great tussocks growing in pasture-land are left, as large boulders might be, on account of the labour required to remove their dense matted roots'.[33] Westropp referred to cliffs east of Portruckagh that were 'sheeted with masses of Osmunda wherever a stream descends their faces'.[34] Today it is a scarce and less stately plant, surviving mainly in drains.

A hundred years ago bracken was found mainly on the margins of the cultivated land. Today it is widespread on many areas that were cultivated a hundred years ago, sometimes overwhelming land that should be productive grassland. It also forms extensive tracts on drier parts of the commonage. Its eradication and control are a major objective of a Walsh Fellowship research project currently under way.

The fungi

At the time of the Original Survey almost nothing was known of the fungi of Clare Island. Sir Henry Hawley and his team spent a week collecting on the island every year between 1909 and 1911, during which time they recorded 283 species, 135 of which were additions to the Irish Census Catalogue, including eight that were new to the flora of Britain and Ireland. Hawley described one new species as *Candelospora ilicicola,* a microfungus living on the fallen leaves of holly; he named it *Candelospora* because its spore-producing structures, the conidia, 'stand up side by side like so many candles'.[35]

Unlike the original survey of fungi undertaken by Sir Henry Hawley and Carleton Rea, mycological endeavour in the New Survey of Clare Island focused on a particular group of fungi, namely those found in grassland. Such fungi have attracted much attention in recent years as valuable indicators of natural and semi-natural unimproved grasslands, typically unfertilised and of low nutrient status.[36] Indeed, analysis of a community of grassland fungi can provide a more robust evaluation of endangered grassland habitats of conservation significance than vascular flora.[37]

The principal components of this fungal community are the waxcaps (*Hygrocybe* species), particularly attractive mushrooms that are often brightly coloured in hues of red, orange, yellow or pink (Figure 7.16). Other fungi that belong to the assemblage are the pinkgills (*Entoloma* species), clavarioid fungi (*Clavaria* and related genera, also known as fairy clubs) and the earth tongues (Geoglossaceae).

Grassland fungi occur widely all over the commonage on Clare Island and in fields with abandoned but well-grazed lazy beds. The latter is surprising, given the extreme disturbance involved in their formation and management, but the key factors that allow for the occurrence of fungi in lazy beds are the absence of fertiliser application, good drainage, an appropriate grazing regime resulting in a low sward and the length of time since cultivation was abandoned. Some species colonise new sites quickly, but others may take many years or even decades before conditions are suitable for them. It is ironic that the richness of the grassland mycoflora should be due in large measure to human interference, and in particular to the survival of an earlier agricultural landscape.

Although Hawley is likely to have missed the main fruiting season for grassland fungi (normally late October to early November), he nevertheless recorded fifteen species of waxcaps, along with

Table 7.2

Grassland target species found on Clare Island November 2008

Species	Type
Clavulinopsis corniculata	C
Clavulinopsis fusiformis	C
Clavulinopsis laeticolor	C
Entoloma conferendum	E
Entoloma papillatum	E
Geoglossum atropurpureum	G
Geoglossum fallax	G
Geoglossum glutinosum	G
Microglossum olivaceum	G
Trichoglossum walteri	G
Hygrocybe aurantiosplendens	H
Hygrocybe calyptriformis	H
Hygrocybe cantharellus	H
Hygrocybe ceracea	H
Hygrocybe chlorophana	H
Hygrocybe coccinea	H
Hygrocybe conica var. *conica*	H
Hygrocybe fornicata	H
Hygrocybe insipida	H
Hygrocybe irrigata	H
Hygrocybe laeta var. *flava*	H
Hygrocybe laeta var. *laeta*	H
Hygrocybe miniata	H
Hygrocybe mucronella	H
Hygrocybe nitrata	H
Hygrocybe persistens var. *persistens*	H
Hygrocybe pratensis var. *pratensis*	H
Hygrocybe psittacina var. *psittacina*	H
Hygrocybe punicea	H
Hygrocybe quieta	H
Hygrocybe reidii	H
Hygrocybe russocoriacea	H
Hygrocybe splendidissima	H
Hygrocybe virginea var. *ochraceopallida*	H
Hygrocybe virginea var. *virginea*	H

C = Clavariaceae; E = *Entoloma;* G = Geoglossaceae; H = *Hygrocybe*

New Survey Volume 7, 263

Fig. 7.17 *Stamnaria persooni* (a small ascomycete) growing on horsetails at Loughavullin (Roy Anderson).

twelve fairy club species, eight species of pinkgill fungi and two earth tongues.[38] The most significant waxcap species he found were the unmistakeable and unforgettable bonnet waxcap (*Hygrocybe calyptriformis*) and the blushing waxcap (*Hygrocybe ovina*).

In recent decades the grassland mycoflora of Clare Island has been studied by several mycologists. Reitze ten Cate, who described Clare Island as 'a paradise for waxcaps', visited in 1992, when he recorded fourteen species of *Hygrocybe*. Roland McHugh made several visits in the early 1990s, recording 25 *Hygrocybe* species (including varieties), three clavarioids, one earth tongue and six *Entoloma* species. During a three-day visit in November 2008 for the New Survey, David Mitchel recorded 25 waxcaps (including varieties) (Table 7.2).[39]

The number of waxcap species recorded for Clare Island now stands at 26, making it the second richest site in Ireland, surpassed only by the Curragh of Kildare which has 32 species (but the Curragh, it should be noted, has been studied far more intensively). The number of fairy club species makes Clare Island the richest site for this group in Ireland by a considerable margin.[40] The occurrence of five fairy club species, six earth tongues and eight pinkgill mushrooms makes the grassland on Clare Island an outstanding location for these fungi also.

Grassland sites across Europe have been ranked for their fungal conservation value using various scoring systems based on the total number of grassland fungi present.[41] Discussion continues as to what constitutes the best system, and no one approach is universally accepted. Accordingly, at present it is not prudent to rank Clare Island grasslands relative to others. However, with 26 recorded species of *Hygrocybe*, the island comfortably meets a European criterion for a site to be designated as one of international importance (SSSI status) if it has at least 22 species.[42] In addition to this, two of the fungi recorded on Clare Island (*Hygrocybe calyptriformis* and *Geoglossum atropurpureum*) are among 33 threatened fungi in Europe proposed for inclusion in the Habitats Directive Appendices of the Bern Convention.[43] Both fungi are also on the Red Data List of Threatened British Fungi, listed respectively as 'vulnerable' and 'rare'.[44]

Ireland and Britain have more and better-quality waxcap grasslands than any other location in Europe, and possibly indeed, anywhere else on

Fig. 7.18 Representative slime moulds: (a) *Badhamia utricularis* (Björn S. (CC BY-SA 2.0)); (b) *Trichia botrytis* (Helen Ginger (CC BY-SA 3.0)); (c) *Craterium minutum* (Janet Graham (CC BY 2.0)); (d) *Didymium squamulosum* (Helen Ginger (CC BY-SA 3.0)).

the planet. The grasslands of Clare Island are endangered habitats, and therefore outstanding and precious in both national and international terms. David Mitchel (2013) concluded his excellent chapter in Volume 7 of the New Survey with the claim that 'grassland fungi are therefore probably the most important nature conservation feature on Clare Island, and parts of the island merit consideration for statutory site protection'.

You will see little of the splendour of the grassland fungi if you are a summer visitor to the island, because the main fruiting season is between late summer and early winter. Work on other groups of fungi was also initiated during the New Survey, and is ongoing. There are hundreds of other species of fungi, but most of these also tend to produce their fruit bodies later in the year, and many are so small as to be missed unless carefully looked for (Figure 7.17). The several hundred parasitic species likely to occur on plants remain to be recorded. One of these is a tiny fungus called *Unguicularia aspera*; if you happen to be admiring royal fern in June take a closer look at the dead stems beneath the fronds and you are likely to find the fruiting bodies of this parasitic species. The small mixed woodland at Ballytoohy More has produced a rich mycoflora, and a diverse community of coprophilic fungi has been identified in sheep droppings.

Phylum Myxomycota (slime moulds)

Slime moulds were at one time thought of as a sort of fungi, but are now recognised as a distinct phylum of protoctists.[45] Those most likely to attract attention look just like colourful blobs of jelly on vegetation. This gelatinous mass, ranging in size from microscopic to several inches across and often yellow or orange in colour, is the *plasmodial* stage in the life-history. It consists of a mass of naked protoplasm that is an aggregate of individual cells without walls or separating membranes, in which a rhythmic circulation reflects ceaseless metabolic activity. It is made up of millions of loose nuclei, shuttling back and forth within a reticulum of branching vessels. The plasmodium is the feeding stage, engulfing bacteria, fungi and decaying organic material in amoeboid fashion as it glides over rotting wood or moss. Some species seek their food in the interior of decaying wood; others feed while creeping over the surface of decaying leaves. Two species, *Badhamia utricularis* and *B. nitens*, prey on the living pilei of leathery fungi (Figure 7.18).

The plasmodium is usually hidden away in rotting wood or leaf mould, under bark or in the soil, but when it senses stressful change in its surroundings it moves towards the surface and undergoes a remarkable process of differentiation to form stalked spore-producing structures that resemble the sporangia of fungi. These *sporocarps* contain spores that will subsequently discharge. The sporocarps are very varied in form, often brightly coloured or iridescent and remarkably beautiful.

Each of the tiny spores contained in these elaborately ornamented vessels is enclosed in a spore wall and has a single nucleus.[46] On germinating, *swarm cells* break out of the spores; these are of two types, amoeboid and flagellate, both of which can differentiate into complementary mating types, a fusion of which results in a zygote that then divides over and over to form the plasmodium, which looks like nothing more than slimy scum on wet logs and leaf litter. How much more there may be to see is well-illustrated by an anecdote recounted by Gulielma Lister in her report for the Original Clare Island Survey. She had been searching in vain for two hours in a conifer wood near Westport.

> I happened to pick up a piece of larch bark two inches square that showed on its under surface, besides about eight small snails and some woodlice, a patch of rose-coloured plasmodium the size of a pin's head. After removing the snails and woodlice, I brought the bark indoors and kept it in a moist chamber for a week. By that time there had appeared on its surface 35 sporangia of *Margarita metallica* (from rosy plasmodium), 65 sporangia of *Physarum nutans*, and one sporangium of *Trichia varia*.[47]

Lister's contribution to the Original Survey begins with an introduction to the life-history of slime moulds.[48] This was followed by an outline of previous work on the group in Ireland and an enthusiastic account of what one might expect to find in suitable habitats.

> Mixed woods and plantations … should form good hunting-grounds for Mycetozoa at any time of the year in our climate. Here old logs and the under-side of fallen branches should be searched for many species of Physarum, for the brown feathery tufts of Stemonites, the dark pin-like sporangia of Comatricha, the clustered yellow balls of Trichia, or the crimson, buff, or grey clubs of Arcyria. On decaying stumps of Conifers appear troops of yellow or nut-brown sporangia of Cribraria, whose minute pores sift out through the delicate network of the sporangium-walls.
>
> … Mixed leaves often yield sporangia of Physarum, Craterium, Didymium, and Lamproderma in countless numbers. Decaying leaves of Holly, Ivy, and Rhododendron are feeding-grounds for the plasmodium of many species. Here we may find the pearl-like sporangia of *Margarita metallica*, the small, bright yellow-spored form of *Trichia Botrytis*, and the dark, white-stalked columns of *Diachaea leucopoda*, while the snowy heads of *Didymium squamulosum*, one of the most abundant species of Mycetozoa throughout the world, are seldom absent.
>
> … About stems of heather, on moor-land, as well as in the woods, one may find clusters of the glossy brown sporangia of Leocarpus, or the inconspicuous chestnut-brown fruits of the rare *Diderma simplex*, a species that is probably often overlooked. Amongst Sphagnum and rushes occur the sessile pink clustered sporangia of *Badhamia lilacina*, and the scattered grey head of Lepidoderma, both species that are more conspciuous in their yellow plasmodium stage. On moss and hepatics growing in sheltered places on steep wet rocks, are found the shining iridescent sporangia of *Lamproderma columbinum*, and the somewhat similar-looking globose-headed variety of *Badhamia rubiginosa*. It is in such places that *Diderma lucidum* resides, whose scarlet sporangia, mounted on black stalks, glistening among cushions of moss, would reward many an hour of patient search.[49]

Miss Lister spent nine days in Westport with Mr and Mrs Carleton Rea in October 1911, collecting 31 species in the district, and was 'impressed with the rich field such a neighbourhood presented for anyone who could search there at all seasons'.[50] However, she does not appear to have visited Clare Island at all, so that the only two records we have for the island are two species collected by Sir H.C. Hawley the previous year (*Badhamia conglomeratum* and *Craterium minutum*). All in all, Lister's contribution is little more than a general—albeit important—overview of the Irish Myxomycota in 1912. At present the count of Irish species has reached 228, with 69 of those recorded in West Mayo.[51] The figure for Clare Island is 24, comprising 14 field collections and 10 species obtained by culturing oak bark in Petri dishes.[52]

Lichens

One group of fungi that is spectacularly on display throughout the year is the lichens, of which Clare Island has a rich flora growing in a wide variety of habitats, most noticeably of the species that grow on rocks and soil. The island's lichen flora is also highly significant in biogeographical terms because of its extreme westerly location.[53]

Lichens evolved early in the history of life on land due to the highly successful strategy of entering into a symbiotic partnership between fungi and algae and/or cyanobacteria, the latter providing photosynthetic products in return for the protection provided by a thallus made up of fungal tissue. This enables them to grow in situations neither partner could colonise independently, very often in situations too extreme for other kinds of vegetation.

Bushy (*fruticose*) lichens are the most familiar type. Other lichens are irregular, frilly plates, rather like certain seaweeds or leafy liverworts; such *thallose* lichens are particularly common on tree bark and rocks. There are also many kinds of *crustose* lichens, which live in the surface tissue of trees or the outer layers of rock. The crustose lichens growing on rocks often present a map-like appearance where the dark borders defining individual specimens look like the borders of the different countries (Figure 7.19). Certain kinds (especially species of *Cladonia*) have fruiting structures (podetia) that look like miniature golf clubs or tees. In the days when the islanders wove their own cloth, lichens were used for making dyes; a superior yellow dye was made from the rock lichen *Parmelia saxatilis* (*scra cloch*), and a saffron dye from *féitóg liath* (*Ramalina siliquosa*), a shrubby lichen that is common on rocks above high-water mark. Beth Moran (the weaver of Ballytoohy) still uses plant dyes made from lichens, bog myrtle and furze flowers.

In her study of the lichen flora of Clare Island Annie Lorrain Smith was helped by Matilda Knowles, whom she described as having 'exceptional ability as observer and collector'.[54] In total 154 taxa (134 by modern taxonomic interpretation) were recorded, including eight species new to the Irish flora.[55] During a visit in 1990, Mark

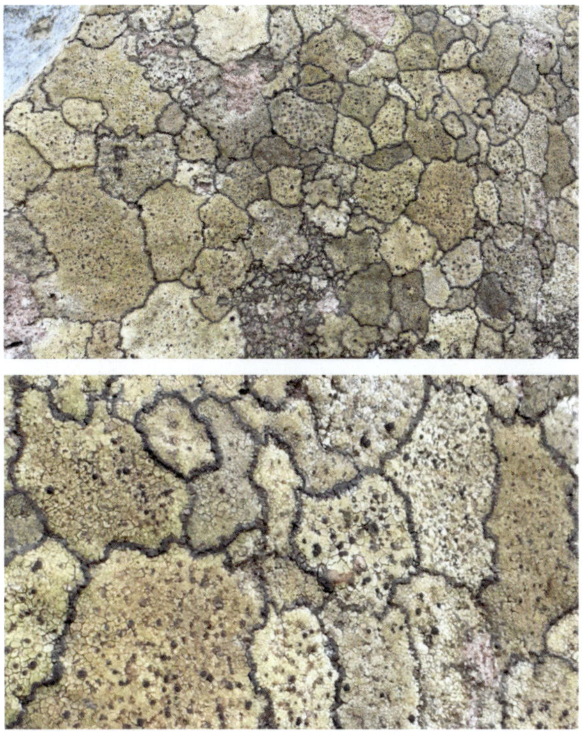

Fig. 7.19 Crustose lichens on rocks of the Maum Formation in Maum. The prolific growth of crustose lichens on rocks—in many cases continuous maybe for thousands of years—does have a down side for the geologist, because it often makes it next to impossible to identify rocks from their surface appearance alone. The rocks in this case are bright red.

Fig. 7.20 Lichens at the shore.
'They clothe the rocks with a sort and variegated colouring, or lend a darker shade to the frowning cliffs by the sea. Everywhere they add their share to nature's infinite variety' (Annie Lorrain Smith (1911), Lichens. Original Clare Island Survey 14, 6).

Fig. 7.21 'The boulders near the sea were gay with the yellow *Physcia parietina* or *Lecanora lobulata*.'.

Seaward, David Richardson and Howard Fox recorded 170 taxa, and a further 168 taxa were added the following year as a result of a visit by the British Lichen Society, bringing the total number so far recorded to 355 lichen taxa (plus 12 lichenicolous and 6 non-lichenised fungi).[56] The report for the New Survey by Seaward and Richardson (2013) includes an annotated checklist, as well as a critical revision of the earlier records.[57]

The lichen flora is seen at its most dramatic and spectacular near the coast (Figures 7.1, 7.20, 7.21), where there is a distinctive and easily followed zonation elegantly described by Annie Smith in her 1911 account:

> The number of species on our list gives but little idea of the vast lichen growths that are to be found on the western shores. The exposed rocks were everywhere covered with sheets of leafy Parmeliae or densely coated with mingling crustaceous forms which competed for every inch of space, while the peaty soil yielded fine growths of Cladonia and Stereocaulon, and spreading over the grass were various species of Peltigera…
>
> The rocks bordering the sea and the great cliffs of the north-west shore of Clare Island are black with an unbroken growth of *Verrucaria maura*. *Lichina pygmaea*, a small, dark, shrubby lichen, is also abundant on some of the rocks washed by high tides; and within reach of the spray *Ramalina scopulorum* [= *R. siliquosa*] and *R. cuspidata* found room for their grey tufts of strap-shaped branches. The boulders near the shore were gay with the yellow *Physcia* [= *Xanthoria*] *parietina* or *Lecanora lobulata* [= *Caloplaca marina*]. Further away the rocks were covered with great white patches of *Lecanora* [= *Ochrolechia*] *parella*, with its large, crowded apothecia; and *Lecanora* [= *Tephromela*] *atra*, with a grey, irregular crust and outstanding black fruits, was very noticeable and very abundant. Further inland still, these two dominant species were partly replaced by *Lecidea rivulosa* [= *Fuscidea cyathoides*], with its thick crust of varying shades of purple brown.[58]

Although the limited extent of woodland restricts the development of arboreal species, lichen diversity can be locally luxuriant, as in the case of the small wood at Lassau, where common species of *Ramalina, Usnea, Parmelia, Physcia, Lecanora* and *Pertusaria* are conspicuous. Smith (1911) recorded *Opegrapha atra* and *Graphina anguina* as being especially notable; both of these species are still to be found on tree bark, the former locally frequent, but the latter only occasionally.

The much higher species count recorded in the detailed surveys of the 1990s is mainly if not entirely due to the greater collecting effort, which has extended our knowledge considerably. A significant element in the difference between the Original and New Surveys, however, is the failure to rediscover many of the interesting and conspicuous corticolous foliose species noted in Smith (1911), such as *Lobaria pulmonaria, Melanelia exasperata, Parmotrema reticulatum, Platismatia glauca* and the fruticose *Usnea florida*. Lichens are extremely sensitive to man-made and natural environmental changes, hence their use as monitors of air pollution and climate change, their sensitivity in this regard varying from species to species.

The terrestrial fauna of Clare Island

It can be difficult to believe that Clare Island is home to more than a thousand different kinds of animals. You see so few as you walk around most of the time; now and then you might spot a group of scouting choughs against the skyline above Knockmore, the occasional rabbit or hare if you are on the lookout for them. It is difficult to appreciate that animals occur in distinct communities, as do the plants of the island. The fact that the majority of the animals are small explains this in part, but we also have to remember that except for the few top predators most species occupy niches somewhere in the middle of food chains (which are linked together to form more complex food webs). Given this, although most animals have to move around in search of food now and then, they hide out of plain sight for much of the time to avoid being eaten or to escape the worst of the weather. Where they occur in large colonies they have developed a peculiar way of life that enables them to do so in security, as with seabirds on inaccessible cliffs, ants in their underground cities or barnacles on a rocky shore.

Animals are herbivores (eating plants), predators (eating only animals) or omnivores (eating dead matter, plants and animals). The distribution of herbivores depends on which plants they eat, and some are very specific in what they require. Their distribution is therefore dictated by the location of the plant communities on which they feed. Omnivores have a wider distribution, but they too have their habitat preferences. Predators, especially larger ones, may kill animals in particular locations, but have a wider choice of location when it comes to resting, reproduction, and so on. Some animals are seasonal in their occurrence, or have biphasic life cycles in which the adults are more readily found than juveniles (e.g. flies and butterflies). All this makes for difficulties in trying to describe the fauna of any locality. These problems were well recognised and recorded by animal biologists in both the Original and the New Surveys.

The Original Clare Island Survey was organised in a taxonomic fashion, with experts in each area looking at their own particular groups and primarily concerned with drawing up lists of species for them. With few exceptions, all the major groups of marine and terrestrial organisms were covered in this way, and in the case of insects each of the main orders was looked at separately.[59] The detailed taxonomic treatment accorded several groups in the New Survey makes detailed comparison between the two surveys—and evaluation of the differences—possible: but this is by no means the case for many others.

Study of the terrestrial fauna in the New Survey divides the flora and fauna into dry-land and freshwater components, and this approach is broadly retained in the pages that follow. However, it should be remembered that several groups of insects spend their early life in the water before emerging onto dry land as adults.

Phylum Annelida
Oligochaetes and leeches

The vast majority of the 150,000 species of annelids known are marine creatures, gloriously represented by the colourful and wonderfully diverse many-segmented bristle-worms (Polychaeta) we met on pages 109–12. The other class of annelids comprises two sub-classes: the parasitic and predatory leeches (Hirudinea) and worms 'proper', which include the large and familiar earthworms (megadriles) and the much smaller worms collectively known as microdriles, which include the tubificids or tubeworms (Tubificidae) that are often so prodigiously abundant in polluted environments, and the enchytraeids (Enchytraeidae), which are among the most important and widespread soil animals. The most important annelids from our human viewpoint are the terrestrial earthworms (family Lumbricidae), whose role in the maintenance of soil fertility it would be hard to overestimate.

The Original Survey was rather dismissive of Clare Island as a habitat for oligochaetes, Rowland Southern summing up his brief account by saying 'There is nothing distinctive of the Oligochaete fauna of Clare Island except its poverty.'[60] This he attributed to the absence of trees, and the general exposure of the island, which made much of it liable to desiccation during periods of drought. He recorded 34 species, but admitted that his account 'must be regarded as very incomplete for all groups'.[61] Fifteen of the species were enchytraeids, and fourteen were lumbricids; the remaining five are

Fig. 7.22 *Lumbricus terrestris* (© David Pérez (DPC), Wikimedia Commons, License CC BY-SA 4.0).

naiads (Naiadae: two species), two tubificids (two species) and a single lumbriculid (Lumbriculidae).[62]

Eight of these species live in the intertidal zone and three are freshwater species. Except for one marine species (*Grania maricola*: see page 112) all the others are soil dwellers. Southern described one of his marine enchytraeids as a new species, *Enchytraeus cliarensis*.[63] His account of the earthworms is particularly brief, partly because the ground was very dry after a period of prolonged drought at the time of his visit in May 1909, 'and earthworms were practically impossible to obtain, except in the gardens near the village'.[64] Earthworms were not studied during the New Survey, but recent work on Achill suggests that the earthworm fauna of fossil cultivation ridges is richer than in other grasslands.[65] It is likely that the relationship between earthworms and grassland biodiversity on the island will receive close attention in the years ahead.

Clare Island's earthworms

There have been considerable changes in earthworm taxonomy since Southern's day. One of his species is no longer considered valid, and most have different names today.

Eiseniella tetraedra
Aporrectodea rosea (Southern's *Eisenia rosea*)
Aporrectodea caliginosa (Grey Worm) (Southern's *Helodrilus* (*Allolobophora*) *caliginosus*)
Allolobophora chlorotica (Green Worm) (Southern's *Helodrilus* (*A.*) *chloroticus*)
Dendrodrilus rubidus (Southern's *Helodrilus* (*Dendrobaena*) *rubidus*, var. *subrubicunda*)
Satchellius mammalis (Little Tree Worm) (Southern's *Helodrilus* (*D.*) *mammalis*)
Dendrobaena octaedra (Southern's *Heodrilus* (*D.*) *octaedrus*)
Allolobophoridella eiseni (Southern's *Helodrilus* (*Bimastus*) *Eiseni*)
Octolasion tyrtaeum
Lumbricus castaneus (Chestnut Worm)
Lumbricus festivus (Ruddy Worm)
Lumbricus rubellus (Red Worm)
Lumbricus terrestris (Common Earthworm)

Phylum Mollusca

Land and freshwater molluscs

The terrestrial and freshwater mollusc fauna of the country is well known, yet the only molluscs familiar to most people are the garden snail and the various slug species whose depredations draw attention to themselves. There are dozens of other species, but most are small and more or less sedentary, and active mainly at night. The number of land and freshwater mollusc species in Ireland stands at 165 at present.

Land and freshwater molluscs were studied for the Original Survey by Arthur Wilson Stelfox who, while extending his area of study to all of West Mayo, spent twelve days on Clare Island. Robert J. Welch also spent several days on the island and passed his notes on to Stelfox, who was also helped by a considerable number of other naturalists who supplied information and specimens.[66] All but twenty out of a total of ninety species found on the mainland occur on the islands, which had only four species that had not yet been recorded from the mainland.

There was considerable interest among Irish naturalists at the time of the Old Survey in 'the intermingling and overlapping of northern and southern—Arctic and Lusitanian—species along the west coast of Ireland',[67] and although Nathaniel Colgan in his review of the marine molluscan fauna felt that 'from a distributional point of view this branch of the subject is perhaps the least interesting and instructive',[68] he did think there was a hint of such 'intermingling' in the group, more apparent in the littoral fauna of the mainland, where he considered six out of the 46 species he identified to be 'austral', compared to two out of 38 on the islands.[69]

A similar 'intermingling and overlapping' can be recognised in the land molluscan fauna. Stelfox considered these two (Arctic and Lusitanian) groups together as an older 'western fauna'. He believed some of these species were here before the Ice Age, but had survived only in certain isolated 'places of retreat', whereas elsewhere they had been overwhelmed by a newer 'eastern fauna'.[70] Stelfox was greatly impressed by Clare Island in this regard: 'Perhaps nowhere in Europe is there a place in which the aboriginal fauna can be better studied.'[71] One of the things that delighted him most was the luxuriance of the molluscan fauna on the sea-cliffs of the island, which he considered to be an outstanding example of such a 'place of retreat'.

> If there existed any place of retreat during the Ice Age on the west coast of Ireland, the great sea-cliff of Knockmore would strike one as being a most probable refuge; facing, as it does, away from the mainland and towards the temperate waters of the Atlantic, the ice descending from the mainland could not surely have affected the conditions prevailing upon the cliff-face to any great extent.[72]

The great sea-cliff at Knockmore (the 'real headquarters' of the land-shells) harboured no fewer than 27 of the 42 terrestrial species found on the island.[73]

> Shells appeared to be most abundant at about 1000 feet. At this elevation every handful of moss yielded dozens of specimens, each patch of Luzula was crowded with *Hygromia fusca*, and if the day were damp hundreds of *Arion ater*—all var. *aterrima*—might be seen crawling upon the vegetation. This slug is reported by Welch and Praeger to show great partiality for the flowers of *Silene acaulis*. I also observed it feeding on this plant, as well as on the flowers of *Jasione montana*, but its customary food seemed to be sheep-droppings, or decaying vegetable matter.[74]

This is a good example of the way Stelfox did more than just make a list of names. He provided detailed notes on local distribution, and tried to record the habits of individual species when opportunity offered (he was particularly exercised about the 'native' character of species). Others, he observed, might consider these to be 'useless and unnecessary notes':

> As, however, there is a probability that, at some future time, this district will be re-surveyed, my aim has been to give those who will be in charge of this second survey as much information as possible, in order that any changes which may have taken place between now and then may be the more easily observed.'[75]

Stelfox was particularly intrigued by the abundance of *Lymnaea truncatula* on Knockmore, where it 'swarms on the bare rock-faces of the cliff and is

Fig. 7.23 (a) The endangered plaited snail (*Spermodiella lamellata*) occurs only among the woodrush stands of the cliffs. It is very susceptible to drought, and its range in Ireland has contracted dramatically in recent years, probably as a direct result of climate change (Roy Anderson).
(b) Juvenile *Limax cinereoniger* (Roy Anderson).
(c) *Leiostyla anglica* (Roy Anderson).

frequently associated among the moss and other vegetation with what would generally be regarded as "woodland" species'. This *Limnaea* normally lives on the margins of streams and ponds.

So many species, many in such abundance! But where are they? Why don't we see them? You must remember that snails and slugs are more or less defenceless creatures; most of them wait until nightfall to begin their foraging. One night in July 1911 Stelfox and Praeger set out under cover of darkness (not for the first time) to see if they could discover the food preferences of that beautiful slug of nocturnal habits, *Limax cinereoniger.*

> ... but we were unable to find them eating anything. We observed, however, that whereas all the specimens of *Limax arborum* were seen on rocks, evidently feeding on lichens or algae, *Limax cinereoniger* was always crawling on the short grass which grows between the boulders and the clumps of taller vegetation.[76]

The non-marine molluscs were studied for the New Survey by Roy Anderson, who recorded a total of 56 species. This is close to Stelfox's total of 57, but there are very significant differences, mainly due to two factors: the absence from the modern list of several of the sea-cliff species that so delighted Stelfox, and the inclusion of several invasive species that have arrived since the original survey: three non-native slugs, a freshwater gastropod, two Mediterranean-Atlantic dune snails and one snail of disturbed ground.[77] Two of Stelfox's aquatic species are also absent from the modern list, but three new pea mussels have taken their place, one of which (*Pisidium lilleborgi*) is red-listed for Ireland. It is interesting that most of the new alien slug species recorded were found only in the vicinity of Granuaile's Tower at the quay, but they are probably widespread around houses further afield.[78]

Detailed comparison and evaluation of the differences between the two lists are however complicated by the inevitable advance in scientific study, which has seen multiple taxonomic revisions, particularly involving cryptic species. For example the *Lymnaea palustris* known to science in 1911 has subsequently been split into several segregate species, the only one of which is widespread in Ireland being *Lymnaea fuscus*; *Lymnaea palustris sensu stricto* has not been confirmed here.[79]

The absence from Anderson's list of the important woodland species recorded by Stelfox does not mean that these have disappeared from the island, but rather that their sea-cliff habitat is less accessible than in 1911. Re-examination of this relict woodland fauna when circumstances permit is considered to be an important task for the future. Although the main area where the old woodland fauna recorded in 1911 was inaccessible, Anderson drew particular attention to the fauna of common woodland species whose occurrence on the island appears to be confined to the stands of woodrush on steep slopes and cliffs, which provide a sort of Lilliputian substitute for the wet woods in which they evolved. They are absent even from the secondary woodland at Portlea. 'In the end', he wrote, 'it may be the relic woodland fauna of Clare Island living in an inaccessible habitat which is woodland in name only which will outlive the native forest fauna in most of the rest of Ireland'.[80]

Phylum Arthropoda
Class Arachnida (spiders and harvestmen)

Araneae (spiders)

Most of us are acquainted with no more than three or four different kinds of spiders. Moreover, that acquaintance is of the most casual sort, and this is the way most of us probably want it to stay. But as is the case with every group of creatures we meet, that would quickly change if we knew them better! Even the common names we give these few familiar spiders are a measure of how vague our appreciation of them is: garden spider, water spider, house spider. The only reason we have names for them at all is that the ways of these few spiders happen to connect with our own in a casual way, but never enough to make real enemies of us, because their activities do no significant harm to our interests. Although our vague vocabulary knows only one 'garden spider', there are several dozen other spider species in your garden as well, most of them smaller than the one we have a common name for, each going about its hidden way of making a living in the garden world. Their unappreciated lives are generally beneficial because they are insect predators.

Although the Prospectus for the Original Survey claimed that 'The report when finished will provide the most complete survey of the Fauna and Flora of a definite area which has ever been made', the species lists compiled for many—perhaps most—groups were very incomplete, and often acknowledged as such. As Denis Robert Pack-Beresford reminds us in relation to spiders, 'it is evident that in order to make anything like a complete survey of the species inhabiting any given area, it would be necessary to live and collect in that area, not only all the year round, but for many years in succession'.[81] Sensibly, he claimed that his spider tally for Clare Island was no more than 'a list of the species which are to be found fairly commonly there'.[82]

Pack-Beresford found that the spiders on the island are not different in any significant way from those of the mainland. Based on his own four-day visit in 1909 and records and specimens provided by others, he compiled a list of 106 species for Clare Island and the adjacent mainland. Seventy-one of these occurred on Clare Island itself; as for the other 35, 'as far as I know there is no particular reason why they should not [also occur]'.[83] Only three of these 106 were 'new to the Irish list', and one of these was found on Clare Island only, the very rare *Diplocephalus castaneipes* (now *Monocephalus castaneipes*). In addition to these three, he listed 31 as 'rare', 22 of which were recorded on Clare Island, many as single specimens.[84]

The presence or absence of particular species of spiders is decided by the range of habitats available. Each part of the island has its characteristic fauna. In and around the lakes and pools we will find the water spider (*Argyroneta*) and the raft spider (*Dolomedes*); indoors the house spider (*Tegenaria*) and the daddy long-legs spider

Fig. 7.24 (a) *Xysticus cristatus* (Roy Anderson); (b) The cupboard spider (*Steatoda grossa*) (Algirdas (CC BY-SA 3.0)).

(*Pholcus*). Few spiders are to be found where conditions are at their most extreme, but Pack-Beresford reported *Segestria senoculata* and *Textrix denticulata* 'in enormous numbers' on the cliffs exposed to the Atlantic gales in the western part of the island. *Segestria* is able to thrive here because it builds a tubular retreat in the rocks where it can keep out of the weather. *Textrix* (a very distinctive little spider with peculiar mating habits) makes a silken bivouac for itself under rocks. In sheltered nooks facing south or west a number of spiders that prefer less exposed situations are found, including *Dysdera crocata*, which has enormous jaws for cracking the hard carapaces of the woodlice on which it feeds.[85] It hunts by night, spending its days under stones in sheltered places. The jumping spider *Heliophanus cupreus* favours similar situations. On the higher ground there is a different fauna again, with hardy species such as *Tmeticus prudens* (now *Centromerus prudens*).

The number of species recorded in the course of the New Survey (from 2562 specimens identified) was 141, nearly twice as many as were recorded by Pack-Beresford.[86]

Adding this number to the fourteen species found by Pack-Beresford but not recorded during the resurvey brings the total spider fauna of the Island to 155, which is nearly 37% of the known Irish fauna; 20 of the species recorded in 2002 are new county records for Mayo. The fact that a new species was recorded for every 18.1 specimens collected suggests a high probability that more species would be recorded under a still more intensive sampling regime.

Although he spent much more time than his predecessor did, Myles Nolan attributed the absence of the fourteen missing Pack-Beresford species from his list to 'the limited and rather sporadic' nature of the resurvey, much more time was spent by him on the New Survey. Several different collecting methods were employed—pitfall traps, Malaise nets and water traps set in association with these, and by hand collecting—and an attempt was made to interpret the variation between the different methods, and between catches from different locations. The difference between the two totals is partly the result of more intensive sampling for the resurvey, and the closer attention paid to habitat associations, although Nolan also points out (*pers. comm.*) that even the most intensive resurveys rarely collect the exact same species as an earlier survey.

At the same time, 19 of the 34 species recorded only on the mainland during the Original Survey were found on Clare Island during the resurvey, showing that there was, indeed, 'no particular reason why they should not [also occur]'. So it is not surprising that only fourteen avoided detection in 2002.

The way in which different species assemblages are associated with different habitats is demonstrated by the difference between the spider fauna taken in Malaise tray-traps in Portlea woodland and that taken in the reedmarsh around Lough Avullin. Five species accounted for more than 75% of the specimens taken in the wood, but less than 14% of the island total. Of the 33 species recorded for Lough Avullin four accounted for over 80% of the total, with one species (*Hypomma bituberculatum*) making up more than 50%. And there was no overlap between the five commonest species in the wood and the four commonest reedmarsh species.[87]

Table 7.3
Species occurring at >3% in pitfall traps (1081) showing also the number and % of total specimens recorded

Species	Specimens in pitfalls	% total pitfalls	total specimens	% total
Trochosa terricola	132	12.21	149	5.81
Pachygnatha degeeri	121	11.19	133	5.19
Pardosa pullata	96	8.88	121	4.72
Tiso vagans	96	8.88	99	3.86
Pirata piraticus	70	6.47	90	3.51
Oedothorax fuscus	64	5.92	81	3.16
Oedothorax retusus	57	5.27	61	2.38
Pardosa amentata	43	3.97	51	1.99
Walckenaeria vigilax	38	3.51	40	1.56
Totals	717	66.3	825	32.18

The six most common species captured in the pitfall traps were among the species most abundant overall. The nine species most abundantly seen in pitfall traps (Table 7.4) are broadly reflective of the open grassland habitats that dominate the island: species not often encountered in either woodland or wetland habitats. Of the 114 species taken by hand collecting, 45 were captured only in this way, but only four of these were present at an abundance greater than 3% of the total. The more interesting species were also generally taken by hand collecting. Both of these factors are a reflection of the wider variety of niches that can be investigated in this way. At the same time, the absence from Clare Island of numerous species commonly associated elsewhere with such habitats as bogs, fens and alder scrub does suggest a somewhat impoverished fauna. Similarly, although the spider fauna of the wood at Portlea is quite typical it is also clearly truncated.[88]

Dicephalus castaneipes (Pack-Beresford's *Monocephalus castaneipes*) is especially characteristic of the moss cover on tree trunks, and rocks on the woodland floor. Pack-Beresford's record was based on a single specimen sent to him, 'amongst a collection made by Mr. H. Wallis Kew'.[89]

The difference in sampling intensity between the two surveys makes it difficult to isolate climate change from other factors responsible for change in the island's recorded spider fauna. The increasingly familiar daddy long-legs spider (*Pholcus phalangioides*) was largely confined to the southern half of Ireland in the1970s, but is now found all over the country; it had probably just arrived on Clare Island before the resurvey of 2002. Nolan does not consider that this can be attributed to climate change. The species lives in houses and other buildings that are artificially heated, and its spread can be attributed to the prevalence of better home heating today. Much the same argument can be made in relation to *Steatoda grossa*.

Order Opiliones (harvestmen)

Harvestmen are not simply spiders with long legs: their biology is quite different. They have no silk or venom glands, but they do have stink glands that produce highly repulsive defensive chemicals. There are thought to be as many as 10,000 species worldwide (although less than half that have so far been described), and there is much greater variety of form and colour among tropical species than our Irish representatives might lead one to expect. Their stilt-like legs enable harvestmen to manoeuvre through a jungle of tall stems and leaves with great agility.[90] Many are omnivorous, and indeed not averse to scavenging, but most also hunt and ambush their prey by night, hiding away during the day.

Pack-Beresford recorded seven species of harvestmen on Clare Island, all but one of which were found in the course of Martin Cawley's more intensive and systematic new survey in 2002, which listed ten species. A notable absentee is *Leiobunum Blackwallii*, which is widespread on the mainland. The alien species *Dicranopalpus ramosus*, which arrived in Europe from North Africa in the late 1940s and has been spreading across Ireland in recent years, has not yet reached the island but its arrival is probably a matter of time.

Fig. 7.25 *Neobisium* (Gilles San Martin (CC BY-SA 2.0)).

Order Pseudoscorpiones (false scorpions)

False scorpions are tiny creatures, few of them as big as a centimetre long and most only a fraction of this, but they are every bit as fascinating as the fearsome larger arachnids after which they are named. Their formidable pincers do indeed resemble those of scorpions, but the stinging tail is absent. They are among the most ancient of terrestrial animals, fossils going back to the Devonian period of earth history 380 million years ago.

Like spiders, false scorpions have the ability to make silk, but the silk glands open at the front end, whereas in spiders they open at the back. Unlike spiders, they do not use silk to trap prey, but to make silken chambers in which to brood their eggs, for moulting, and for hibernation. Some 3,380 species are known worldwide, of which seventeen or so live in Ireland. Some are not uncommon, but often overlooked because of their small size—even indoors, where they can be very useful because they eat the larvae of clothes moths and carpet beetles, mites and other small creatures.

In spite of careful search made by Wallis Kew in June 1910, only three species of false scorpions were found on Clare Island, and only one of these appears to be common.[91] This is the unusual little *Neobisium maritimum*: unusual among false scorpions in that it lives between the tide marks. At low tide it is to be found sheltering in narrow fissures in rocks covered in wrack, barnacles and limpets. It is often in the company of a little group of other invertebrates with which it is characteristically associated: the blue springtail *Anurida maritima*, a small (2mm) yellow carabid with a large head (*Aëpus marinus*), and a little black rove beetle (*Micralymna brevipenne*). The false scorpion can survive being submerged twice daily because an air bubble remains trapped in its nest when the tide is in; oxygen diffuses into this from the water, and carbon dioxide diffuses out.

Among the *Neobisium* observed by Wallis Kew at Gubanooneen were several females in their brood nests 'and bearing the embryo mass', as well as two moulting-nests containing the cast skins of the animal.[92] The other two species found in small numbers on the island were *Obisium muscorum*, a common and widely-distributed species that lives under stones, in leaf litter, and so on, and *Chthonius tetrachelatus*, which also lives in such places, more especially in maritime and mountainous districts.

Nothing of significance was added during the New Survey, only two of Wallis Kew's three species being recorded again, each in only two of Clare Island's 27 square kilometres.[93]

Subclass Acari (mites and ticks)

Mites are found in every sort of habitat apart from the open ocean. They are among the most ancient of terrestrial animals, with forms as old as 380 million years. They have learned to exploit every conceivable way of making a living. There are predatory mites, parasitic mites, herbivores of every variety, scavengers, detritus feeders, fungivores. So successful have they been at exploiting ecological opportunity that they are found in the most unlikely micro-habitats imaginable. An almost typical example—typical in being so extraordinarily particular—is *Dicrocheles*, which lives in the ears of noctuid moths; one ear only, so as not to impair the hearing of its host unduly, resulting in the untimely demise of both moth and mite. There are mites that flourish in dust, and mites that (famously and familiarly) inhabit the follicles of our eyelashes.

There are over than 45,000 species, ranging in size from a full stop to a grape. Many are of great medical and agricultural importance because of the diseases their specialised activities cause to humans and domestic animals (such as Lyme disease, encephalitis and Varroa), and to crops, both in the field and in storage. The great majority however, live lives that have nothing to do with us, going about their business unseen and unknown, until acarologists shine the light of their microscopes and diligent investigations on their fascinating lives. The mites and other small creatures found in soil play a major role in the decomposition of soil organic matter, and as selective grazers of soil fungi. A 100g sample of rich forest soil may contain as many as 500 mites. Although the number of species described worldwide is no more than 40–50,000, the actual number is thought to be over a million.

Relatively little attention was paid to the study of mites during the Original Clare Island Survey ('except as regards one or two of these groups'); the time devoted by James Halbert to insects and other invertebrates meant that 'little time was available for collecting the Acarina'. In spite of his admission that it was 'very incomplete', an impressive list of 186 species was compiled for the district as a whole (not counting the eighty species of freshwater mites), including three new genera, 22 new species and five new varieties. At least ninety species on

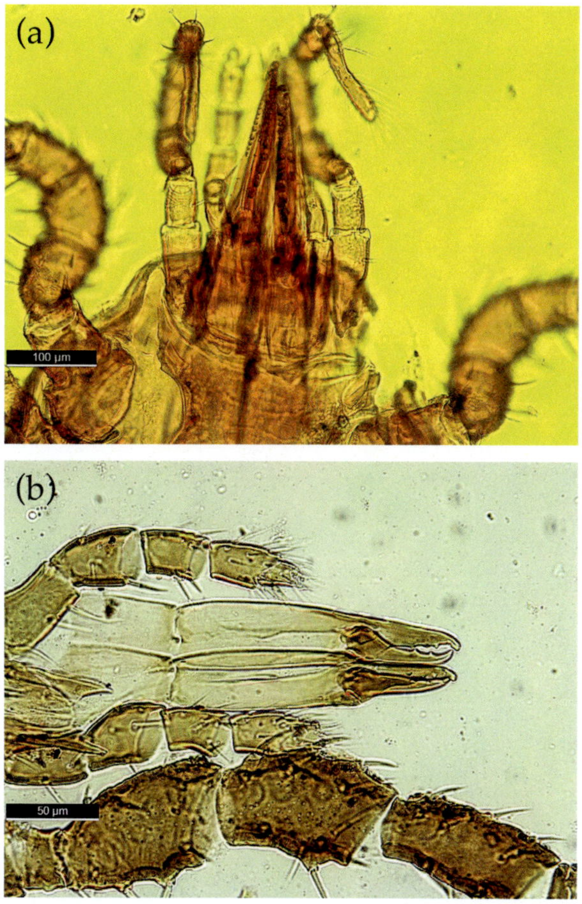

Fig. 7.26a Two predatory species of Mesostigmata showing the variety of forms of chelicerae (a) *Longicheles mandibularis*, (b) *Cosmolaelaps claviger* (Tom Bolger).

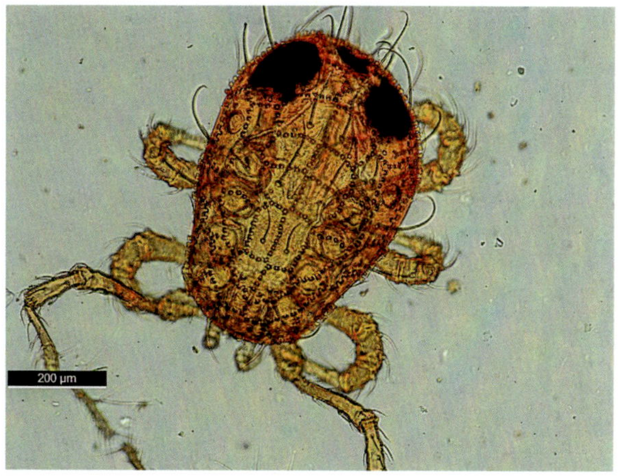

Fig. 7.26b *Epicrius mollis* a species from the Order Mesostigmata found on the island in the New Survey but not in the Original Survey (Tom Bolger).

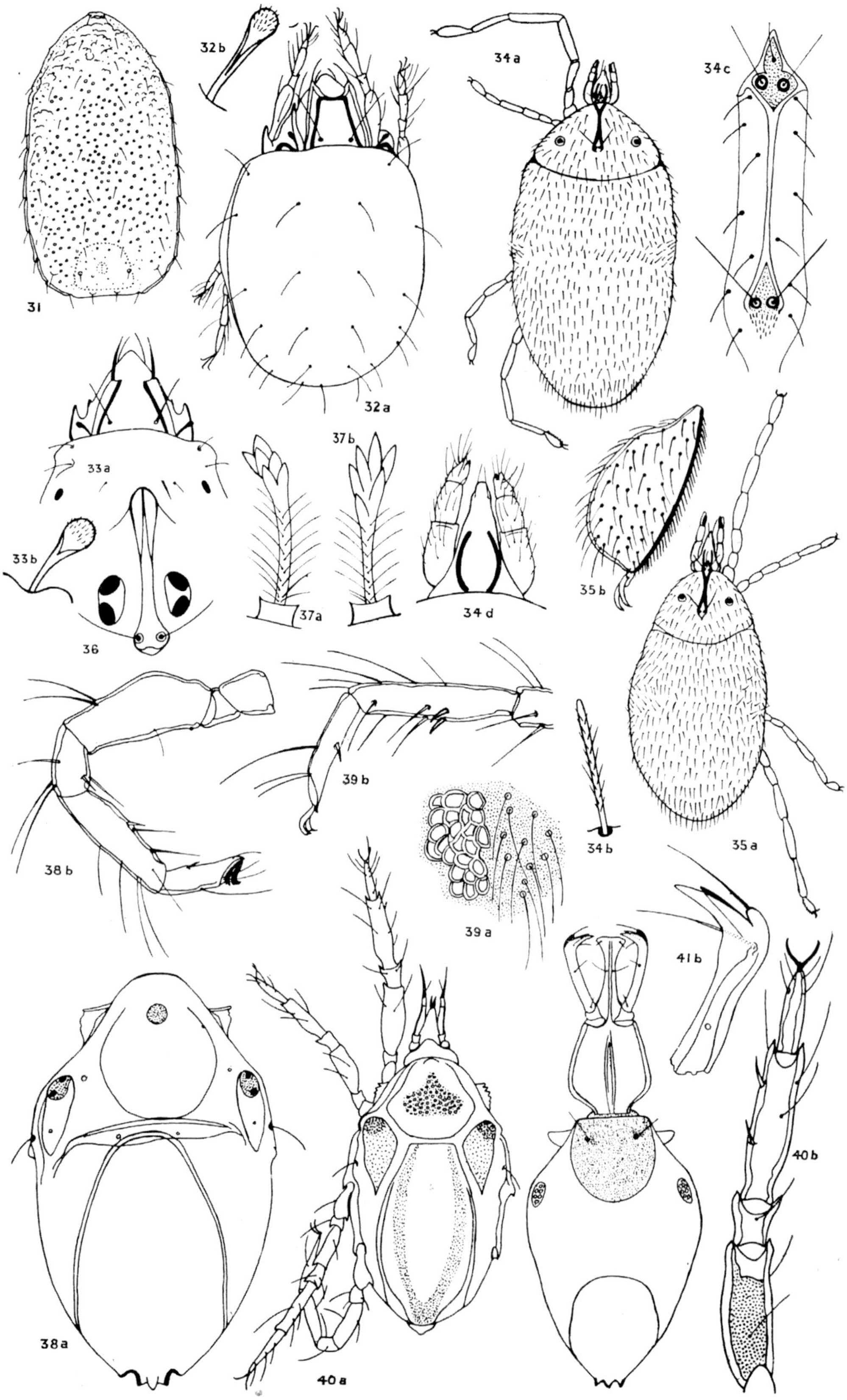

Fig. 7.27 Halbert's mites: *Original Survey* 39 ii, Plate 8.

31 *Sciulus minutus* sp. nov.
32 *Oribata alpina* sp. nov.
33 *Notaspis venustus* Berl.
34 *Rhyncholophus tardus* sp. nov.
35 *Rhyncholophus sabulosus* sp. nov.
36 *Thrombidium valgum* George
37 *Thrombidium ramosum* George
38 *Halacarus Southerni* sp. nov.
39 *Halacarus areolatus* sp. nov.
40 *Halacarus lamellosus* Lohmann var. *septentrionalis* nov.
41 *Scaptognathus Trouessarti* sp. nov.

Fig. 7.28 (a) Sea-slater (*Ligia oceanica*) (Roy Anderson); (b) *Oniscus asellus* (Roy Anderson).

the list were new to both Britain and Ireland, and at least 156 were previously unrecorded for Ireland.[94] Halbert's report runs to 136 pages (with eight plates), but it is largely devoted to matters of taxonomic detail, and showed little interest in focusing on Clare Island in itself. Sixty-seven species are given specific Clare Island records, but there can be no doubt that many others would have been found with a more intensive search.

Halbert thought the gamasid mites the most interesting.[95] These are largely predatory mites, found in a wide range of habitats; the majority of those on Clare Island were found in moss and under bark. Many are parasitic, but apart from two kinds found on field mice he made no attempt to collect these. They are relatively large as mites go (200 microns up to 4mm). Among the most interesting and specialised of the gamasid mites are those that live in ants' nests, one beautiful species of which (*Urotrachytes formicarius*) is common on Clare Island.

The only mite with which most people are familiar is the common tick (*Ixodes ricinus*), which was and is abundant everywhere, 'especially amongst grass and heather in wooded districts'.[96] Ticks constitute a small but very important sub-order of acarines—important because they are blood-sucking ectoparasites of vertebrates, and the most important vectors of human and animal disease after mosquitoes.

The only representative of the Sarcoptoidea[97] recorded for Clare Island was *Hyadesia fusca*, which was found among the coralline algae in rock pools, in the company of sea mites of the genus *Rhombognathus*; this was the first occurrence of this marine species on the coast of either Britain or Ireland.[98] Oribatid mites are strongly sclerotised forms usually found in leaf litter and soil.[99] One of their many talents is the ability to fold themselves up to a fraction of their size as occasion demands. They are rather small mites, from 0.2 to 1.5mm.

Thrombidid mites[100] are the most diverse order, both in terms of species diversity and in structure: in spite of which they 'were but little collected, except for the aquatic kinds. Indeed this part of the list is a record of casually collected species often represented by single specimens.'[101] A number of truly marine forms were recorded (in the family Bdellidae). These species retire into crevices in the rocks or under stones when the tide is in. Other species live under stones and seaweed at or near high water mark. Although reference is made to the zonation of mites on the seashore, little attention was paid to this topic (beyond admitting that 'much observation is needed').[102] Halbert was among the first to realise the existence of 'an undoubted sphagnum fauna' among mites, and provided a short list of fifteen or so species.[103]

For the New Survey Thomas Bolger of UCD's School of Biology and Environmental Science collected mites during several visits to Clare Island between 1994 and 2001. He used soil-coring and suction sampling methods, whereas Halbert simply turned over stones and searched the ground by eye, which meant that many fast-moving species, as well as those that live below ground, were missed.

Bolger collected 64 species of oribatid and gamasid mites compared to Halbert's 60, but only 30 were common to both surveys; the combined total between the two collections is 94 species.[104] The startling difference between the two lists is due mainly to the different methods employed during the two surveys, but it still emphasises the incomplete nature of both surveys and the need for further work. Although the great majority of the species are common, three of the newly-recorded mites were first records for Ireland, but this is more a reflection of how poorly the Irish acarine fauna is known than of anything unusual about the Island's mites.

Order Isopoda (isopods)

Woodlice are the terrestrial branch of an order of crustaceans that have their headquarters in the sea; they are the only group in the order to have successfully made the transition to dry land in the course of their evolution. They are familiar to everybody, but few of us realise how very abundant they are because they stay hidden during the day in order to conserve body moisture: their aquatic ancestry remains with them to that extent.

At the time of Nevin Foster's review for the Original Survey 27 species of woodlice had been recorded in Ireland, nine of which occurred on the island itself, none of them more than usually out of the ordinary (although *Porcellio pictus* did merit comment as 'even invading the beds in several of our stopping places in the district').[105] The large woodlice that inhabit the upper shore (and which are most active after dark) are sea-slaters (*Ligia oceanica*). Shine a torch and they scuttle for shelter as fast as their woodlice legs will carry them. The ubiquitous water hog-louse (*Asellus aquaticus*), the only freshwater representative, occurs in all the small lakes of the island, as it did a hundred years ago.

All of Foster's nine species were recorded by Martin Cawley in 2002, along with three new species, two of which (*Haplophthalmus mengii* and *Trichoniscus pygmaeus*) Foster had expected to find but did not, even though 'the special attention of Mr. Stelfox and myself was directed to its discovery', and 'we have succeeded in finding it in every other Irish county in which we have searched for it'.[106] Cawley believes that four of his twelve species may have arrived with human assistance, including one new arrival, the rosy woodlouse (*Androniscus dentiger*). The absence of *Porcellionides cingendus*, which occurs on Inishbofin, is puzzling.

Class Myriapoda (centipedes and millipedes)

Praeger and his colleagues on the Organising Committee were unable to enlist the services of a recognised myriapod expert, so they asked the Rev. William F. Johnson (whose main field of expertise was beetles) to take on the task. With the help of others he compiled a list of 13 species for the wider Clew Bay area, out of the 34 then known in Ireland, eleven of which were on the island and two were species hitherto unrecorded in Ireland. Of these eleven species eight were centipedes and three were millipedes.

There are two orders of myriapods, readily distinguished from each other by the number of pairs of legs on the segments of the trunk: two pairs in the millipedes (Diplopoda) and just a single pair in the centipedes (Chilopoda). Both are highly successful groups, with 2,500 centipede species and up to 10,000 millipedes worldwide. Most centipedes are predators (many look like diminutive dragons), hunting under stones, bark and logs and in other dark underworlds for insects and other invertebrates, which they grab with their venomous jaws; most millipedes are herbivores, feeding on decaying leaves and other dead plant material. Both groups are poisonous, but whereas centipedes produce a neurotoxin which they inject with their impressive fangs, millipedes have a pair of poison glands on each trunk segment, which makes them highly distasteful to would-be predators.

Diplopoda

Martin Cawley's survey of 2002 recorded ten millipede species, six more than Johnson. Most of the new species were probably present a hundred years ago; it is hardly surprising that Johnson and his helpers missed some since 'none of those who collected had special knowledge of the habits and habitats of these creatures, and all were engaged in other work'.[107] However, his assumption that the presence of millipedes is evidence for Clare Island's former connection with the mainland 'since these creatures could not make their way across the sea' was somewhat naïve. Several of the species are associated with human habitation and are most likely to have reached the island with people. Two other alien species whose range on the mainland is expanding are likely to join them soon (*Leptoiulus belgicus* and *Cylindroiulus londinensis*).[108]

Chilopoda

The two new species found on Clare Island by Johnson were *Lamcytes emarginatus* and *Hydroschendyla submarina*.[109] *Lamcytes emarginatus* lives on the banks of small streams, and is reported to be able to make brief underwater forays. *Hydroschendyla submarina*

Fig. 7.29 (a) *Lithobius variegatus* (Roy Anderson); (b) *Hydroschendyla submarina* (Anthony Barber).

is an intertidal species.[110] The fact that *Hydroschendyla submarina* was not seen during the New Survey suggests that the naturalists who collected centipedes for Johnson were not quite as inexperienced as he may have led us to believe, because there has only been one subsequent sighting of this rare species in Ireland (Figure 7.29).

Among the newly recorded species is *Cryptops truncorum*, an inconspicuous centipede probably overlooked by the early naturalists, and *C. hortensis* which is larger and more active and may be a recent arrival. These two species are particularly associated with human habitation and probably arrived by boat.[111]

Phylum Tardigrada (water bears)

Few creatures better illustrate the truth of Linnaeus' maxim that 'nature is at its most-to-be-marvelled-at in the tiniest of creatures' than tardigrades.[112] Most of them are not much more than 0.1mm long, but they have the most extraordinary qualities.[113] They are sometimes referred to as 'water piglets', but we will retain the name coined for them by Johann August Ephraim Goeze, privileged to be the first to behold one of these wonderful creatures, and (thinking its rolling, shambling gait when on unfamiliar ground was somewhat bear-like) described it as *ein kleiner Wasserbar* (a little water bear).

Fig. 7.30 Coloured scanning electron micrograph (SEM) of a tardigrade (*Paramacrobiotus*) in moss. Magnification x555 (Science Photo Library).

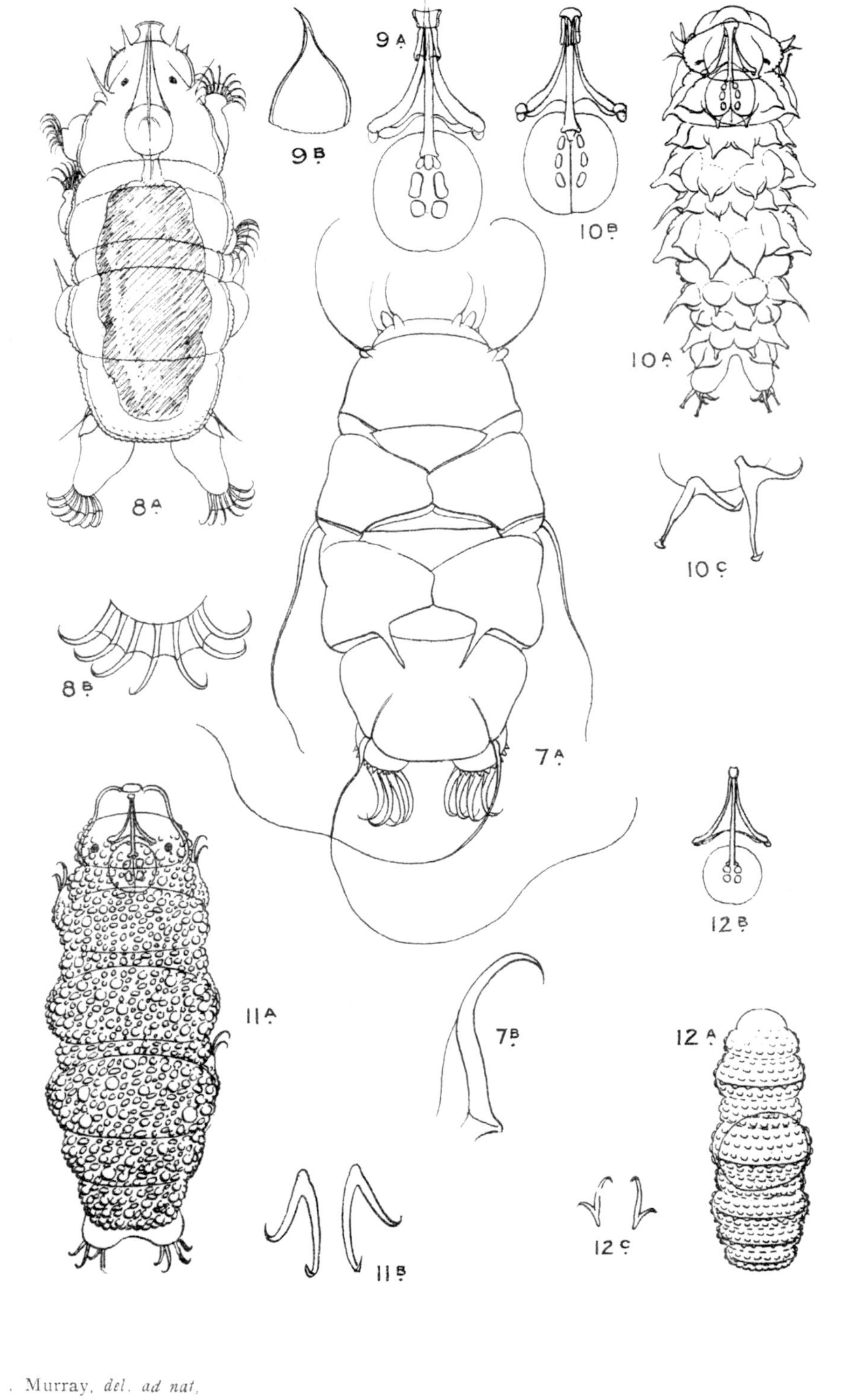

Fig. 7.31a Murray's tardigrades: *Original Survey* 37, Plate II, 10-12.

10 *Macrobiotus papillifer* Murray; variety with large papillae (B: teeth and pharynx; C: claws.

11 *Macrobiotus scabrosus* sp. nov.; B: claws.

12 *Macrobiotus ornatus* var. *verrucosus* Richters; A: skin with egg; B: teeth and pharynx; C: claws.

They can live anywhere, and in favourable situations can be tremendously abundant (as many as 25,000 in a litre of sediment). Some live in freshwater, a few in the sea, others in soil, many among decaying leaves, in rot holes, in moss or liverwort forests. A few are parasitic. Clogged eave-chutes are often a good place to look for them. They live on bacteria or minute algae, or tiny animals, which they pierce with a pair of calcified stylets and then use their muscular pharynx to pump fluids into their gut.

They come in a great variety of shapes, and many colours: red, brown, blue, purple, yellow, orange, olive, pink or translucent (their colour depends partly on what they are eating). All have four pairs of stumpy legs that end in claws or adhesive disks which they move by hydrostatic pressure. Nobody knows how many there are. The best current textbook estimate is 1,200 described species and subspecies.

Colin Tudge in his wonderful book *The Variety of Life* tells us that if the world is ever laid waste by ecological disaster or collision with yet another asteroid the tardigrades are good candidates for survival.[114] They can withstand extreme desiccation for centuries—perhaps for millennia; they can endure temperatures as low as –272°C or as high as 151°C and levels of radiation that would be a fatal dose for humans. When they are active water accounts for 85% of their body weight; if their abode is overtaken by severe drought they can enter a state of suspended animation where this drops to 3%, and remain in this 'tun' state for a hundred years. In conditions less extreme they transform into thick-walled 'cysts' in which state they can comfortably wait out the drought, hatching to take up where they left off.

Prior to the Clare Island Survey nobody had looked for water bears in Ireland. At that time the leading authority on the group in the English-speaking world on the microscopic fauna associated with bryophytes was James Murray.[115] In March 1911 he spent a week collecting in the district—on the island and facing mainland—and he also examined material supplied by others. He recorded 35 species, nearly all of them 'firsts' for this country, 11 new to Britain and Ireland, and 4 new to science.

Thirteen or fourteen species were recorded on Clare Island, a number of them very common, and

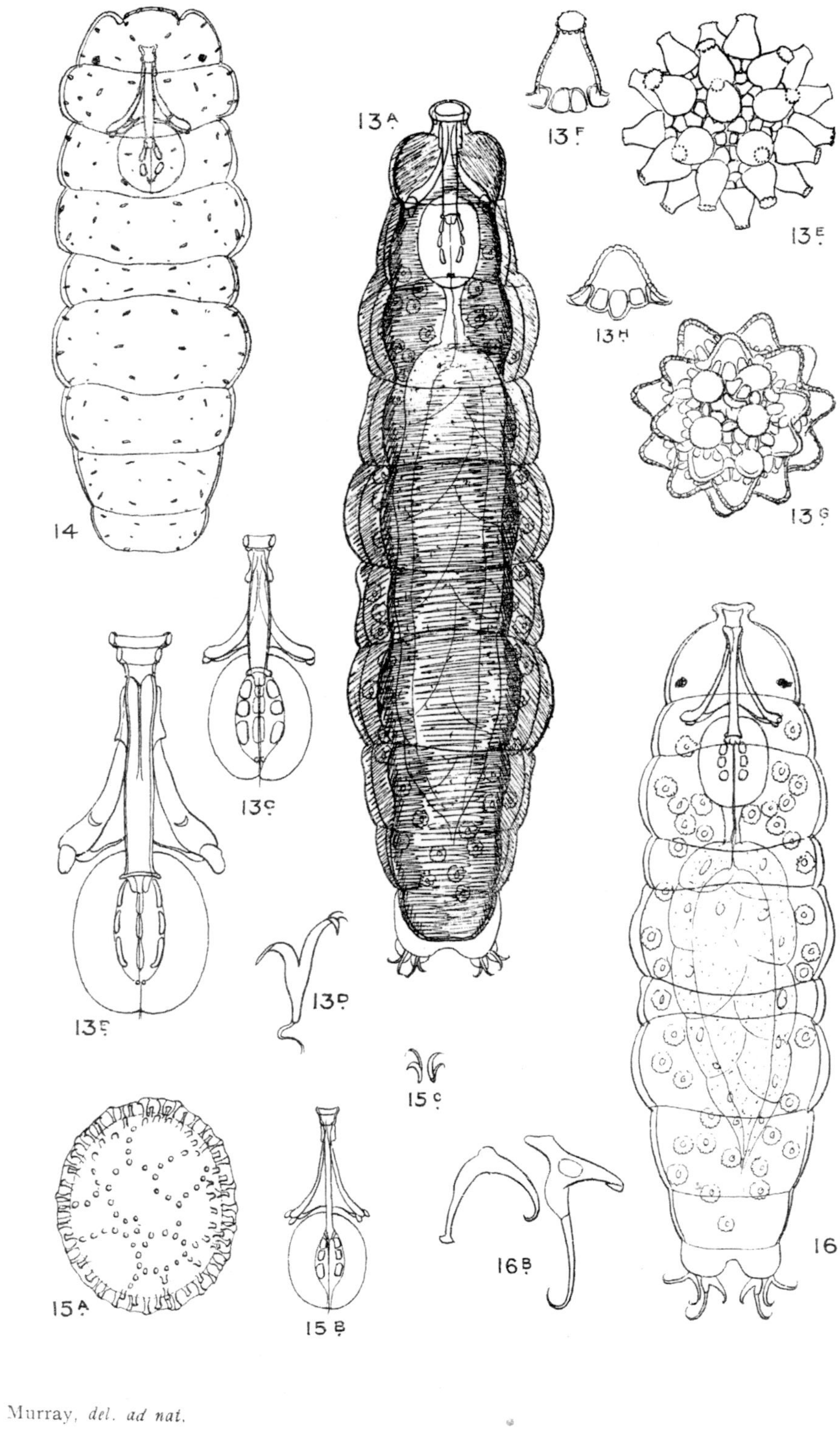

Fig. 7.31b Murray's tardigrades: *Original Survey* 37, Plate III, 13.

Macrobiotus richtersii sp. nov. (B: teeth and pharynx; C: teeth and pharynx of young in the egg; D: claws; E: egg; F: one process of the egg; G: variety of the egg; H: one process of the egg).

three species were seen only here. A special variety of *Echiniscus gladiator* (var. *exarmatus*) was found 'at all elevations.' *Macrobiotus crenulatus*, which was found in some numbers on Knockmore, is one of half a dozen Arctic species; another is *Diphascon chilenense*. One species, which he identified as *Macrobiotus richtersii* , was found in the saltmarsh at Kinnacorra, and is a good example of the extraordinary distribution of some species of water bears. Although hitherto undescribed, this 'species' was known to occur in Samoa, South Africa, Scotland, Uganda—and Clare Island.[116] Murray believed there was something of a distinct 'Canadian element' to the fauna, 'this Atlantic distribution due to some real community of conditions'.[117]

Very little work has been done on tardigrades since Murray's day; indeed, his contribution to the Original Survey still accounts for some 70% of all records for the phylum in Ireland to the present day. A contribution by Erica DeMilio and her colleagues at NUIG for the New Survey in 2013 found that the great majority of the specimens examined (89 out of 131) belonged to the *Macrobiotus hufelandi* species group. Ten other species were identified, two of them new Irish records.[118]

Direct comparison between the results of the two surveys is difficult because of the differences in the number of samples and uncertainty about the quantity of material gathered and examined by Murray at each of his three Clare Island locations (compared with the New Survey's eight). The differences between the two surveys are most likely due to the patchy distribution of species and the small number of samples collected.[119]

Subphylum Hexapoda
Class Entognatha
Sub-class Collembola

Though hidden from everyday eyes and on that account largely unnoticed by us, the fauna of soil is among the most species-diverse on land; indeed, below-ground diversity greatly exceeds that in plain sight. Springtails are the largest of several groups of wingless arthropods long considered to be insects, but now relegated to a distinct class (Collembola). They are highly successful and enormously abundant, but unfamiliar to most people because they are very small creatures (seldom more than 5mm long), and the worlds they inhabit are hidden from

Fig. 7.32 (a) *Petrobius maritimus* (Jymm, Wikimedia Commons).
(b) *Anurida maritima* (under-surface); a: papilla of ventral tube. 'It appears to be gregarious, and when the tide is high, to shelter itself against the commotions of the water in chinks of the rocks and other positions of advantage. When the tide is out the Insects apparently delight to congregate in masses on the surface of the rock pools. This *Anurida* can endure prolonged immersion; but ... it is, when submerged, usually completely covered with a coat of air so that the water does not touch it.' (*Cambridge Natural History*, 1895, 194-5).

Table 7.4: A summary and comment on the species of Collembola recorded by G.H. Carpenter for the Original Survey (Tom Bolger).

Name used in original survey (Carpenter 1913)	**Name used in checklist of Irish species (Bolger 1986)**	**Nomenclature according to Hopkin (2000)**	**Comments on status in Ireland**
Achorutes armatus (Niciolet)	*Hypogastrura armata* (Nicolet)	*Ceratophysella armata* (Nicolet)	Widespread species occurring in a variety of habitats.
Achorutes longispina Tullberg	*Hypogastrura longispina* Tullberg (sp. dub.)	*Ceratophysella longispina* (Tullberg)	The only Irish record of this species is from herbage near the beach at Portakilly.
Achorutes viaticus (Linn.) Tullberg	*Hypogastrura viatica* (Tullberg)	*Hypogastrura viatica* (Tullberg)	Widespread species occurring in a variety of habitats and often in very large aggregates.
Xenylla maritima Tullberg	*Xenylla maritima* Tullberg	*Xenylla maritima* Tullberg	Appears to be widespread on seashores.
Anurida maritima (Guerin)	*Anurida maritima* (Guerin)	*Anurida maritima* (Guerin)	Very widespread and abundant species on the seashore.
Folsomia quadrioculata (Tullberg)	*Folsomia quadrioculata* (Tullberg)	*Folsomia quadrioculata* (Tullberg)	A very widespread and abundant species in soil and litter.
Isotoma sensibilis (Tullberg)	*Isotoma sensibilis* (Tullberg)	??	Very widespread and abundant species.
Isotoma olivacea (Tullberg), var. *grisescens* Schäffer	*Isotoma olivacea* (Tullberg)	See comments	Widespread and abundant species but now considered by some to be a species complex.
Isotoma viridis (Bourlet)	*Isotoma viridis* (Bourlet)	See comments	This is one of the most widespread and common species of Collembola. However, it has recently been split into a number of species, two of which occur commonly on Clare Island.
Isotomurus palustris (Müller)	*Isotomurus palustris* (Müller)	*Isotomurus palustris* (Müller)	Very widespread and abundant species.
Tomocerus minor (Lubbock)	*Tomocerus minor* (Lubbock)	*Tomocerus minor* (Lubbock)	Very widespread and abundant species.
Tomocerus longicornis (Müller)	*Tomocerus longicornis* (Müller)	*Tomocerus longicornis* (Müller)	Very widespread and abundant species.
Entomobrya Nicoletii (Lubbock)	*Entomobrya nicoleti* (Lubbock)	*Entomobrya nicoleti* (Lubbock)	There have only been four records of this species in Ireland but this may be due to limited sampling.
Lepidocyrtus cyaneus (Tullberg)	*Lepidocyrtus cyaneus* (Tullberg)	*Lepidocyrtus cyaneus* (Tullberg)	Very widespread and abundant species.
Orchesella cincta (Linn.)	*Orchesella cincta* (Linn.)	*Orchesella cincta* (Linn.)	Widespread species occurring on surfaces of soil and vegetation.
Orchesella villosa (Geoffroy)	*Orchesella villosa* (Geoffroy)	*Orchesella villosa* (Geoffroy)	Only five Irish records but they are from disparate habitats.
Sminthurides aquaticus (Bourlet)	*Sminthurides aquaticus* (Bourlet)	*Sminthurides aquaticus* (Bourlet)	Carpenter says that this occurs in Dublin but the only real Irish record is from Clare Island where it was found in enormous numbers on a pool behind the storm beach at Kinnacorra.
Dicyrtomina minuta (O. Fabr.)	*Dicyrtomina minuta* (O. Fabr.)	*Dicyrtomina minuta* (O. Fabr.)	Very widespread and abundant species.

Table 7.5: Species recorded during 2001–2002 which had not previously been recorded on Clare Island (Tom Bolger)

Neanura muscorum (Templeton)	Very widespread and abundant species.
Xenylla humicola (O. Fabricius)	Has previously been recorded from coastal sites in Antrim, Dublin, Meath and West Cork. In this survey it was found in the plantain sward at the west end of the island.
Protaphorura fucifera (Börner)	Very widespread and abundant species.
Folsomia candida Willem	Very widespread and abundant species.
Isotoma anglicana (Lubbock)	This species was common on the island and was probably recorded by Carpenter as *Isotoma viridis*.
Isotoma notabilis (Schäeffer)	Very widespread and abundant species.
Isotomiella minor (Schäeffer)	Very widespread and abundant species.

view. For the most part they live in soil and among litter, where they play a major role in the decomposition of soil organic matter, and as selective grazers of soil fungi. More than 7,500 species have been described, but springtails have been around for as long as there has been soil and litter to exploit. They are very ancient creatures, belonging to lineages that go back as far as the Devonian period of earth history 400 million years ago.

Bristletails are another ancient lineage, and although represented on Clare Island by only two species, one of these is the most widespread wingless insect on the island. This is the shore bristletail *Petrobius maritimus*, which is ubiquitous all along the coast, between the tides and for some distance above high-water mark. Those that live between the tides retire into sheltered cracks and crannies when they are under water; when the tide is out populous assemblies congregate on the surface of rock pools. It is most unusual among insects in being so at home at the edge of the sea, but if it is submerged it stays dry because it remains coated in a protective coat of air trapped on its hairy surface (Figure 7.32).

George Carpenter spent a week collecting apterygotes on Clare Island in July 1911, and he was also supplied with specimens by various colleagues. Of the 18 species of springtails recorded by him, the most common were *Isotoma viridis*, *Tomocerus minor* and *Tomocerus longicornis* (see Tables 7.4, 7.5). Although his tally of these primitive wingless insects is little more than an attenuated list of forms dominant on the mainland, it is remarkable for its generally northern and American aspect. All but six of the eighteen springtails on the island also inhabit arctic or sub-arctic countries, and all but three also occur in North America or Greenland. The shore-haunting bristletail *Petrobius* also has American affinities. On the other hand, a few species were considered to be members of 'the southern or Mediterranean faunistic group'.

Carpenter took this as further support—albeit tenuous, as he was careful to point out—for the theory of 'a western Continental land linking the present Irish area with south-western Europe, and with America by way of the Arctic regions. At the same time the presence of the springtails recorded from Puffins' and Gulls' nests on the Bills Rock, warns us that we must admit the possibility of the accidental transport of these small insects across sea-channels by the agency of birds'.[120]

Whereas Carpenter collected his springtails simply by turning over stones and searching the ground by eye, Bolger used suction sampling and soil-coring. [121] Carpenter's somewhat superficial approach meant that many fast-moving species, as well as those that live below ground, would be missed. It is hardly surprising then, that the results of the two Surveys were very different. Bolger recorded 24 species compared to Carpenter's 18: but only 4 species were common to both, making a total of 38. All are common and widespread elsewhere in Ireland.

Fig. 7.33 House cricket (Petr Gebelt, (CC BY 3.0)).

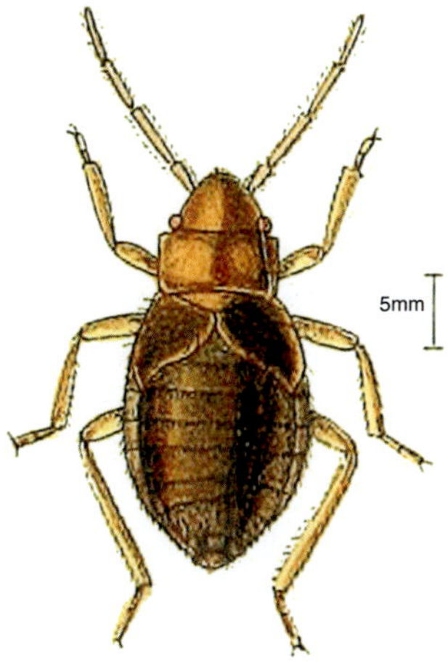

Fig. 7.34 *Aëpophilus bonnairei.* From *The Hemiptera Heteroptera of the British Islands* (1892) by E. Saunders.

Class Insecta

Order Orthoptera

Grasshoppers and crickets are the main insects assigned to the Orthoptera. Their principal economic importance is that they include the locusts that can cause such damage to crops (locusts are really just big grasshoppers). The great reduction in the area of species-diverse grassland over the last half century or so has seen a great decline in the diversity and abundance of these insects. The singing of stridulating grasshoppers is one of the unforgettable hallmarks of these wonderful habitats, and its scarcely noticed disappearance from common experience means one more chord lost from the symphony of earlier summers.

Two grasshopper species (the common green grasshopper *Omocestus viridulus* and *Stenobothrus bicolor*), and one groundhopper (*Tetrix bipunctata*) were recorded on Clare Island by George Carpenter during the Original Survey. The house cricket (*Acheta domesticus*), not uncommon in Praeger's day, is no longer found on the island.[122]

Order Dermaptera (earwigs) and Order Dictyoptera (cockroaches)

In Carpenter's day earwigs and cockroaches were classified under Orthoptera, but are now assigned to different orders. The common earwig (*Forficula auricularia*) was then widespread on the island, and is still common today, but the common cockroach (*Blatta orientalis*), which was at that time 'found in the outhouse on Clare Island which served the naturalists as a laboratory' has disappeared.[123]

Order Hemiptera (bugs)

In everyday discourse we often use the word 'bug' for any sort of creepy-crawly, just as we use the word 'fly' for any insect with wings. These words have a more precise meaning for the naturalist, however; 'bugs' are a particular order of insects characterised by mouthparts that have become adapted for sucking the juices of plants (and sometimes of animals, mainly arthropods). It is an enormous and truly successful group, with some 70,000 species described so far.

The order comprises three groups. Cicadas and froghoppers (suborder Auchenorrhyncha) account for nearly half the total, but there are nearly as many Prosorrhyncha, the suborder that includes the insects to whom the name Hemiptera properly applies. Hemiptera means 'half-wings', and in the Prosorrhyncha one part of each forewing is horny and tough, the other part membranous. At rest the hard part folds protectively over the soft part. The third group of Hemiptera is the Stenorrhyncha, and although this is the smallest of the three (with 12,500 species), it is the most familiar, because it includes the aphids or greenflies all too familiar to anybody with a garden, but actually nearly every species of plant or plant group has its own distinct aphid or aphid group, specifically adapted to feed on them.

James Halbert wrote the section on Hemiptera for the Original Survey. In common with so many others, he considered that his list gave nothing more than 'a fair idea of the prevalent forms of Hemiptera actually occurring in the Clare Island neighbourhood', the result of a few short visits made during the three years previous to his report. These excursions were made in June and early July, so that species reaching maturity later in the season are likely to have been missed, and no attempt was made to collect aphids.[124]

Of the 172 species of Hemiptera recorded, 66 occur on Clare Island itself (including 26 Homoptera: froghoppers and leafhoppers), mainly in the eastern part of the island. Most of the species are common and widely distributed, but with the usual leavening of rare and local species, including the mixture of 'northern' and 'southern' forms found in all the insect orders in the west of Ireland. In terms of distribution the most interesting species on the list is the rare *Aëpophilus Bonnairei*, which lives between tide-marks in cavities bored in rocks by *Saxicava*. On the continent this species is known only from the French and Galician coasts, and in Britain is largely confined to the coast of Devon and Cornwall.[125]

Dicyphus pallicornis, which is common everywhere—but only on foxglove–is a striking example of the specialised feeding habits of many Hemiptera. *Orthesia (Arctorthesia) cataphracta* is a scale insect (whitefly) that occurs commonly in moss throughout the district, especially on the mountains, as on Knockmore. A notable feature of the list is the complete absence of larger shield bugs (family Pentatomidae).

Order Coleoptera (terrestrial)

The report on terrestrial beetles was compiled by W.F. Johnson and J.N. Halbert, and was based not only on their own field work between 1909 and 1911, but on the records of the several other coleopterists who had explored the district before them. For the broader 'Clare Island district' they recorded 521 species (not including the 90 species of water beetles, which were the subject of a separate report), but emphasised (as did several others) that 'this cannot be regarded as anything like a complete list of the beetles of the district. Such a list could only be compiled by resident collectors working all the year round for several years in succession'.[126] Johnson and Halbert recognised 'a western trend' in several species, and evidence of 'a more northern distribution' that was much more in evidence than 'a distinctively southern origin'.[127]

The number of species recorded on the island was 238, and although a few of these had not been recorded on the mainland, 'there is no reason why they should not occur there'. The groups best represented were Carabidae (ground beetles), Staphylinidae (rove beetles) and Aphodii (dung beetles). The authors were struck by 'the very frequent occurrence' of such species as *Carabus clathratus*, *Staphylinus caesareus* and *Otiorhynchus blandus* on the island. *Carabus clathratus* is a beautiful northern ground beetle, brassy to greenish in colour, each of its wing cases with three rows of large metallic pores lodged between three distinctively raised ribs. *Staphylinus caesareus* is a somewhat rare northern species of rove beetle (13–16mm). *Otiorhynchus blandus* is a boreal coastal weevil.

Among other more 'northerly' species on the island are *Patrobus assimilis* and *Tachinus proximus* (both found on the summit of Knockmore; the first Irish record of the latter). *Patrobus assimilis* is a wingless black ground beetle (7–8mm) with red legs and antennae. *Tachinus proximus* is a small rove beetle (5–6mm). A few interesting species live on

Fig. 7.35 (a) *Carabus clathratus* (Roy Anderson); (b) *Cychrus caraboides*, which is one of Clare Island's woodland carabids, feeds on slugs and snails (Roy Anderson).

the shore, including *Aepopsis robinii*, a tiny and very local carabid species (2mm) that lives in crevices in the rocks between tide-marks, and *Oxytelus maritimus* (a small rove beetle) that is found in decaying seaweed.

Four dung beetles belonging to the genus *Aphodius* were recorded on the island, their habitat described rather more delicately than would be the norm today; *Aphodius fimentarius* was taken *'in stercore bovino'*, whereas *Aphodius putridus*, *Aphodius ater* and *Aphodius lapponum* were found *'in stercore ovino'*, the last in great numbers.[128]

As an example of how beetle populations could change with the seasons Johnson and Halbert pointed out that whereas *Aphodius contaminatus* was exceedingly abundant everywhere on the island from the sea shore to the top of Knockmore in September, it could not be found anywhere at all in June. On the other hand, *Aphodius scybalarius* was found in June but not in September, and two other species, *Aphodius lapponum* and *Aphodius ater*, were much more abundant in June than in September.[129]

The only group of land beetles studied for the New Survey were the ground beetles (Carabidae), 66 species of which are now known from Clare Island; 50 of these were recorded by Johnson and Halbert in 1912 and 50 were recorded by Roy Anderson and Stephen McCormack in 2002/3; but out of that total of 66, 16 species are unique to each list. The fauna is described by the authors of the New Survey as an impoverished subset of that found across Ireland as a whole, reflecting the isolation of the island ecosystem, its small size and the restricted range of habitats.[130]

Although the difference between the two lists is striking, it is thought to be due to random factors and the use of pitfall trapping alongside hand collecting in the recent survey rather than the result of such drivers of change as the advent of alien species or climate change.

The larger ground beetles continue to be well represented; this may reflect the absence of foxes, which include them in their diet when nothing else is available. Woodland species continue to be poorly represented, which is not surprising in view of the limited extent of suitable habitat. The newly added species include seven that are typical of acidic, exposed upland environments. Their absence from the old list is surprising; we would expect them to have been widespread a hundred years ago, so it is more than likely that they were simply overlooked. Four other species on the new list absent from the Original Survey list occur in wetlands of one kind or another. It is impossible to say whether these two groups are new arrivals or are more common today than they were in the past. On the other hand, the absence from the New Survey list of four *Bembidion* species characteristic of riparian habitats is unexpected.

Order Lepidoptera (butterflies and moths)

Butterflies belong to a small number of families in the insect order Lepidoptera, the vast majority of whose members are moths. About 15,000 butterflies have been described, less than 10% out of a total of 165,000 species worldwide. In Britain and Ireland there are about 2,500 species of moths in fifty families, of which five are butterflies. We are much more familiar with butterflies though, because they fly by day and are often large and colourful. Moths, by contrast, are mainly nocturnal and often rather drab-coloured and small—or very small. Over 1,600 species are assigned to the informal subdivsion of the 'microlepidoptera'—the smaller moths—in Britain and Ireland.

Moths are a very important faunal group, between them depending on a wide variety of plants in the habitat and occupying a wide variety of niches. Every single one of these species has a way of life that is different from each of the others, exploiting between them a really wide range of food sources. There is scarcely a species of plant that is not eaten by a host of invertebrates. Some of these herbivores are plurivorous, which means they feed on several different species of plants, but others have developed an almost exclusive dependence on a particular family, or in some cases a particular species, which is all very well as long as the host plant is abundant, but not so if it becomes rare or extinct.

Most moths are nocturnal, remaining out of sight and motionless during the day. Their wing patterns are superbly adapted in this regard, enabling them to practically disappear when they come to rest. The behaviour of adult moths during the day is also very different from the day-flying butterflies, because they are very reluctant to fly during the day, even if disturbed. A few families (particularly tiger and cinnabar moths) deter predators by being downright distasteful.

Caterpillars too have evolved all sorts of protective devices. They are equally adept at the art of camouflage, with their predominant green—the colour of the foliage in which most of them live—being no more than a first step. In some species the colours are brilliantly elaborate, making the creature appear dazzlingly conspicuous once it is removed from the background for which it has evolved. The great family of the geometrids (which are readily identified by their loopy way of walking) are superb twig mimics. Others simply stay out of view; tortricids for example spin webs of silk, drawing leaves or flowers into a rolled-up covering around themselves. Some of the larger ones adopt a fierce posture (the puss moth and elephant hawk-moth are wonderful examples). Some caterpillars are very distasteful; the woolly bear caterpillars of tiger moths for example, which you often see scooting across a track looking for somewhere to pupate.

Micromoths are often very specific in their choice of host plants. There are some wonderful examples among the leaf-miners: insects that spend their larval lives feeding on the nutritious tissue of leaves, sheltered within by the protective roof and floor of the epidermis, which is made of tough cellulose. *Relatively* sheltered only however, because countless parasites have learned how to breach this barrier to get at the now defenceless creature inside. Adult micromoths are seldom noticed because they are so small, but where they congregate they are much more likely to catch your attention.

William Francis de Vismes Kane, who prepared the report on Lepidoptera for the Original Survey with the help of Thomas Greer and Bonaparte Wyse, was disappointed with his tally of 217 species for the broader Clare Island area, which he attributed in part to the limited extent of habitat suited to large numbers of moth and butterfly species; 'the enormous tracts of very wet bog, peculiarly barren of herbage and heather, that stretch for miles in every direction in Co. Mayo, seem, in my experience, peculiarly devoid of Lepidopterous life'. And then there was the weather: '... with the exception of the summer of 1911, the frequent rainfall, the high winds that so frequently prevail, and the chilly fall in temperature toward evening that characterized the western coast even in the hot summer of 1911, and proves so disastrous to the flight of moths, are too well known to the entomologist in that part of Ireland'. The absence from his list of the numerous species that emerge from March to May and the large host of autumnal species attracted to flowering ivy in October and November was to be attributed to 'the impossibility of getting research carried on at these periods of the year, in so remote a district, characterized by so considerable a rainfall'. (His visits took place in June and July).[131] Another thing that added to his list of disappointments was the failure of the lighthouse lamp to attract the night-flying moths he was counting on; the high winds on the lofty cliffs on which the lighthouse stands put paid to this.[132]

For all that, he found the absence of certain species that *should* be on the list as remarkable as the presence of many rare and local species. He gave as an example the complete absence of Dianthoeciae—various species of *Dianthoecia*—even though its food plant (sea campion, *Silene maritima*) is plentiful. Kane was particularly struck by the number of species on his list that are natives of Finland: 126 out of the total of 217.

Fieldwork for the New Survey by Ken Bond of University College Cork in 2002–06 and 2016 from late spring to early autumn, combined with extensive light-trapping, has increased the tally of species for Lepidoptera more than three-fold. A total of 282 species were recorded during the New Survey, including all but 25 of the species recorded in the Original Survey, bringing the total number of species ever recorded on the island to 307. [133]

This great increase can be attributed in large measure to the much more intensive schedule of the New Survey, especially the use of modern light-traps, combined with the greater availability of identification guides to the micromoths. The striking increase in the number of noctuids in particular is likely due to the increased efficiency of the mercury-vapour lamps used.[134] Seventeen species were singled out for particular comment because of their special conservation significance, including a tortricid moth that was recorded for the first time. This is *Epinotia sordidana*, which feeds on alder, a shrub that is quite scarce on the island (Figure 7.36).

A significant factor is probably habitat change. The disappearance of the small blue butterfly (*Cupido minimus*) is certainly due to the loss of its food plant, kidney vetch (*Anthyllis vulneraria*), which in Praeger's day was frequent on the island. Overall, the groups that showed the greatest decline fall into three categories: those of unimproved herb-rich grassland, of well-devel-

Fig. 7.36 Some Clare Island Lepidoptera.

(a) The green hairstreak (*Callophrys rubi*) is a butterfly of local distribution thought to feed on bilberry.

(b) One of Clare Island's many species of micromoths: *Ochsenheimeria taurella* is a rare grass-feeding species whose apparent decline is probably due to the loss of unimproved grassland.

(c) *Pyraustra purpuralis* feeds mainly on wild thyme.

(d) *Micropteryx calthella* is unusual among moths in having functional jaws; it feeds on pollen.

(d) *Epinotia sordidana*.

Photos: (a-d) Patrick Clements, (e) James K. Lindsay (CC BY-SA 3.0).

oped heathland (which were the most interesting groups at the time of the Original Survey), and of wetland. Significant losses in the first category included the Grayling (*Hipparchia (Satyrus) semele*), and such moths as the Grass Rivulet (*Perizoma (Emmelesia) albulata*), *Dichrorampha plumbana* and *Stenoptilia bipunctidactyla (Misaeseoptilus plagiodactylus)*, none of which were seen during the New Survey.[135] On the other hand, the New Survey recorded several localised species not seen during the Original Survey, such as *Pempeliella dilutella*, *Phycodes saxicola* and Pretty Pinion (*Perizoma blandiata*).

Heathland species that were not rediscovered during the New Survey included *Macrothylacia (Bombyx) rubi* (Fox Moth), *Aphelia viburnana*, *Eupoecilia angustana*, *Phragmatobia fuliginosa* and *Diacrisia sannio*.[136] Wetland species which appear to have declined or disappeared include *Platyptilia isodactyla*, *Argyroplace palustrana* and *Argyroplace olivana*. On the other hand, species associated with woodland and scrub have been recorded in good numbers, mainly in and around the Portlea-Lassau area, reflecting the unimpeded development of the wood here, which at the time of the Original Survey would have been in existence for little more than 65 years (see pages 272–4).

Trichoptera (caddis-flies) are reviewed below in this chapter in the section that looks at life in freshwater.

Order Hymenoptera (bees, wasps, ants and sawflies)

The only Hymenoptera most people are familiar with are a few of the larger species: honey bees and bumblebees, the wasps that annoy us on summer picnics, and a few species of ants. There are in fact something like 175,000 ants, bees and wasps (aculeates) (including 25,000 species of bees), each in its own way as complex and fascinating as the honey bee whose biology and ecology have been studied so closely.

The report on Hymenoptera for the Original Survey was prepared by the great English naturalist Claude Morley, who spent a week on the island in July 1910 in the company of a large party of other naturalists, having spent a week previous to this collecting on the mainland around Louisburgh. His report also incorporated the work of several other earlier collectors of Hymenoptera.

What seems to have struck him more than anything else was the absence of anything distinctive about the hymenopteran fauna: 'here were mountains of over 2,000 feet, the sea-shores and geological formation and climate were totally different, yet the fauna bore none of that individual stamp one so confidently anticipated in so distinct a locality'.[137] He compiled a list of 324 species, among which were 'very few with claims upon our interest'. From our somewhat different point of view of course, they all have a claim on our interest: each has its own unique story to tell.

Very little was known about Irish parasitic Hymenoptera in particular a hundred years ago, so it is hardly surprising that about a hundred of the species on Morley's list were new Irish records, although only six of the aculeate Hymenoptera were new. Nor was the experienced Morley altogether disappointed. There were many scarce species on the list of parasitic Hymenoptera; '*Pimpla ovivora*, *Glypta nigrina*, and *Mesoleius maculicollis* are extremely rare everywhere, and I have seen no other *Cremastus albipennis* and *Rhogas bicolor* than those from Clare Island.'[138] On Clare Island itself there were 132 species, including 5 ants, 2 fossorial wasps, 3 social wasps (*Odynerus*), 17 bees including 8 bumblebees; 9 sawflies, 1 fire wasp, 55 ichneumons, 21 braconids, 7 chalcids, 9 proctotrupids and 3 cynipids. The most remarkable absentees were the Honey Bee (*Apis mellifera*) and Common Wasp (*Vespula vulgaris*): 'The islanders know them only through visits to the mainland; to

Fig. 7.37 (a) Moss carder bee (*Bombus muscorum*); (b) nests of *Lasius flavus* on a slope in Capnagower; (c) the nest of *Lasius flavus* showing the galleries, and workers with pupae (Photos: John Breen).

Table 7.6: Species of Social Hymenoptera recorded by Morley (1911), their current nomenclature, and status in the present study. Items highlighted in bold are changes from the original Clare Island Survey (John Breen).

Morley (1911) record	**Current nomenclature of species recorded in 1911**	**New Survey**
Apidae		
Bombus smithianus White and *Bombus venustus* Smith	*Bombus muscorum* (Linnaeus, 1758)	*Bombus muscorum* (Linnaeus, 1758)
Bombus agrorum Fab.	*Bombus pascuorum* (Scopoli, 1763)	*Bombus pascuorum* (Scopoli, 1763)
Bombus derhamellus Kirb.	*Bombus ruderarius* (Müller, 1776)	*Bombus ruderarius* (Müller, 1776)
Bombus jonellus Kirb.	*Bombus jonellus* (Kirby, 1802)	*Bombus jonellus* (Kirby, 1802)
Bombus hortorum L.	*Bombus hortorum* (Linnaeus, 1761)	*Bombus hortorum* (Linnaeus, 1761)
Bombus latreillellus Kirb. var. *distinguendus* Mor.	*Bombus distinguendus* Morawitz 1869	Not found in present survey
Bombus latreillellus Kirb. 'type form'	*Bombus subterraneus* (Linnaeus, 1758) This record was dismissed as 'some error' by Stelfox (1927; p. 282). No other Irish records exist of this species.	
		Bombus cryptarum (Fabricius,1775)
		Bombus lucorum (Linnaeus, 1761)
		Bombus magnus Vogt, 1911
		Apis mellifera Linnaeus, 1758
Vespidae		
None		*Dolichovespula sylvestris* (Scopoli,1763)
		Vespula rufa (Linnaeus,1758)
		Vespula vulgaris (Linnaeus,1758)
Formicidae		
Formica fusca Latr.	*Formica lemani* Bondroit, 1917	*Formica lemani* Bondroit, 1917
Lasius flavus DeG.	*Lasius flavus* (Fabricius, 1781)	*Lasius flavus* (Fabricius, 1781)
Lasius niger Linn.	*Lasius platythorax* Seifert, 1991	*Lasius platythorax* Seifert, 1991
Myrmica ruginodis Nyl.	*Myrmica ruginodis* Nylander, 1846	*Myrmica ruginodis* Nylander, 1846
Myrmica scabrinodis Nyl.	*Myrmica scabrinodis* Nylander, 1846	*Myrmica scabrinodis* Nylander, 1846

the island children they are quite unfamiliar.'[139] Honeybees have however been re-introduced to the island in recent years.

Sawflies are the most ancient group of Hymenoptera; they are rather wasp-like insects that often form galls on the plants on which their larvae feed. 200 million years or so ago one group of sawflies gave rise to the Apocrita: the ants, bees and wasps, all of which have a common ancestor. The larvae look very like caterpillars for the most part, and they feed on a great variety of wild plants (and some crops), many being quite specific in their choice of host. They lack the narrow waist of the Apocrita, and have a saw-like ovipositor at the tip of the abdomen which they use to cut into plants in order to deposit their eggs.

Nearly all ichneumons and braconids are *parasitoids*, feeding as larvae on the living bodies of other insects which the mother wasp has first immobilised by stinging, and which die when the little wasp grub has finished with them. Most are very small insects, seldom noticed by anybody except entomologists (and very few of those indeed!), even though they play a very important role; many are useful in pest control, and they are considered to be among the most valuable biological indicator groups. Chalcids are also very small, and mainly parasites or hyperparasites of other insects—and of great economic importance on this account. They prey mainly on butterflies and moths, bugs and flies, and many are parasites of other Hymenoptera. Cynipids also form galls, but they are much smaller insects than sawflies.

The social Hymenoptera were intensively studied for the New Survey by John Breen and his team over a fifteen-year period (2002–16), but no attention has been paid to the other groups since Morley's study in 1910–11. He recorded five of Morley's bumblebees. Morley's *Bombus latreillellus* var. *distinguendus* (*B. distinguendus* today) was not re-recorded, and his record of *B. latreillellus* 'type form' (*B. subterraneus*) is thought to be a mistake. The absence of *B. lucorum* / *B. terrestris* from Morley's list is surprising, since it is one of the more common bumblebees on the island today. However, although these white-tailed bees with black and yellow bands are easy to recognise as a group, they are very difficult to assign to species on morphological grounds. They are now recognised as a cryptic complex of three species, which can be reliably distinguished only by their DNA barcodes. Using this technique, specimens from Clare Island examined by Robert Paxton at the Martin-Luther University in Halle-Wittenberg could be assigned to three species—*B. cryptarum*, *B. lucorum* and *B. magnus*—bringing the total bumblebee tally for Clare Island to nine (Table 7.6).[140]

Two species of social wasps were regularly encountered during the New Survey: the Tree Wasp (*Dolichovespula sylvestris*) and the Red Wasp (*Vespula rufa*). Colonies of both species are small however, and easily overlooked. Workers of the Common Wasp (*Vespula vulgaris*) were spotted on just one occasion during the survey.

Clare Island still harbours the same five species of ants that were present in Morley's day. The Yellow Meadow Ant (*Lasius flavus*) is abundant, its large mound nests a conspicuous feature of older fields (Figure 7.37). Nests of the *Myrmica* species are usually found under stones and decaying logs. The New Survey measured the effects of the activity of *Lasius* ants on soil chemistry and found that soil pH was higher and electrical conductivity reduced in the nest mound compared with the adjacent soil; potassium was much higher in the nest, but phosphate lower.[141] The nature and extent of nematode parasitism in *Lasius* was also looked at during the survey. Both queens and males of *L. flavus* were found to be infected by *Pheromermis myrmecophila*, infection tending to be heavier in the queens. In *L. platythorax* only queens were infected.[142]

Order Diptera (two-winged or true flies)

The term 'true flies' is sometimes used to distinguish such two-winged insects as houseflies and mosquitoes from all the other 'flies' we are familiar with—butterflies, caddisflies, stoneflies, dragonflies—all of which have four wings (although why this should make them less 'true' is not at all clear!). However this may be, in the course of their evolution the hind-wings of these flies have been transformed into whirring stabilisers called halteres, and these has given them unparalleled mastery of flight. They are an enormously successful group of insects, with about 150,000 species known and hundreds of thousands more awaiting discovery and description.

Percy Grimshaw was enlisted to prepare the account of two-winged flies for the Original Survey. This was surely an extremely daunting task, given that nothing whatsoever was known of the dipteran fauna of Clare Island, and little enough of Ireland as a whole. Added to this was the challenge of an order of insects of such complexity and diversity. Moreover, Grimshaw himself was only able to spend a few days on the island in July 1910 and a further fortnight in the area around Westport a year later. Nevertheless, he was able to supplement his relatively brief visit to the area through examination of the collections made by other naturalists.

Just how little was then known about two-winged flies can be gauged from the fact that of the 519 species in some 45 families identified in the course of the Old Survey, 30% were new to the Irish fauna (and five to Britain). Of this number, 211 species were recorded on Clare Island, 43 of which were found on Clare Island only. On the other hand, 308 were recorded on the mainland only; 168 were common to both. There is little that is unusual or distinctive about the Clare Island list. Although

Fig. 7.38 *Tipula scripta* (Roy Anderson).

there might appear to be 'a large percentage of peculiarity' about the fact that 43% of the then known Irish total of 1,050 species occurred in the broader Clare Island district, 13% had not been recorded *outside* of it, 'whether this is due to insufficiency of knowledge or to an actual feature in distribution it is not yet possible to say'.[143]

Non-biting midges (chironomids)

Chironomids are among the most successful of aquatic insects in terms of species diversity and abundance. The adult flies are familiar to most people as the swarms of dancing midges above streams and other freshwaters, which are the males awaiting the arrival of females. The juvenile stages of these little flies—and so, much the greater part of their lives, for their adult life lasts no longer than a week or two—is spent in watery places: freshwater bodies of every conceivable size, and such wet terrestrial habitats as rotten wood and other decaying vegetable matter, dung and moss. The great diversity and numerical success of the group is indicative of the great range of microhabitats they have learned to exploit over the course of their evolution. There are several thousand species, of which about a thousand are European. Under favourable conditions they can often reach population densities of several tens of thousands per square metre.

Many chironomids live in tubes of silk fortified with salivary secretions through which they maintain a current of oxygenating water, from which food particles are strained through a net of the salivary muslin, the net and its contents being swallowed at regular short intervals. Apart from these detritus feeders there are many herbivores—grazers of algae—in the family. One sub-family, the Tanypodinae, consists for the most part of carnivores. A few have much more specialised requirements; *Eukiefferiella ancyla*, for example, lives nowhere except in the mantle cavity of the freshwater limpet (*Ancyla fluviatilis*).[144] Of particular interest on Clare Island is the presence of several saline-tolerant shoreline species.

The number of two-winged flies recorded in Ireland today—around 3,400—is nearly three times what it was in Grimshaw's day. Grimshaw recorded 39 species of Chironomidae altogether, eight of which occurred on Clare Island itself.[145] The number of chironomids now known from Clare Island as a result of Declan Murray's work for the New Survey is about 4% of the Irish total for the family, *sixteen* times what it was in the Original Survey: a quarter of the mainland total.[146]

The chironomids are a particularly good example of how increased sampling efforts and modern methods of study of complex and little-known groups can greatly increase the number of species. Since Grimshaw's day the taxonomy of the Chironomidae has undergone radical restructuring (one result of which has been a halving of species), and new techniques have been developed, most notably the use of pupal exuviae (the skin cast by the adult as it emerges from the water) in which a wealth of vital taxonomic detail can be discerned, along with meticulous slide preparation and detailed critical microscope examination of adult morphological features and external genitalia.[147] In his research for the New Survey Declan Murray recorded no fewer than 129 species, and 241 taxa for the mainland near Clare Island [148]

We can be sure that for many other faunal groups—on Clare Island or anywhere else in Ireland—a comprehensive programme that takes full account of habitat diversity and seasonality, carried out over several years, would indeed

Fig. 7.39 (a) A typical chironomid (adult) (*Chironomus plumosus*) (© entomart); (b) a typical chironomid larva (Frank Fox (CC BY-SA 3.0 DE)); (c) stream foam is a rich source of chironomid exuviae (Declan Murray); (d) pupal exuviae washed from stream foam (Declan Murray).

demonstrate the truth of Grimshaw's view that 'our knowledge of the Dipterous fauna of Ireland is very far behind', and that 'with prolonged periods of collecting any conclusions now formed might be entirely upset'.[149] Advances in knowledge, and taxonomic revision since the days of the Original Survey make it difficult for a non-specialist to get a sense of how limited this might be. The biting midges (family Ceratopogonidae) are one further example of how much remains to be discovered. Of the 171 species recorded from Great Britain 78 occur in Ireland, although 'the majority, if not all, of the 93 additional species found in Britain should occur here'.[150] On Clare Island only *four* species have been recorded, all by Grimshaw in 1912.[151] Not a single species has been recorded in modern times, a fact that will utterly astonish the island's inhabitants!

Snail-killing flies (Schiomyzidae)

The only other group of flies that has been studied in any detail in recent years is the snail-killing or marsh flies (Schiomyzidae), the majority of whose larvae feed on various molluscs. Grimshaw recorded six species, and a further ten were discovered during research in 2002, captured mainly by sweep-netting over the course of three visits. Of these sixteen species, ten are new records for the island (although six of the ten had been recorded by Grimshaw on the mainland). On the other hand, three of the species recorded by him were not seen in 2002.

Three species accounted for 60% of the total catch of 66 adult flies: *Ilione lineata* whose larvae feed on pea mussels, *Pherbina cortyleti* which eats various marsh snails, and *Tetanocera ferruginea*, which is a predator of water snails.[152]

Class Aves (birds)
Introduction

Mammals are large, earth-bound creatures, ill-equipped to maintain themselves on a small island as densely populated as Clare Island was in pre-Famine times. As a consequence, the island no longer harbours foxes, stoats or pine martens, let alone the wolves or deer of a more distant yesterday. With birds it is a different story, because of their mastery of the air. Where their ways of life on a small island overlap closely with ours, their fortunes rise and fall in response to our changing management of the land.

Fig. 7.40a Wheatear (*Oenanthe oenanthe*) (William Clarke).

Fig. 7.40b 'One of the most magnificent of the Island's land birds is the Rock Dove, 'absolutely magnificent looking' flint coloured acrobats, whose displays on the cliffs of the north coast are breath-taking (A. King (2008), quoting Tom Kelly). The cliffs are also home to the dove's natural predator, the Peregrine Falcon, which Praeger described as Ireland's 'master of the air.' (William Clarke).

No group of animals on Clare Island—or anywhere else in Ireland for that matter—has received more attention than birds during both the Original and New Surveys, and indeed independently of these. Although the report for the Original Survey was prepared by Richard Ussher, he saw himself merely 'as one among several observers and as compiler of our joint observations, coupled with records from every available source.'[153] Between them they listed some 91 species for Clare Island, but after careful study concluded that 'none of the Clare Island birds showed any distinction from those from other parts of Ireland.'[154] However, no other group of animals is more immediately reflective of the ways in which the island environment has changed over the past century, on Clare Island just as on islands all around the coast.

A survey carried out by C.S. Lloyd for the Royal Irish Academy in the early 1980s found that ninespecies had been lost since 1911: but a further seven were new arrivals; another five had been lost since 1945, but there were four new arrivals.[155] Birds were intensively studied for the New Survey by several ornithologists under the direction of Thomas Kelly of UCC, who also reviewed the results of the numerous expeditions of other naturalists to the island by ornithologists down the years; unfortunately, differences in the timing of visits, as well as in approach and methodology between different surveys, make detailed comparison difficult. Around a hundred different species altogether have now been recorded on the island, 59 of which have been recorded as breeding at one time or another, the 'equilibrium number' being 24.[156]

Land birds

The New Survey also included a study of the equilibrium number of the island's land bird species and of the 'apparent turnover' in species numbers since the findings of the Original Survey. The equilibrium number is the number at the point where the curves for extinctions and immigrations meet. This was estimated to be 24, which is four fewer than was calculated by Ussher in 1912. While the equilibrium number will (broadly speaking) remain the same over time, the *species composition* will change: this is what 'turnover' means. Of the 35 land bird species known to have bred at least once on Clare Island over the 104-year interval between the Original Survey and recent surveys, seven have become extinct and there have been six new arrivals. The analysis found that the longer the interval between surveys the lower the rate of 'apparent turnover' tends to be.[157]

Table 7.7: Breeding landbird species on Clare Island

Species	Scientific name
Blackbird	*Turdus merula*
Chaffinch	*Fringilla coelebs*
Chough	*Pyrrhocorax pyrrhocorax*
Dunnock	*Prunella modularis*
Hooded crow	*Corvus cornix*
House sparrow	*Passer domesticus*
Kestrel	*Falco tinnunculus*
Lesser redpoll	*Carduelis flammea*
Linnet	*Carduelis cannabina*
Magpie	*Pica pica*
Meadow pipit	*Anthus pratensis*
Pheasant	*Phasianus colchicus*
Pied wagtail	*Motacilla alba*
Raven	*Corvus corax*
Reed bunting	*Emberiza schoeniclus*
Robin	*Erithacus rubecula*
Sedge warbler *	*Acrocephalus schoenobaenus*
Skylark	*Alauda arvensis*
Song thrush	*Turdus philomelos*
Starling	*Sturnus vulgaris*
Stonechat	*Saxicola torquata*
Swallow *	*Hirundo rustica*
Wheatear *	*Oenanthe oenanthe*
Willow warbler *	*Phylloscopus trochilus*
Wren	*Troglodytes troglodytes*
Tree sparrow	*Passer montanus*
Rock dove	*Columba livia*
Rock pipit	*Anthus petrosus*

* migrant species

Broadly speaking there are two communities of land birds: a garden community found in the vicinity of human settlement where such species as Robin (*Erithacus rubecola*) and Blackbird (*Turdus merula*) are ubiquitous; and a 'grassland community' in the open grassland that covers most of the island, dominated by a large population of breeding Wheatears (*Oenanthe oenanthe*), along with Meadow Pipits (*Anthus pratensis*) and Skylarks (*Alauda arvensis*). At the time of Ussher's survey Clare Island could 'happily be called a sanctuary of the chough' (*Pyrrhocorax pyrrhocorax*);[158] the bird is still a common sight on the grasslands of the coast and commonage, as are the Raven (*Corvus corax*) and Hooded Crow (*Corvus cornix*).

A total of 28 breeding landbird species were recorded during the New Survey, which also made a pioneering attempt to estimate their numerical abundance (Table 7.7).[159] The total landbird population of ≥8000 is dominated by generalist species (such as Wren, Dunnock, Robin and Blackbird) and species characteristic of open habitats. The two most abundant species on the island are two open-habitat species: Wheatear (≥1700 individuals) and Meadow Pipit (≥1500 individuals); the other open-habitat species—Skylark—is much less abundant with only ≥200 individuals. The great majority of species feed on invertebrates; there are only three herbivorous species (Rock Dove, Linnet, Lesser Redpoll) and three vertebrate predators (Peregrine Falcon, Kestrel, Sparrowhawk). Double-brooded species with clutches of 3–5 eggs were found to predominate.

Fig. 7.40c Chough (*Pyrrhocorax pyrrhocorax*) (William Clarke).

Only four migrant landbird species breed on the island (Swallow, Wheatear, Willow and Sedge Warbler), which is a very small proportion compared to the mainland. There are few hole-nesting woodland species: no tits, no Treecreeper or Goldcrest. There are only two *Carduelis* species (Linnet and Lesser Redpoll) and one *Emberiza* (Reed Bunting). There are four crow species, but the Rook is absent (see page 187). The recently arrived Chaffinch is the only finch. The total landbird biomass is estimated to be in the region of ≥386kg or ≥24kg per km^2.

Water and wetland

The absence of woodland and wetland of any extent is a major factor in accounting for the relative paucity of species. In spite of the scarcity of reedbeds and similar habitats however, the Reed Bunting (*Emberiza schoeniclus*) is common, and the Sedge Warbler (*Acrocephalus schoenobaenus*) is also present.[160] More intensive study might discover the presence of several less easily recognisable warbler species that are regularly seen on Inishbofin, such as *Hippolais*, the unstreaked *Acrocephalus* and such scarce *Phylloscopus* species as the Yellow Browed Warbler (*Phylloscopus inornatus*). A number of scarce passage migrants might be added by more

Table 7.8: Waterfowl species

Oystercatcher *Haematopus ostrelagus* Linnaeus
Ringed Plover *Charadrius hiaticula* Linnaeus
Lapwing *Vanellus vanellus* (Linnaeus)
Common Snipe *Gallinago gallinago* (Linnaeus)
Common Sandpiper *Actitis hypoleucos* (Linnaeus)
Mallard *Anas platyrhynchos* Linnaeus
Water Rail *Rallus aquaticus* (Linnaeus)
Moorhen *Gallinula chloropus* (Linnaeus).
[The Dipper *Cinclus cinclus* (Linnaeus) *hibernicus*]

careful and sustained attention in the autumn months.[161]

The Original Survey recorded eight species of waterfowl. The five wading bird species recorded at that time are still here, apparently in much the same small numbers and thinly scattered over the island.[162] The reduction in the area of suitable wetland habitat (only two of Praeger's five 'lakes' remain today) has not resulted in notable decline, nor on the other hand has the absence from Clare Island of the mainland's predatory mammals seen an increase.[163] Snipe appears to have been more common in the past, although the wide extent of blanket bog and associated grassy and small-sedge habitats still provide ideal conditions for breeding today. The close-cropped grazed areas continue to be suitable habitat for Lapwing and Ringed Plover. Areas of short grassland close to the shore (such as the *Plantago* sward) are ideal for nesting Oystercatcher.

The Mallard is still the only species of duck on the island, and was also probably more common at the time of the Original Survey. The Moorhen is still here but not often seen. However, the Water Rail, a common bird a hundred years ago, has gone; its disappearance is all the more striking in view of the transformation of Lough Avullin from open-water lake to the impenetrable reedswamp favoured by the species. Although there have been occasional sightings of the Dipper over the years, extensive searches have failed to find nests.

Seabirds

The seabirds that nest on the cliffs that fringe much of the island retain their fierce and wonderful independence, secure in those cliff-bound fastnesses, 'where the rows of tiny dots that stand for Puffins, Razorbills, Guillemots, and Gulls rise tier upon tier to a height of a thousand feet, absolutely unapproachable'.[164] The Atlantic cliffs of the north-west are however no longer the great sanctuary they once were for Kittiwake ('breeds in great colonies'), Razorbill ('There is no more remarkable colony in Ireland than on the great cliff of Clare Island, where the Razorbills breed up to 1000 feet'), Guillemot ('considerable colonies'), and Puffin ('Considerable colonies extend along the upper parts of the Clare Island cliffs among and above those of Razorbills').[165]

> When I was a gasúr I used to go out near the cliffs. There were so many birds. They were like a great orchestra playing. They are not there now. It's a silent place. I don't know what happened because there's not much pollution around here. The Razor Bill and Guillemot were along the cliff like bottles on a shelf, like penguins, like little nuns. I often ate their eggs. They have a very fishy taste. They used to lay their eggs on the rock. I always thought it extraordinary how the egg remained on the shelf of rock. They laid only one egg, of the most beautiful colours.[166]

Such anecdotal accounts notwithstanding, the marine avifauna has experienced a remarkable increase in species richness: possibly as high as 75%, from the 9 species at the time of the Original Survey to 14 today.[167] The seabird colony on Clare Island is still one of the largest in Ireland, with a total biomass that may be as high as 10.7 metric tons and consuming up to 3 metric tons of food a day. More than half the species are mainly pelagic, feeding mainly offshore.[168] Recent surveys confirm a large total population of seabirds, with 14 recorded breeding species and an estimated total population of approximately 8,000 pairs: although it is difficult to determine numbers accurately because of the inaccessibility of some of the cliffs.[169]

The Fulmar (*Fulmarus glacialis*) was absent from Clare Island in Ussher's day, but he cautioned that it should be looked for: sure enough, it had arrived by 1935. Although it has experienced a startling decline of over 50% in recent years, the Fulmar is still the most abundant seabird on the island, numbering several thousand birds in what is the third largest colony in Ireland.[170]

Kittiwake and Guillemot are the most numerous species after the Fulmar. Three other seabirds that have arrived since the Original Survey are Gannet (*Morus bassanus*), Great Cormorant (*Phalacrocorax*

Fig. 7.40d Fulmar (*Fulmarus glacialis*) (William Clarke).

Fig. 7.40e Gannet (*Morus bassanus*) (William Clarke).

carbo carbo) and Lesser Black-backed Gull (*Larus fuscus*). The cliffed coastline is now a Special Protection Area under the EU Birds Directive.

Among the most dramatic changes in the island's avifauna has been the establishment of a colony of Gannets. Less than five pairs were present over the period between 1978 and 2004, but since then the colony has grown substantially, to at least 65 pairs by 2011. In 2014 the number of apparently occupied sites (AOS) was thought to be between 250 and 300.[171] The increase cannot be attributed to change in the immediate island environment; over the ten-year period leading up to 2015 the Irish Gannet population as a whole increased by nearly a third, for reasons that are not at all clear, but European fisheries policy on discarding may be a key factor.

The Razorbill has also increased, but the Guillemot colony appears to have declined since 1999. There has also been a dramatic decline in the number of Common Gulls, which only began to breed in the early 1940s, whereas Greater and Lesser Black-backed Gulls show large increases.[172] The most dramatic change in recent years has been the arrival of the Great Skua in 2015; it bred for the first time in 2016. The Skua is a notorious predator of seabirds, and is likely to have major impact in the years ahead.

On the other hand, the small colony of Arctic Terns (*Sterna paradisaea*) that lived in 1912 'on the detached rocky western end of Clare Island' is no longer there. It is possible that the Storm Petrel (*Hydrobates pelagicus*) and Manx Shearwater (*Puffinus puffinus*) are also present, but small populations could easily be missed because they are only out and about at night, spending the day in subterranean burrows.

The change in the fortunes of many species—such as Shag, Fulmar and Kittiwake—is tied to change in the wider marine environment, although it is difficult to pin these down in terms of cause and effect. The Fulmar, for example, is a pelagic species that ranges over the open waters where it is exposed to such potential hazards as oil contamination, long-line fisheries and microplastics pollution. It is thought that the increase in stormy weather over recent years may be a factor in the decline of numbers of Shag, which seems especially vulnerable in this regard.[173]

Until late in the nineteenth century the same unassailable cliffs were the home of the Golden Eagle (*Aquila chrysaetos*). A mounted specimen shot on Clare Island by a Captain Boxer (RN) in 1877 is on display in the Natural History Museum in Dublin. The last formal record (in the *Irish Naturalists' Journal*) seems to have been on Knockmore in 1891.[174] In spite of extensive persecution waged against it in the nineteenth century, the White-tailed Sea Eagle (*Haliaeetus albicilla*) survived on Clare Island between 1867 and 1870, even taking lambs; they are also reported to have 'lifted geese from the cottiers's very doors due to no guns being allowed on the island'.[175] The last Sea Eagle seen on the island (in 1935) was an injured bird, a victim of misfortune elsewhere:[176]

> One summer evening, when children were at play on the beach in front of our house, a very large bird landed in their midst. The children scattered in all directions in fear of the huge bird that had dropped from the sky into their playground. My watching father, with concern for the frightened children, rushed to the strand and, discovering that the unfamiliar bird carried a severe injury, he returned to the house, reached for his shotgun and did what he thought was the correct thing to do.'[177]

*****

The change in the avifauna is a dramatic reflection of environmental change generally. Many changes reflect change on the island itself more immediately, while others relate to change in a wider world. Birds that were once common have gone or become much less common, especially those that depended on the traditional pattern of mixed farming or on the more extensive heather cover of earlier times.

Perhaps the most significant cause of change is in the pattern of agriculture. There is no longer the patchwork of arable fields with their winter gleanings that drew Thrushes, Blackbirds and other birds from the mainland in such numbers at one time.[178] The once 'very common' Corncrake has vanished.[179] The House Sparrow, a common resident in the years of the Original Survey, no longer has the thatched roofs that were the norm a hundred years ago in which to nest.[180] By 1945 the Yellowhammer, in Ussher's day 'all over the island', had gone, along with the Corn Bunting, both victims of the changes in farm practice.[181]

As the heather disappeared, the once-common Red Grouse followed. When Jackie O'Grady of

Gurteen was growing up (during the1930s and 40s) he could walk the hills 'where the heather was knee-high, and it was not unusual to flush six or seven brace of woodcock and a half dozen hares in a while of an evening. With the advent of government subsidies, the sheep population grew and those same grounds are bare'. His grandfather remembered a time in his youth when 'there was an abundance of red grouse to be found in the long heather, on the hills and bogs throughout the island, and that the landlord at that time took shooting parties to the island on organised shoots. Sadly, there are no longer red grouse to be seen in Clare Island and, indeed, they are now all but extinct on the mainland as well'.[182]

It is impossible to distinguish any impact on the avifauna as a result of climate change that can be isolated from other more immediate influences, especially from changes in the pattern of agriculture, and other changes resulting from the human presence. The island's populations of farm and garden birds are very small. This makes them very vulnerable to changing conditions, and is largely responsible for the way some species seem to come and go.

The Rook

The absence from Clare Island of the Rook, one of Ireland's commonest and most widely-distributed birds, is surprising, but a study carried out by Thomas Kelly and Patrick Smiddy as part of the New Survey has shown that this is simply because the island is too small, as indeed are all of Ireland's offshore islands.[183] The Rook is a highly social bird, living in colonies that increase in security and reproductive success as they get bigger; the minimum size for a rookery is probably around 16 nests. The colony needs a considerable area of grassland for foraging however, which small offshore islands are unable to provide. It also needs trees for its rookeries, and there are virtually no trees on Clare Island (those in the wood at Lassau are scarcely of sufficient stature).

The threshold island area seems to be in the order of 38,000ha, which is nearly 25 times larger than Clare Island's 1,600ha (16km^2). However, what matters more than area alone is that the combined area of grassland and woodland habitat needs to be large enough to meet the bird's nutritional and behavioural needs.[184]

The New Survey of Clare Island's birds also provides the opportunity to broaden the focus of attention, and through a careful review of the published information try to identify patterns and trends in the populations of breeding landbirds on Ireland's offshore islands as a whole. The Birds Volume once again highlights the significance of the species-area relationship. The presence or absence, and the real extent, of such important habitats as woodland and wetland are also shown to be key determinants of bird species richness. An interesting fact about Clare Island from the study is that it is the only one of Ireland's 29 islands from which the Dipper has been recorded.[185]

Other vertebrates

Mammals

The animals most likely to come into conflict with humans on a small island are mammals, particularly if they are useful or if they cause damage, but also simply because they are larger than most other animals—they are more easily seen and take up more space. On the other hand the smaller size of invertebrates gives them vastly greater opportunity to exploit microhabitats and smaller areas. It is no surprise therefore that Clare Island had only two native vertebrate species in the nineteenth century, and—not coincidentally—two of the smallest: the pigmy shrew (*Sorex minutus*) and wood mouse (*Apodemus sylvaticus*). Rabbits and hares had disappeared and were only re-introduced in the early years of the twentieth century. In 1906 half a dozen hares were introduced from Murrisk by J.J. McCabe, and a similar number of rabbits the following year.

By the time of the Original Survey the number of hares was estimated at around a hundred. Brown rat and house mouse are both introduced species, confined to the vicinity of human settlements. Gerald Barrett-Hamilton, who studied the mammals of the island for the Original Survey, was informed that only four bats had been seen on the island during the previous ten or twelve years. Otters and grey seals were plentiful, but both have the protection of inaccessible seas in evading persecution.[186] On one occasion, as he was swimming out to identify a plant he had spotted on an offshore rock, Praeger found himself escorted by an honour guard of grey seals: 'The silent presence of those six-foot carnivores

Fig. 7.41 (a) Pygmy shrew and (b) wood mouse. From Millais (1904-1906), *The Mammals of Britain and Ireland.*

Table 7.9: Terrestrial and amphibious Irish mammal species recorded on Clare Island

Mammal species	Present	Absent
House mouse *Mus domesticus*	X	
Wood mouse *Apodemus sylvaticus*	X	
Bank vole *Myodes glareolus*		X
Brown rat *Rattus norvegicus*	X	
Black rat *Rattus rattus*		X
Red squirrel *Sciurus vulgaris*		X
Grey squirrel *Sciurus carolinensis*		X
Red fox *Vulpes vulpes*		X
Stoat *Mustela erminea*		X
Badger *Meles meles*		X
Pine Marten *Martes martes*		X
Mink *Mustela vison*		X
Otter *Lutra lutra*	X	
Grey seal *Halichoerus grypus*	X	
Harbour seal *Phoca vitulina*	X	
Irish hare *Lepus timidus hibernicus*	X	
Brown hare *Lepus europaeus*		X
Rabbit *Oryctolagus cuniculus*	X	
Red deer *Cervus elaphus*		X
Sika deer *Cervus Nippon*		X
Fallow deer *Dama dama*		X
Feral goat *Capra hircus*		X
Pygmy shrew *Sorex minutus*	X	
Hedgehog *Erinaceous europaeus*		X
Daubenton's bat *Myotis daubentonii*	X	
Brown long-eared bat *Plecotus auritus*	X	
Natterer's bat *Myotis nattereri*	X	

gave one a tingly feeling about the toes, as they rose up, and looked intently, and silently sank again'.[187] The mammals might therefore not appear to have much to offer in the way of significant information, and the fact that the wood mice on the island are indistinguishable from those found everywhere on the mainland indicates that the island has only recently been severed from the mainland.[188]

Twelve species of terrestrial mammals (including the otter and two seal species) were recorded during the course of the New Survey by Colin Lawton and Kate McAney.[189] Half of these are either species native to Ireland or very early (Mesolithic) arrivals (Table 7.9). This list is very similar to that of Gerald Barrett-Hamilton's list from the Original Survey, the main difference being the greater attention paid by the New Survey to bats and seals.

Although they were recorded by Barrett-Hamilton and have been caught on several occasions since, house mice were not recorded during the New Survey. This may be because they have adapted to home ranges away from houses, which is not unusual on islands. Brown rats maintain a more tenacious presence around human settlements. Very few wood mice were captured by Barrett-Hamilton during the Original Survey, and they are still scarce, with only five taken in Longworth traps over 200 trap nights during the survey. This is because there is little optimal habitat for them on the island, although these five were exceptionally large individuals by mainland standards. It appears that Clare Island's wood mice follow a general pattern of island wood mice being larger than those on the mainland.[190]

Bats were surveyed on two occasions: in July 1993 and July 2002. During the more recent of the two surveys, seven brown long-eared bats and a Daubenton's bat were caught in mist nets, and echolocation signals of both *Myotis* and *Plecotus* bats were recorded.[191] The predatory impact on wildlife of the island's ubiquitous feral cats remains unquantified. The hedgehog is unknown on the island.

Amphibians and reptiles

The common lizard (*Zoothoca vivipara*) also managed to survive the high tide of human impact on Clare Island during the nineteenth century. Its presence on the island and the fact that the island population is indistinguishable from lizards on the mainland further indicates a recent land connection with the mainland. The absence of frogs and smooth newts from the Island was and remains something of a puzzle ... [192]

Freshwater flora and fauna

> Yes, a stagnant pool, though but a few feet wide, hatched by the sun, is an immense world, an exhaustible mine of observation to the studious man and a marvel to the child. All these creatures of the water: what are they all doing here? I do not know. And I stare at them for ever so long, held by the incomprehensible mystery of the waters. The worlds of the ponds are so vast, I should lose myself in their immensities, where life swarms freely in the sun. Like the ocean, they are infinite in their fruitfulness.
>
> J. Henri Fabre

Freshwater algae

Although algae never enjoy the prominence on land they attain in the seas and oceans of the world, which they have dominated since the early days of life on earth, they nonetheless play a vital role in the ecology of freshwater habitats. Algae are not a discrete group, in the way mosses, ferns or flowering plants are. They are a very diverse assemblage encompassing numerous major groups of living things (phyla): only some of which can properly be described as 'plants'. In fact, it is hardly correct to refer to them as plants; only three groups, the green algae or chlorophytes, the red algae (rhodophytes), and a small group of rare freshwater algae known as glaucophytes have close affinities with plants. The so-called blue-green algae (cyanophytes), for long lumped together with 'other' algae, are actually photosynthetic bacteria (see page 193).

In no other part of Ireland has the algal flora been looked at in anything like the detail of Clare Island, where its amazing diversity was studied systematically by a team of specialists over a 15-year period for the New Survey. Well over 700 species have been recorded, making it (for its size) one of few known 'hotspots' of freshwater algal diversity anywhere in the world. The New Survey volume devoted to freshwater algae contains a detailed descriptive catalogue of species unmatched for level of detail in any other part of Ireland: desmids, diatoms, stoneworts, cyanophytes, subaerial algae and others.[193] There is considerable disparity between the Original and New Surveys, which makes it difficult to make reliable comparisons. The disparity is believed to be due mainly to differences in sampling strategy and preferences, and insufficient attention to the important factors of seasonality and small-scale patterns of variation (especially among the desmids).

It is impossible to convey the wonder of the lives of algae in words; some of them you might not even recognise as living things. The little gelatinous olive-green sausages and spheres you sometimes see floating in Creggan Lough and in various other ponds are colonies of a protozoan peritrich ciliate, *Ophrydium versatile* (Figure 7.42). The little black cushions you often see on water-splashed rocks close to waterfalls are clusters of the blue-green alga *Rivularia beccariana*. That slimy green coating on sloping peat surfaces is a colony of blue-green algae, often accompanied by olive-green or yellowish cushions of a green alga, *Coccomyxa ellipsoidea*. The purple coating that develops on bare peat surfaces is a dense mat made up of the finest cotton-like threads of the green alga *Zygogonium ericetorum*, whose chlorophyll green is masked by purple pigment. Those little brown patches you see on stones at the edge

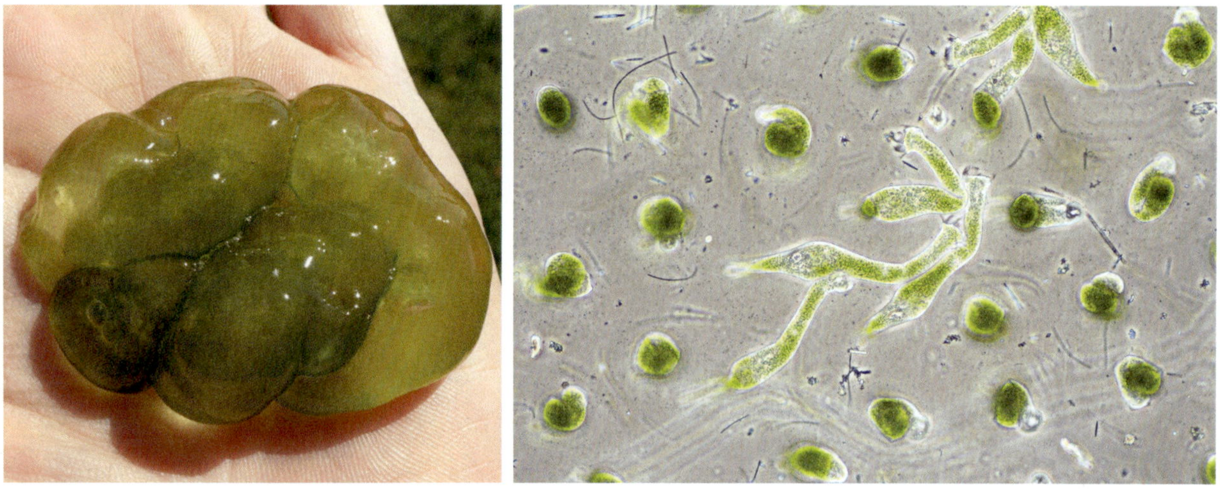

Fig. 7.42 (a) *Ophrydium versatile* colony (Jakob Fahr (CC BY-NC 4.0)); (b) *Ophrydium versatile* cells form symbiocenosis with the alga *Chlorella vulgaris* (Laboratory "Microcosmos" (CC BY-SA 4.0)).

of undisturbed streams are likely to be clusters of the blue-green alga *Dichothrix*.

The survey of freshwater algae for the Old Survey was carried out by William West and his son George Stephen West, who were the foremost authorities on the freshwater algae of Britain and Ireland at the time. The Wests visited the area on two occasions, at Whitsuntide in 1910 and 1911, and several collaborators also provided them with material, not only from Clare Island itself, but from Achill Island and the mainland round by Curraun and Westport to Louisburgh and Mweelrea, and inland as far as Castlebar. They spent 'considerably more than a thousand hours' on microscopic examination of the material, and identified 887 species, 254 varieties and 46 forms from the district as a whole, 769 species, 230 varieties and 40 forms of them from freshwater habitats. Of these, some 550 species of freshwater algae were recorded on Clare Island itself. However, some of the original names have changed as species concepts have been revised and taxonomy updated.

The island abounds in habitats suitable for freshwater algae. There are innumerable bogholes and peaty pools, ponds and ditches, and the island is drained by several streams. But although the only algae you will see in a river or lake on casual inspection are the larger macroalgal forms, which often form extensive tangles in the water, every one of these is a miniature ecosystem, home to numerous microscopic algae of many kinds, of great importance because they are the foundation of the food chain here, the producer organisms on which everything else depends.

In Praeger's day there were four lakes on the island: Lough Avullin, and three lakes in the gap that separates Knockmore from Knocknaveen. Of these only Leinapollbauty and Creggan Lough remain, and the only one with open water is Creggan; Leinapollbauty is little more than a large pond 20m across and less than a metre deep.[194] There are floating mats of vegetation in the others, and there is every likelihood that Creggan and Leinapollbauty will deteriorate over time unless there is active management to maintain these hotspots of algal diversity.[195] A fourth lake in the group (Loughnabéasta) had already gone by the time of the Original Survey. These pools are the most species-rich freshwater habitats on the island; several of the desmids found here have never been recorded elsewhere in Ireland.

A hundred years ago Lough Avullin was a body of open water up to 2.5m deep, fringed with reeds and bulrushes. Even at the time of the earlier Survey it was only half its former size 'owing to the deepening of its outflow'.[196] Today it is all reedmarsh (see page 272). Praeger never mentioned Poirtín Fuinch, the little lake in a drumlin hollow in the commonage between Maum, Capnagower and Glen townlands (Figure 8.5). The reason for this is that it did not exist at the time. Several other new lakes have also appeared since then: the lakelet in Toormore sometimes mistakenly identified as Loughanaphuca, and the little lake in Breacail, east of Doon Cove.[197] All of these formed as a result of turf-cutting (see page 271). Lough Avullin too is man-made, constructed no more than a century earlier by diverting the course of the Dorreen River in order to provide a

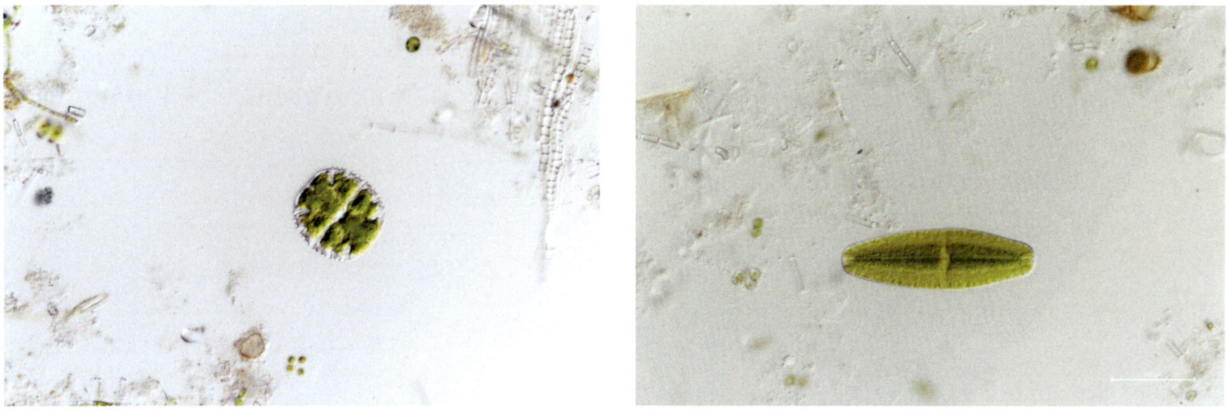

Fig. 7.43 Two desmids from one of the sphagnum pools that are widespread on the island (Bryan Kennedy).

non-spate water supply for the newly constructed mill at Maum (see page 272).

The algal flora in the small permanent bog pools that are so widespread on the commonages (including areas of old turbary) is dominated by a distinctive group of green algae known as desmids (order Desmidiales), often accompanied by three other groups: the flagellated dinoflagellates and euglenophytes and another group of green algae known as the Volvocales.[198] Bright green mats or slimy masses of filamentous algae (mainly *Spirogyra, Mougeotia* and *Zygnema*) are often found in ponds and pools, developing very rapidly after dry spells, along with green filaments of *Microspora*. In the numerous streams the filamentous green algae *Ulotrix, Oedogonium* and *Microspora* grow attached to rocks and other surfaces. They are often especially noticeable on the lips and vertical faces of waterfalls, intertangled with the mosses that are often so abundant in such places. The only identifiable red algae known from the island were found in the River Dorreen where *Batrachospermum turfosum* was well established in the upper reaches and occurred together with *Lemanea fucina* and *Torularia atrum* further downstream.[199]

Freshwater seepage areas or 'flushes' often contain diverse desmid assemblages. Such 'flushes' are most evident in the low cliffs of the south-eastern shore, and are easily recognised from a distance by the greenish-brown spongy mats of entangled filaments belonging to *Spirogyra, Mougeotia* and *Zygnema*. In the splash zone these give way to the green seaweed *Ulva intestinalis* (prosaically known to some as 'Gut Weed'). Rocks in seepage areas and intermittently flowing streams, especially in the south-west of the island, are also often covered with mats of green filamentous zygnematalean algae and the dark-brown gelatinous masses of various blue-green algae (e.g. *Dichothrix orsiniana, Rivularia beccariana* and *Nostoc verrucosum*).

Desmids

As a result of the work of David Williamson, one of the foremost living authorities on desmids, who identified material collected by David John and his colleagues in 2001 and 2002, we now have a comprehensive account of the desmids of the island. His account is essentially a very detailed checklist, accompanied by drawings of all the species.[200] Desmids (Desmidiales) are an enormously diverse and abundant group of beautiful green algae. They are at their best in old pools that are little disturbed from one year to the next, and are especially characteristic of the plankton of bog pools. The number of desmids recorded from Clare Island now stands at 310. The Wests, who were world authorities on desmids in their day, recorded 182 species and 37 subspecies during the Clare Island Survey.[201]

A great number of the desmids recorded by the Wests have not been seen since; in fact, 122 of the Wests' 219 taxa have not been recorded again from the island. Equally surprising, 91 species that *are* present today, many of them common species, including some of the most obvious and strikingly recognisable desmids (such as *Micrasterias crux-melitensis* and *Micrasterias papillifera*) are absent from their records. The difference could be due in part to site disappearance, and in particular the loss or reduction in the open water of lakes present a hundred years ago. Another factor could be the temporary flushing out of habitats due to flooding.

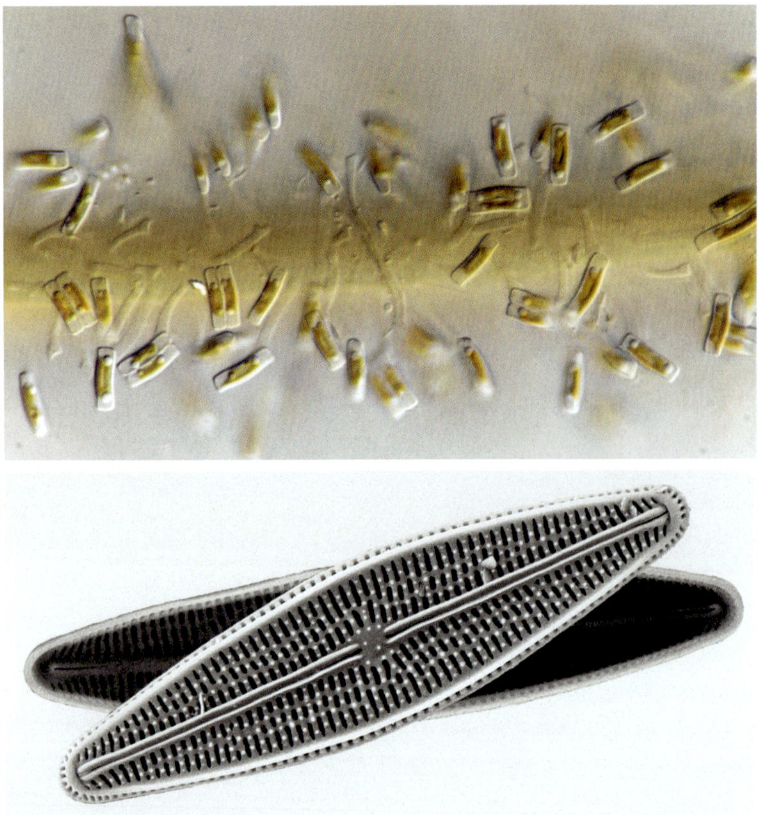

Fig. 7.44 Diatoms
(a) Living epiphytic cells of *Achnanthidium minutissimum*, a stalked benthic diatom common in the rivers of Clare Island. The individual valves and their brown-coloured chloroplasts are visible (Bryan Kennedy).
(b) The two halves (valves) of a diatom frustule (*Brachysira praegeri*, recently described from the Connemara area adjacent to Clare Island), viewed with scanning electron microscope (oxidised to remove all of the living contents) (Bryan Kennedy).

Fig. 7.45 Stonewort (*Chara*). Note the typical reproductive organs: male antheridia (red) and female archegonia (brown) (Christian Fischer (CC BY-SA 3.0)).

Diatoms

Diatoms are unicellular organisms that are enclosed in what for all the world look like ornate glass boxes, with a bottom and a lid on top: they are among the most beautiful organisms occurring in the plankton or in benthic biofilms in rivers and lakes. The distinctive pores and slits in these boxes allow the passage of nutrients into the cell and of waste products out.

They occur in freshwater of every sort—streams, puddles and loughs, aquatic plants, wet moss and rocks, and at the bottom of bog pools—in fact, wherever it is wet.[202] The richest habitats on the island are where there is a regular supply of saltwater or sea spray (such as suitable sites on cliff faces, or in streams and pools close to the sea).[203] Of the 281 species recorded from Clare Island to date, 77 are species recorded by the Wests (among their tally of 181) that have not been seen since his day.

Stoneworts

Stoneworts (charophytes) are a specialised group of green algae, of which two genera occur on the island. They look a bit like aquatic horsetails, the genus *Chara* with unbranched whorls, *Nitella* with branched whorls. They are typically associated with lime-rich habitats (one of their most distinctive features is the way they are encrusted with lime), so it is not surprising that they are poorly represented on Clare Island, which is not obliging in this regard. Although charophyte nomenclature has changed somewhat in the interval, the same three species we find today were present a hundred years ago, and in many of the same sites, especially Lough Leinapollbauty and Creggan, and in a number of boggy moorland pools and streams, especially in the east of the island.[204] They are very sensitive to nutrient enrichment, but this is unlikely to have affected them significantly here.

Blue-green algae (Cyanophyta / Cyanobacteria)

Cyanobacteria are an important part of the flora of all fast-flowing streams on the island, and of the loughs. These organisms played a key role in the evolution of life on planet earth when, around two billion years ago they developed the ability to use the energy of sunlight to split water molecules. This led to the build-up of free oxygen in what had previously been an essentially oxygen-free atmosphere, thereby preparing for the development of life on land. For a long time the free oxygen produced in this way reacted with reduced iron compounds to form massive iron oxide deposits (which are still mined commercially). In time other bacteria developed the ability to use this atmospheric oxygen in their respiratory metabolism. The chloroplasts in eukaryotic cells are thought to have arisen at a later stage through the endosymbiotic incorporation of cyanobacteria.

The detailed account of the blue-green algae prepared for the New Survey was based on 100 samples collected in various locations in July 2004. Sixty-eight species were recorded from terrestrial and freshwater habitats, and 18 from marine and brackish sites, resulting in a total of 75 species overall when overlap between the two is taken into account. When earlier records are added (after critical evaluation) the total is 118 species, one third of the cyanophyte flora of Britain and Ireland.

Both Creggan and Leinapollbauty have quite a diverse blue-green algal flora. The group is also well represented in streams with rocky substrates, except where these are subject to periodic flooding, which has a scouring effect on the algae. Among the most common species here are *Rivularia beccariana*, *Dichothrix orsiniana*, *Nostoc verrucosum* and *Phormidium autumnale*. The cyanophyte flora on wet rocks beside streams is often visually striking, with gelatinous masses and brown leathery sheets (the latter dominated by *Schizothrix coriacea*). The downstream stretch of the Owenmore River has a particularly diverse flora. The flora of the stable environment found in the small flushes in the upper reaches of this stream and in the River of

Fig. 7.46 *Batrachospermum turfosum* (Chris Carter, Algaebase [Mike Guiry]).

Fig. 7.47 Iron-encrusted *Batrachospermum* from the upper Dorreen River (Bryan Kennedy).

Rouing are especially well developed, and of particular ecological and conservation value. The high quality of these water bodies undoubtedly provides some of the finest examples of this macroalgal community yet documented in Ireland.

The recent arrival and spread of Chilean Rhubarb (*Gunnera tinctoria*) along streams (at the time of writing largely confined to the east of the island) is thought to represent a particularly serious threat to the blue-green algal flora. Apart from its general shading effect, the stems of *Gunnera* host symbiotic colonies of *Nostoc*. It is feared that this may raise the nitrogen status of the streams, which could bring about reduced populations of nitrogen-fixing blue-green algae (such as the native *Nostoc verrucosum*, *Dichothrix* and *Rivularia*), encouraging an increase in non-nitrogen-fixing algae, such as *Homoeothrix crustacea*, and green algae such as *Chaetophora* and *Gongrosira*.

Of the 51 species recorded by West, 45 are accepted as valid today. A further 67 were added by Whitton in 2004 and another 6 by other recent surveys; 16 of West's valid species have not been found since his day. Of the species found during the New Survey 35 had not been found by West or Cotton. The picture is much the same for marine and brackish species. Thirty of Cotton's marine and brackish species are considered valid today; the New Survey recorded 18 species, but failed to find 20 of Cotton's while finding 11 new ones. In summary:

Total species recorded in 2004 Survey	75
West and Cotton species accepted as valid	67
Valid West and Cotton species not found in 2004 survey	31
Species found in 2004 but not by West or Cotton	35
TOTAL NUMBER OF VALID SPECIES	118

In only a few cases is the difference likely to be due to significant environmental change (for example, the conversion of Lough Avullin to reedmarsh). It is more likely due to different sampling strategies and preferences, and difference in the times of sampling.

Subaerial algae

The New Survey includes a preliminary examination of subaerial algae, which are not well represented on the island because conditions are so exposed. Ironically, they are the most immediately visible algae, though scarcely recognisable as such to the casual observer! For instance, the black, red or green streaks and patches that often discolour damp concrete surfaces are colonies of cyanobacteria and of green algae belonging to the order Trentepohliales. Subaerial algae also occur in other above-ground situations where there is some shelter from direct sunlight and strong winds. The most common of the dozen or so single-celled green algae recorded on the island is *Apatococcus lobatus*, which is found just about everywhere on tree bark and damp wood, often with a variety of other species. The dark red streaks seen on the walls of older cottages (and concrete surfaces elsewhere) are due to the presence of *Trentepohlia iolithus*. The blackish crusts on damp rocks and walls are produced by cyanobacteria of various kinds (inside of Granuaile's Castle is a good example).[205]

Fig. 7.48 *Trentepohlia* (Jason Hollinger (CC BY 2.0)).

West's account of the great quantity of material collected on Clare Island is a checklist for the most part, but he did make a preliminary effort at identifying distinct communities of algae, acknowledging that 'these may be of some interest in these days of ecology-worship'. In one particular pool dominated by *Oedogonium braunii*, he identified eighteen other species of algae, six of which were especially prominent, '*Pediastrum tetras* in great abundance, *Coelastrum sphaericum* frequent, *Lyngbya Rivulariarum* mostly clinging to and around the Oedogonium, *Anabaena inaequalis* mostly agglutinated to the Oedogonium by its sheath, and various forms of *Scenedesmus quadricaudata* with *Tetrapedia reinschiana*'.[206]

In a pool near the lighthouse he identified more than forty species living among the submerged

Sphagnum moss. In a small pool on the summit of Knockmore the dominant *Chrococcus turgidus* (a cyanophyte) was accompanied by sixteen other species. In another pond on top of Knockmore fifteen species lived among *Bulbochaete* and *Oedogonium*, the dominant species of the fifteen being *Closterium moniliferum* and *Netrium digitus*. A pond near the old hotel had more than seventeen species, dominated by *Rhizoclonium hieroglyphicum* and *Cladophora flavescens* (now *C. glomerata*). A different *Cladophora*-filled pond hosted twenty species, among which one diatom (*Cocconeis placentula*) was a particularly abundant epiphyte on the Cladophora.

Among bladderwort (*Utricularia minor*) in Lough Avullin he found over twenty species, four being more or less co-dominant: *Cosmarium rectangulare, Chlorobotrys regularis, Xanthidium smithii* and *Staurastrum aciculiferum*. A pool filled with *Sphagnum plumulosum* had over thirty species; a tuft of *Microspora* had fourteen species; associated with a submerged *Hypnum* species he found over thirty-three species. But algae are also found away from ponds and streams wherever conditions are wet. Among a large patch of *Cladophora glomerata* from the cliffs of Knockmore, West described an association of over twenty species of diatoms, one of which (*Cocconeis pediculus*) was so abundant it almost hid the *Cladophora* from view. In one ditch near the chapel he identified twenty-three species.[207]

Protoctista

Flagellata and Ciliata

If so many small aquatic animals can occur in such enormous numbers, the quantity of tiny creatures upon which they feed must be astronomical beyond computation—indeed it is, and tremendously diverse in parallel. However, because it is so small, most of this Lilliputian world is hidden from everyday eyes, revealed to us only by those of the microscope. The world of the plankton contains both plant and animal forms, although at this level of biological organisation this distinction loses the meaning it acquires in multicellular forms. Many planktonic creatures have plant and animal qualities in the same one-celled body.

For the Clare Island Survey Praeger enlisted the services of John Samuel Dunkerly, who spent ten days on the island and also examined material from further afield, confining his attention to

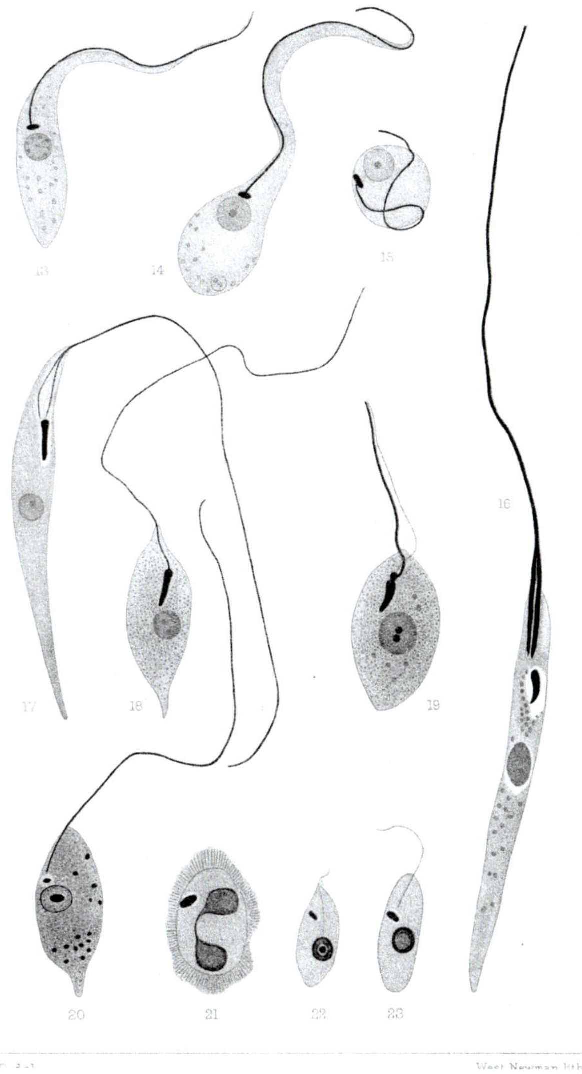

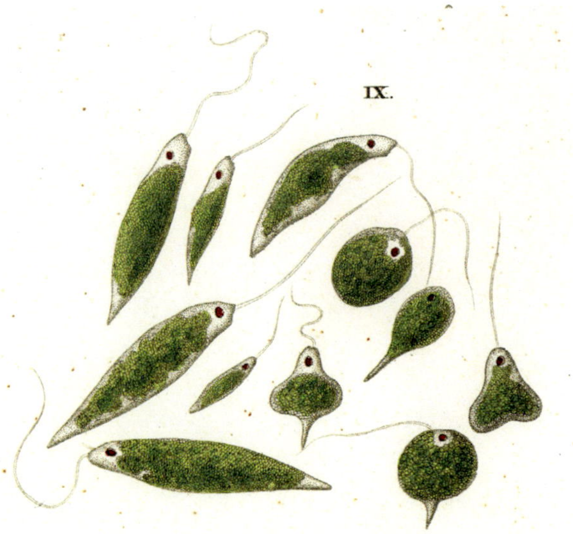

Fig. 7.49
(a) Plate II (forms of *Leptomonas*) from Dunkerley's *Flagellata and Ciliata* (Original Survey 61-62).
(b) *Euglena viridis*, from C.G. Ehrenberg, *Die Infustionsthierhcen* (1838).

freshwater and parasitic forms. Of the five localities for which he provided species lists (for various places around Westport and Louisburgh, as well as Clare Island and Achill), the richest assemblages came from Clare Island, and more particularly from Creggan Lough, to which he devoted most attention: 'The cause of this, however, [being] that this material was extremely rich in the number of species, if not so much so in the number of individuals, and indeed I have seldom examined more interesting material.'[208] Dunkerly used the most advanced microscope and imaging techniques of his day, enabling him to execute fine drawings at a magnification of up to 1,250 times (Figure 7.49a).

Of the 102 forms he recorded, no fewer than 90 were new to Ireland at that time, and a possible 20 new to Britain. Furthermore, he was at pains to point out that this had been 'a very incomplete investigation' because he was unable to take account of seasonal variation or the variation within an individual water body known to be so important in planktonic groups (for example, see desmids, page 191).

Since Dunkerly's day the classification of what at the time were unicellular creatures in two classes of the phylum Protozoa (Flagellata and Ciliata) has been radically revised. The Ciliata are relatively intact as the new phylum Ciliophora, but the Flagellata have been broken up into several new phyla. There has also been substantial taxonomic revision at the lower levels of species, genera and family. However, for our limited purposes, it is enough to retain Dunkerly's classification.

Among the most abundant flagellates—not only in Creggan Lough but in most of the other localities sampled—was *Euglena viridis* (one of several *Euglena* species recorded) (50–70 microns) (Figure 7.49b). This is something of a 'type species' for the dual plant/animal nature of many of these organisms. *Tricholomonas volvocina* (30 microns diameter) is closely related, but looks like nothing so much as a tiny balloon with a long string (the flagellum) attached. *Phacus pleuronectes* is rather similar (25–75 microns), but has a longitudinally striated surface, a curly flagellum and what looks like a little handle (the caudal process) at the other end. *Oikomonas* is rather *Euglena*-like but smaller (20–30 microns) and tends to occur in groups. In *Bicosoeca*, which is close to *Oikomonas*, the zooids are each enclosed in a protective urn called a lorica, which is attached to a substrate by a pedicel. The zooid itself is anchored within the lorica by a slender peduncle (the lorica is about 15–18 microns). In *Peranema trichophorum* the oval body of the zooid is flattened and very flexible, and has a very long flagellum which is vibratile at its tip. All of these species occur in an aquatic jungle of algae or other water plants A related species is *Anisonema acinus* (25 microns) which is pear-shaped and has two flagella, with a trailing flagellum that is very prominent and curls around the anterior end. It seems to be especially common where diatoms are abundant.

Anthophysa vegetans favours more stagnant conditions. It is an extraordinary-looking thing, consisting of a dense, rosette-like cluster of individual zooids, each with its pair of flagella, either attached to a common stalk (the pedicel) and anchored by this to a substrate, or else swimming with a rolling motion through the water. Each zooid measures 5–10 microns. *Chilomonas paramecium* (25–40 microns), which has a lip-like projection beside two subequal flagella, is also very common in stagnant water.

Several species of *Leptomonas* were identified during the study. These are parasites closely allied to the trypanosomes that cause malaria. The various species identified on Clare Island parasitise different insects; for example, *Leptomonas scatophagae* isolated from the common dung fly *Scatophaga stercoraria* and described as a new species.

Vorticella is one of the commonest ciliates—indeed one of the most fascinating of these tiny creatures (Figure 7.50). Each is about 150 microns in size, and they normally occur in clusters, attached to whatever provides secure anchorage—usually algae or other water plants—but they are occasionally found hitchhiking on the bodies of Entomostraca. The body is bell-shaped, and covered with a lid that remains slightly open; two rows of strategically-situated cilia create a little whirlpool that draws water into the mouth of the bell, where any food particles present can be sieved out. At the slightest disturbance it will instantly withdraw by coiling its stem into a tight spiral, and wait until things have quietened down. There are many species, but Dunkerly was unable to identify his Clare Island Vorticellas to species level with certainty.

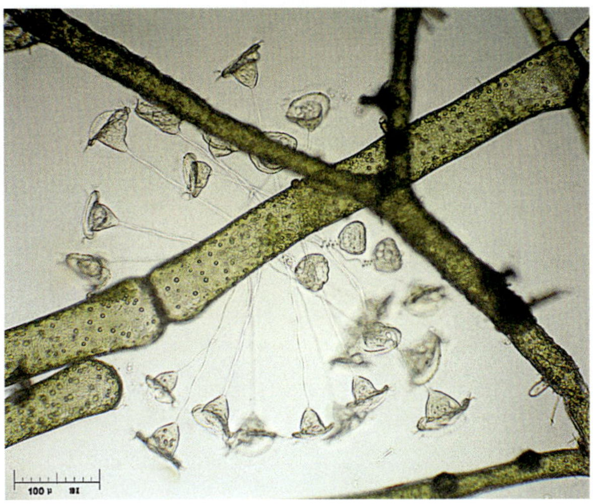

Fig. 7.50
(a) *Vorticella* (C.G. Ehrenberg, *Die Infusionsthierche*, 1838)
(b) *Vorticella* (Giuseppe Vago (CC BY 2.0)).

Phylum Rhizopoda
Class Lobosea
Subclass Gymnamoebia (naked amoebae)
Subclass Testacealobosa (testate amoebae)
The amoeba is tiny, never more than a few hundred microns from one end to the other, an animal that is no more than a single cell. That something so small can *behave* with such apparent purpose is one of the truly astonishing first things you learn when you start to study biology. But it is really only when you get the chance to *observe* this behaviour, to see the amoeba going about its daily business, that you begin truly to appreciate the wonder of it. Such is the power of the microscope in helping us to see things as they really are. Later we learn that 'the' amoeba is many. There are approximately two hundred known species of amoeba in addition to most likely countless others we are not aware of, all different, each exploiting the resources of the earth in a slightly different way from the others.

Even more astonishing than the 'ordinary' amoebas (if there is such a thing) to our sense of what should be possible, are those that build houses for themselves. These amoebas carry their homes around with them just as snails do, and are thereby provided with a measure of shelter and protection. These are structures of quite amazing architectural complexity and precision, constructed from a wide range of materials, but unique to each kind of amoeba, depending on the particular ecological circumstances of its evolutionary apprenticeship. With its arms (pseudopodia) it carefully selects its building material from its watery surroundings, then takes it inside itself where it coats it with 'cement' before moving it outside to where construction is ongoing. Different amoebas have different diet sheets, selected from a range of food items, but most feed on small algae. The amoeba simply flows around and engulfs its prey whole and entire for later digestion. Amoebas reproduce asexually, simply by splitting in two.

For the Original Survey Praeger recruited the services of the great Swiss expert Eugene Penard and of G.H. Wailes, who were together responsible for producing the report.[209] A small amount of material was collected from Clare Island by Penard in July 1909, but no fewer than 56 'gatherings' were made on the island by Wailes between March and June 1911.[210] Material from the adjacent islands and from a range of localities on the mainland was also collected (and supplied by others), but in spite of the diligence with which it was later examined, the authors considered this to be 'little more than a preliminary survey'.[211] Five heliozoans and 133 species and varieties of rhizopods were collected from sphagnum and wet moss on Clare Island, 22 more than on the mainland; 28 fossil species were identified in the peat in which the fossil pine forest is entombed.

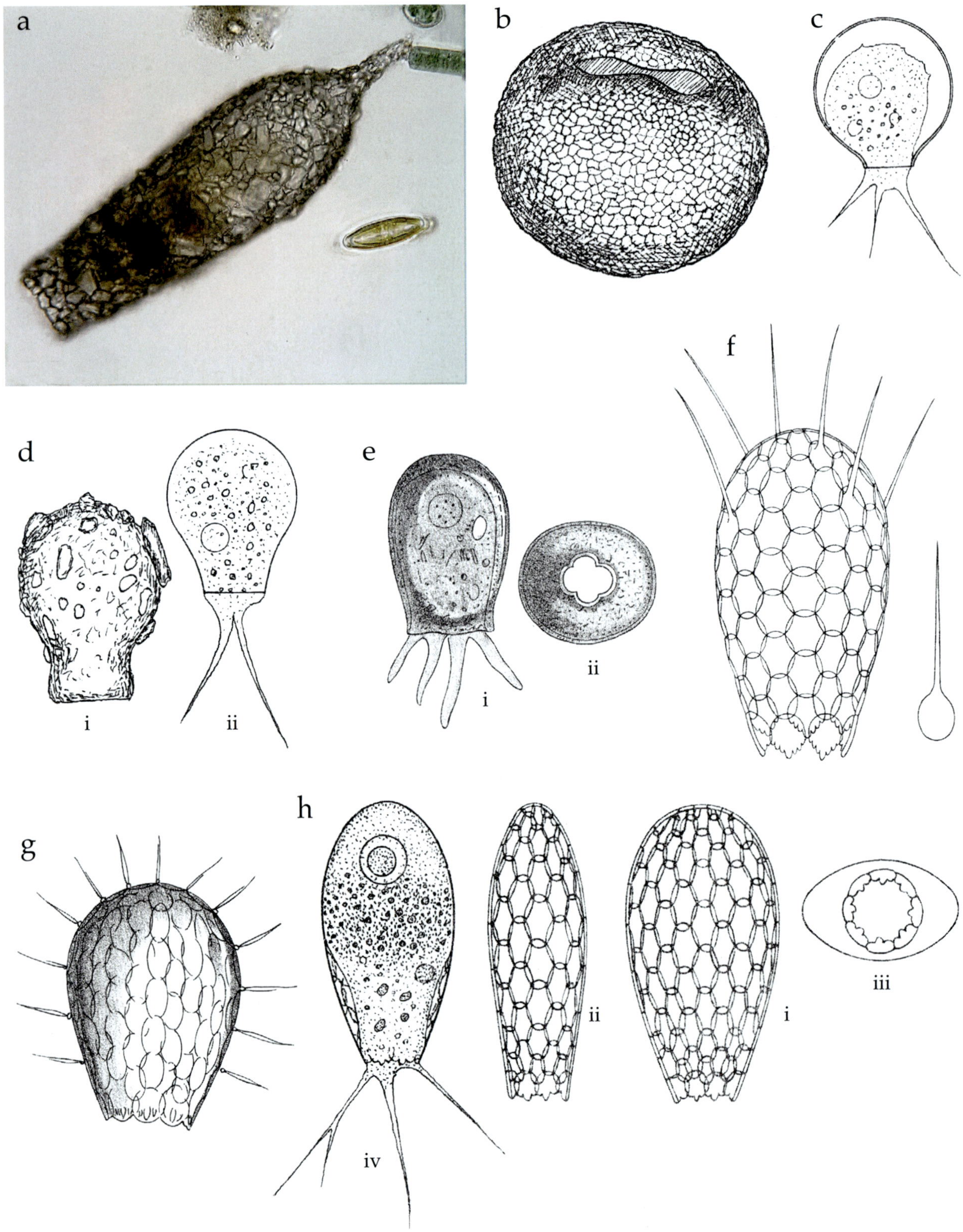

Fig. 7.51 (a) *Difflugia acuminata* (Deuterostome (CC BY-SA 4.0)).
Rhizopods in Wailes and Penard's *Original Survey* Volume 65 (1911)
(b) *Bullinula indica* x300.
(c) *Cryptodifflugia compressa*, broad view of living animal.
(d) *Cryptodifflugia sacculus*; (i) empty test of typical individual, (ii) side view of living animal in a smooth test.
(e) *Difflugia oviformis*; (i) living animal with its pseudopodia extended x420, (ii) buccal view of empty test.
(f) *Euglypha armata*; side view of empty test and one of the scale spines.
(g) *Euglypha compressa*; side view of empty test, x340.
(h) *Euglypha rotunda*; (i, ii, iii) front, side and buccal views of an empty test x900, (iv) front view of living animal x900.

The character of the rhizopod fauna is remarkably similar in different parts of the world: 'probably 90 per cent of the known species of Rhizopoda have a world-wide range'. Because of this, Penard was confident he would see little he had not seen before. However, although 'The essentially cosmopolitan character of the fresh-water Rhizopoda left me in no doubt as to what I should find', and that 'one could not expect to make any very important discovery', he was not disappointed in his hope of finding something new. In any case, 'the rhizopodan fauna of Clare Island was yet undescribed, and the suggested study could not fail to prove one of real interest.'[212]

The number of species found on the island proved to be 'considerable for its small and little diversified area', although the numbers of individuals in many gatherings were below average. The richness of the Clare Island list compared to other areas is however due in part to the fact that it was more intensively studied. Two of the species recorded—*Difflugia oviformis* and *Nebela griseola*—had up to that time only been seen in the Australian Alps. Several forms proved to be new to science: six forms of *Euglypha*, three of *Cyphoderia* and a new *Cryptodifflugia*. Over thirty were very rare: those 'of which few or no records exist of their occurrence in the British Isles'.[213] Among the commonest forms of rhizopods on Clare Island—all represented by several species—are *Amoeba, Arcella, Difflugia, Euglypha, Heleopera, Nebela* and *Phryganella* (Figure 7.51).

The Original Survey recorded six species of *Arcella*, of which the most common species is *Arcella vulgaris* (diameter 80–140 microns). It has a domed, snail-like shell (but plain, without the whorls, a bit like a World War I soldier's helmet). It is yellowy-brown in colour and moves around in a snail-like way but without a front or back. The pseudopodia are long, straight and colourless, and the creature is attached to its shell by delicate threads.

Difflugia oblonga is one of some eighteen species of its genus recorded on Clare Island. The shell is more pear-shaped than *Euglypha*, and covered with sand grains, diatom frustules, and so on, which its pseudopodia have carefully selected from the surroundings. It is found everywhere in the ooze of ponds and lakes, and is very variable in size and form: some of the larger forms are 450 microns long, just about big enough to be seen with the unaided eye. It feeds on filamentous algae and the like, the cell walls of which it perforates and then engulfs the protoplast within. This chlorophyll diet gives it a green colour, but the pseudopodia are invariably colourless.

Euglypha alveolata (length 45–100 microns) is one of the commonest of the ten different species of its genus recorded on the island. The shell is composed of overlapping plates of chitin, which gives the surface a regular hexagonal appearance, and it has one opening around which the scales have notched spines. It is a very common species among algae and mosses in the ooze of ponds and bogholes. Some of the other species (e.g. *Euglypha ciliata, Euglypha compressa*) have needle-like spines on the shell.

Nebela species have pyriform or ovoid shells, somewhat compressed, composed of plates that are round, oval or irregular. *Nebela collaris* and *Nebela lageniformis* are nearly always found where there is abundant *Sphagnum*. Species of *Heleopera* are distinguished by an aperture that is elliptical or linear; one species, *Heleopera rosea*, is very distinctive by virtue of its wine-red colour.

In *Phryganella* a chitinous semi-globose or ovoid shell is densely covered with sand grains, diatom shells and other roofing materials. *Phryganella nidulus* is the commonest of the three species that live in the peaty ooze at the bottom of lakes, ponds and bogholes. The pseudopodia are very slender. There is a distinct ring of large sand grains around the ventral aperture underneath the shell. Large forms are about 90 microns in diameter.

Phylum Actinopoda

Heliozoa (freshwater sun animalcules)

At the time of the Original Survey the Heliozoa were grouped with the rhizopods, but today they are usually placed in the Actinopoda ('ray-feet'). The single-celled body is spherical, but perforated with innumerable tiny pores through which long, slender threads of protoplasm (axopods) are extruded, each strengthened by a bundle of reinforcing microtubules (known as an axoneme). They feed on freshwater plankton. Five kinds of Heliozoa were recorded on Clare Island (there are approximately 100 species known worldwide).

Freshwater sponges

Freshwater sponges merit no more than a footnote. Of the five species recorded in Ireland at the time

Fig. 7.52 *Hydra vulgaris* (Stephanie Guertin (CC BY-SA 3.0).

of the Original Survey only one occurred on Clare Island: *Heteromeyenia ryderi*, which grows as 'soft, more or less lobed, yellowish masses on water-plants, or sometimes in a thin grayish-white film on the under surface of stones'.[214] *H. ryderi* is widely distributed in the west of Ireland, and of particular interest because it is a North American species, known from nowhere else in Europe at the time of Jane Stephens' survey. What is surprising though, is the complete absence from Clare Island of the most widely distributed Irish species, *Spongilla lacustris*.

Cnidaria

The phylum to which sea anemones and jellyfish belong is almost entirely marine, but it does include a small number of freshwater species. *Hydra vulgaris* is the only freshwater representative of the hydrozoa on Clare Island, and is extremely common in most of the western lakes. It varies greatly in colour from place to place, and the Clare Island population is pale orange.

Bryozoans (moss animals)

The vast majority of bryozoans are marine (see page 116), but about 50 species live in freshwater. The only species common in freshwater is the beautiful *Plumatella repens*.

Flatworms

Roland Southern studied flatworms (phylum Platyhelminthes, class Turbellaria) for the Original Survey, but concentrated mainly on the marine forms; only five of his list of fifty were freshwater species, taken from among vegetation in the streams and pools on the island.[215] The five species are *Stenostomum leucops*, *Stenostomum unicolor*, *Typhloplana viridata*, *Planaria alpina* and *Dendrocoelum lacteum*.

Dendrocoelum lacteum (up to 2.5cm) is easily recognisable from its milky colour. *Planaria alpina* is variable in colour and only about half as big, but is interesting as an indicator of the cool water of springs and mountain streams, emigrating upstream if it experiences a change in temperature.[216]

Annelids: Hirudinea (Leeches)

Southern recorded only three freshwater species of annelids. *Lumbriculus variegatus* occurred everywhere in the lakes and in sphagnum pools on the bogs, and the two other species—*Chaetogaster diastrophus* and (in large numbers) *Naias communis*—occurred among weeds from a stream on the island. Naiads are among the commonest of freshwater annelids, and can be recognised by the long bristles they have down the back.[217] They often occur as chains of joined individuals, because they reproduce by budding in this way more often than by laying eggs. These common species are a few centimetres in length.

A dozen or so species of leeches (Hirudinea) were recorded by Southern, but he had very little to say about them. One of the most interesting was *Protoclepsis tessellata*, mature females of which he found 'bearing young on the ventral surface' in August.[218] The young of this species attach themselves to the underside of the parent once they hatch (as many as 215 have been counted). The other common species were *Glossophonia complanata* ('very abundant in the lakes and streams on Clare Island'), *Herpobdella stagnalis* ('in several of the lakes'), *Herpobdella atomaria* ('in a pond near the north cliff of Clare Island, known as the "Light-house Marsh"'), and the horse leech *Haemopis sanguinea* ('very abundant … lurking under stones in every damp place').[219]

Nematodes

We go through our lives unaware of the existence of these practically invisible worms (see page 100), of which there may actually be millions of species, accounting for as much as four-fifths of all the animals on earth! One estimate puts the number of nematodes in an acre of topsoil at three billion, and there might be 90,000 in a rotting apple. Many of the 20,000 species described so far are free-living inhabitants of soil and water, where they are enormously important in recycling the elements needed for plant and animal growth. Others are parasitic, and these often have extremely complex life cycles.[220]

Phylum Rotifera (Rotifers)

Of all the groups of small creatures strange to our eyes that live in freshwater, the one that most enthralled and fascinated microscopists of an earlier generation were the rotifers: the wheel animalcules as they were fondly called. They are the most abundant and cosmopolitan, and indeed the most characteristic of the freshwater zooplankton, in that they are found, with few exceptions, nowhere else. Nor does 'freshwater' just mean streams and ponds, where they do indeed occur in immense numbers and variety: drains and ditches, eave-chutes, rot-holes in trees, stagnant puddles, all have their rotifer communities.

Such is the extreme diversity of form and organisation found among rotifers that it is difficult to characterise them all together in a concise description, but they all have a head end where the most striking feature is an extraordinary ciliated structure called the corona. The cilia of the corona beat in co-ordinated and rhythmic fashion to set up tiny whirlpools that waft prey towards the mouth and (when required) to propel the tiny creature through the water. They also maintain a flow of water over the body that enables it to obtain the oxygen it needs.

At the other end is a structure for attachment called (what else) a 'foot', which has two or more pointy 'toes'. It has a well-developed alimentary canal, a muscular pharyngeal region (the mastax) and wonderfully complex jaws (trophi). It has a simple excretory system but lacks circulatory or respiratory systems. It has a (for its size) prominent brain, with nerves and sense organs. It reproduces sexually, but in most rotifers the male is a minute, degenerate creature lacking an alimentary canal. Many rotifers wear chitinous suits of armour suitably reinforced with arrays of knobs and spines (the lorica).

Most rotifers feed on organic debris and play an important part in freshwater food chains, but their diet also includes bacteria, small algae and invertebrates, and rotifers smaller than themselves. Some are planktonic, others creep over or are attached to vegetation. Our best guess is that there are between 1,800 and 2,000 species, ranging in size from 0.04mm to 2mm. About 50 species are marine, and there are a good number of epizootic or epiphytic forms (for example, Rousellet recorded *Proales parasita* in great abundance in Creggan Lough inhabiting the hollow spheres of the planktonic *Volvox aureus*).[221] There is one group that lives in the tiny, helmet-like leaves of certain liverworts (especially *Frullania*).

One of the most extraordinary things about rotifers is their ability to withstand drying and extremes of temperatures (in a special resting stage). Another remarkable thing is how cosmopolitan they are. Species found in streams in Africa are the same as those in North American streams; species that characterise sphagnum pools in Europe are the same as those in Asia. Some species undergo a cycle of regular change from month to month in a yearly cycle.

Most rotifers belong in the class Monogonta, but there is another species-diverse class known as the Bdelloidea that is especially abundant among sphagnum and other mossy plants. They are ceaselessly active little animals; when not attached they creep about like leeches. Their body is composed of rings that can be drawn one within the other like a telescope when they want to make themselves less conspicuous. Bdelloid rotifers are characterised by a very distinctive corona, consisting of two flat ciliated, stalked discs on either side of the head. These are especially wheel-like, and it was for these in particular that the name 'rotifer' was coined. What also distinguishes them is that they have given up on sex; most reproduce by parthenogenesis because there are no male bdelloids. They comprise an immense number of species, often very similar: and identification is rendered even more difficult by constant changes of form.

Charles Rousselet reported on the non-bdelloid rotifers for the Original Survey. He spent a week on the island in June 1909, but was very careful to point out that his list of species should be regarded as no more than 'a first instalment towards the rotiferous fauna of Clare Island', a complete record being 'a task requiring many months' work at different seasons, and continued for several years'.[222] He recorded sixty species on Clare Island and sixty-five on Achill, attributing any difference between the two lists to nothing

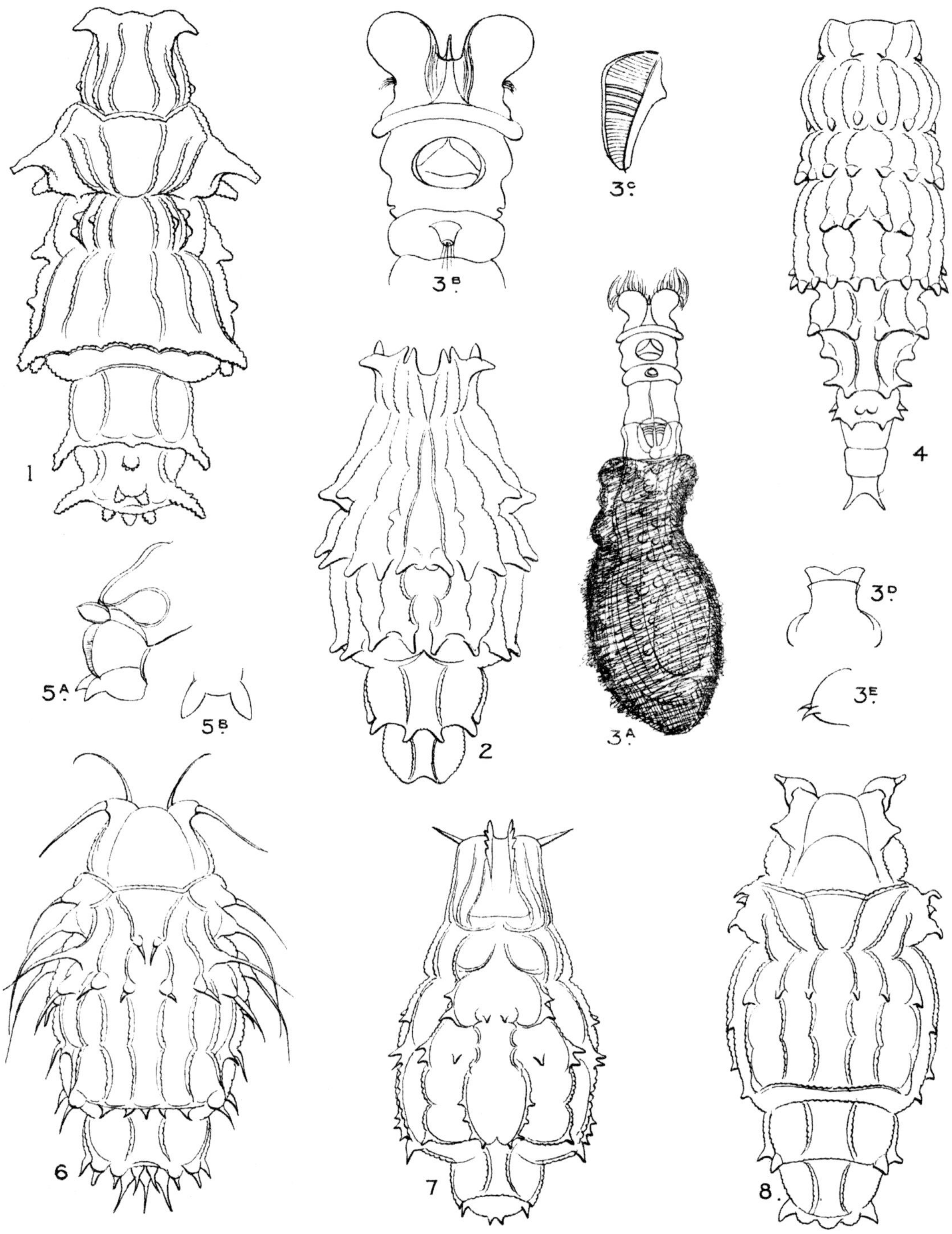

Fig. 7.53 Plate from James Murray's fascicle on bdelloid rotifers (Original Survey 52).

1. *Callidina papillosa* (Thompson): variety with elongated tubercles.
2. *Pleuretra brycei* (Weber): variety having six tubecles on the preanal segment.

3A. Habrotrocha, an undescribed tube-dwelling species.
3B. The same: the head more highly magnified.
3C. The same: jaw.
3D. The same: rostrum.
3E. The same: side-view of foot.

4. *Callidina aculeata* (Milne): trunk and foot, dorsal view.

5A. *Habrotrocha collaris* (Ehr.): side-view of foot, showing process at anus, like that of the tailed species (*H. caudata*, etc.).
5B. The same: spurs.

6. *Callidina multispinosa* (Thompson): long-spined variety.
7. *Dissotrocha spinosa* (Bryce): variety.
8. *Callidina multispinosa* (Thompson): short-spined variety, having lateral hooked spines on the second trunk segment.

more than 'want of sufficient search and opportunity'.[223] It is hardly surprising in view of the cosmopolitan nature of rotifer distribution that the fauna shows no special features, and yet a book could be written about the life of every one of these species.

The bdelloid rotifers were studied for the Original Clare Island Survey by James Murray, who visited the area in 1911, and was also supplied with material by others.[224] He recorded over sixty species, thirty-six of them on Clare Island, nearly all of the sixty new to Ireland; a number of the species recorded are more or less ubiquitous, others more local; one, a 'pretty little species' (*Callidina asperula*) which occurs everywhere on Clare Island, was at that time known only from Canada.[225] Among the most characteristic bdelloids for the region are *Callidina plicata, Callidina quadricornifera, Callidina papillosa, Rotifer longirostris, Habrotrocha angusticollis, Habrotrocha pulchra, Adineta gracilis* and *Adineta barbata. Ceratotrocha cornigera* is also widely distributed over the entire district, apparently more abundant in Ireland than anywhere else it occurs. A number of species of *Habrotrocha* are confined to sphagnum, living inside the outer retort cells of the moss (Figure 7.53).

Anuraea is among the most common of the planktonic rotifers, usually found near the surface. It is a good example of a rotifer in which the lorica is highly developed, composed of armour-like plates that often have warty, verrucose knobs sticking out of them. In *Eucharis*, a group that prefers well-aerated water, the lorica is composed of front and back segments. Like so many rotifers (except bdelloids) it swims with a rotary motion, but in *Eucharis* the play of light upon its highly refractive, glassy lorica gives it a beautiful crystal-like appearance at times. The 'toes' of rotifers are usually short and pointy, but in *Furcularia longiseta* they are bizarrely elongated into stilts (of different lengths). *Proales* forms gall-like structures on aquatic plants, inside which hundreds of young notifers may be nurtured.

Perhaps the most fascinating of all rotifers are those that live in tubular dwellings of their own making, usually attached to water plants. *Limneas* lives in a cylindrical tube fashioned with its body secretions, the sticky nature of which causes particles of debris to adhere to it, giving it a somewhat untidy appearance but doubtless effective as camouflage. In *Melicerta* the tube is made of bricks fashioned from particles of mud gathered from the water, moulded 'by hand', and then arranged with meticulous precision on its growing tower.

Phylum Crustacea
Freshwater Microcrustacea
(Entomostraca)
Class Branchiopoda (water fleas or cladocerans)
Class Copepoda (copepods)
Class Ostracoda (seed shrimps)

Crustaceans are divided into several major and minor classes, the principal ones being Branchiopoda, Ostracoda, Cirripedia and Malacostraca.[226] Branchiopods are most familiarly represented by water fleas (Cladocera): mainly small freshwater creatures, generally much less than three millimetres in length. There are about 850 species, some families of which are important planktonic forms (e.g. Daphniidae); others, more benthic in nature (e.g. Chydoridae), are much more common among vegetation and closer to the shoreline. They are wonderfully complex: 'probably no other animals of so small size have so complex a structure'.[227]

These groups of tiny crustaceans are a vitally important link in freshwater food chains: the link between the microscopic cells of the phytoplankton on which they largely feed, and the innumerable aquatic invertebrates and juvenile fish that feed on them. Each species has its own story, its own unique part to play in the ecology of the pool or stream in which it lives: and this is the aspect that is most hidden from us. Most of what we know is based on animals (often just a few, often indeed just one) plucked from their world.

Copepods are enormously abundant in both marine and freshwater environments, occurring in astronomical numbers in the sea, and in just about every body of water, including puddles and rain gutters. They often dominate the zooplankton and so are among the most important food animals for fishes and other aquatic animals. There are some 13,000 species, typically under 2cm long. The class includes many commensal and parasitic species (some of which are among the weirdest-looking creatures you can imagine).

Chydorus sphaericus

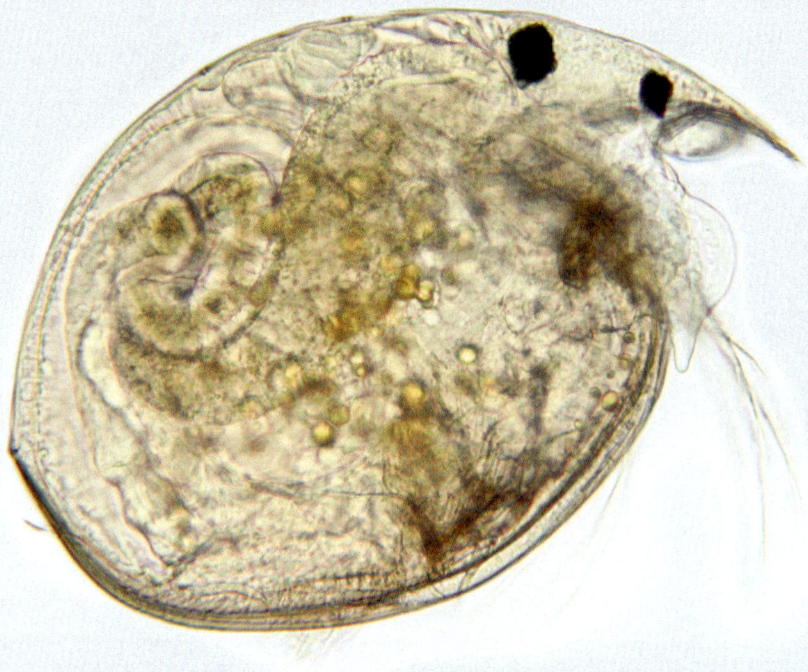

Fig. 7.54 *Chydorus sphaericus* (© Malcolm Storey, www.bioimages.org.uk (CC BY-NC-SA 4.0).

For all their diversity, all Cladocera are built to the same specifications; they all have the same features, differently shaped and proportioned, and you wonder what peculiarly favoured combination of size and shape accounts for the extraordinary success of this little animal, which occurs in just about every lake or pond in the world, in comparison with all the others in its class.

Chydorus sphaericus is a broadly elliptical water flea, the female about 0.3 to 0.5mm long, the male considerably smaller.[1] The body is enclosed in a two-valved (but one-piece) transparent carapace, open at the bottom (which has a diagnostic reticular pattern). The head with its prominent 'beak' (rostrum) is separated from the rest of its body by a rather faint depression known as the cervical sinus. There is a single prominent eye which is ever in motion, looking this way and that. The brain and various principal nerves (including the optical ganglion) and the front part of the alimentary canal are in the head. There are two pairs of appendages on the head: the antennules which have prominent sense-rods (olfactory setae), and the antennae which are used for swimming. In the mouth the algae on which the animal mainly feeds are ground by the teeth and ridges of its strong, chitinous mandibles. These grind away all the time, even when there is nothing to eat. Behind the head is the heart, whose constant pulsation is one of the hypnotic features of the animal. There are no blood vessels as such (although the blood does move in a structured way through various compartments). The intestine is a prominent convoluted tube that also pulsates constantly as food is churned and absorbed by the blood.

In a few families the feet are five pairs of leaf-like structures retained within the shell; these are very complex structures with lots of hairs and setae. These are used to generate a current of water that wafts items of food towards the mouth and maintains a steady flow through the valves, the inside surface of which is the main organ of respiration. In most families however, the first pair of legs has hooks and bristles that enable the animal to clamber about in vegetation. At the back of the body (its 'post-abdomen') two prominent claws protrude between the two valves of the shell. These have an array of spines and teeth of the most varied shape and structure, whose function it is to comb the legs and the frilled margins of the shell and keep them clean. The surface of the body is equipped with an array of sensory hairs which varies greatly in different species and is important in classification.

When there is plenty of food and the future looks rosy the females produce eggs in a special brood chamber in the upper part of the space under the valves. The eggs hatch and develop without fertilisation into replicas of their mother, who retains them in the brood chamber until they are able to fend for themselves. The mother may have several such broods in the season. Sooner or later however, 'true' females and males are produced, mating between which results in the production of eggs of a very different character. These are few in number and when they are shed the mother's shell often comes with them as a protective wrapping. These eggs will not hatch until favourable conditions return once more.

[1] Chydorus was a river of Macedonia mentioned in Herodotus. *Daphnia*, the 'original' water flea, first seen in 1669 by Jan Swammerdam, who described it as *pulex aquaticus arborescens* (the water flea with branching arms) is named after the river goddess Daphne.

Ostracods are tiny, bivalved crustaceans, abundant, ubiquitous and very diverse, and perhaps typically no more than 1mm across. They look for all the world like little pea-shells until you see their bristly antennae poking out to explore. And unlike other shelly look-alikes, their shell is shed every time they moult, and a new one is grown. There are some 5,700 species, most of them benthic crawlers, but some live in freshwater.

D.J. Scourfield undertook the study of the freshwater entomostraca for the Original Clare Island Survey. He and W.F. de Vismes Kane collected in the area in June 1909, and supplemented this work with material supplied by others.[228] They collected from the island itself, and from the mainland inland up to 1.5km from the coast. On Clare Island they collected a total of 54 species: 25 Cladocera, 19 Copepods and 10 Ostracoda. Most of these species are fairly common in Britain and Ireland as a whole, but the list included a number of species new to Ireland at the time, several of them living on the island. Five of the cladocerans, fifteen copepods and possibly one of the ostracods were considered to be first records for Ireland.[229]

However, one of the copepods—a single specimen found in wet moss at the side of a stream flowing from the lighthouse marsh to the coast–was identified as a species new to science, which Scourfield named in honour of Praeger, 'to whom the Clare Island Survey owed its inception and continued inspiration, and to whose patient labour and never-failing enthusiasm and encouragement is due so much of what is known today of the Irish fauna and flora'.[230]

Scourfield considered the Clare Island fauna to be 'mainly noticeable, as perhaps was to have been expected, for negative rather than positive characters'. The main reason for the lower species richness of the island compared to the mainland is probably the absence of larger lakes, but Scourfield believed the absence of certain genera that were common on the mainland (*Diaphanosoma*, *Bosmina*, *Alonopsis* and *Polyphemus*) could only be accounted for by something to do with their means of distribution.[231] A number of forms were thought to specially connect the fauna of the west of Ireland with Scotland and possibly Scandinavia.[232]

Elvira de Eyto, Adriana Trojanowska, Jens Petter Nilssen and Kieran McCarthy studied the Cladocera and Copepoda for the New Survey in 2001, 2002 and 2008.[233] They recorded 37 species of Cladocera and 8 Copepoda, including 14 not on Scourfield's list. They attributed the difference not to a deficiency in sampling during the earlier survey, but to an increase in species richness in the intervening 90 years. However, they found there was no apparent shift in the dominant chydorid taxa; they took this to mean there has been little change in the ecological character of the waters over the last century.[234]

The most species-rich sites were found to be Lough Leinapollbauty with 22 Cladocera, Creggan

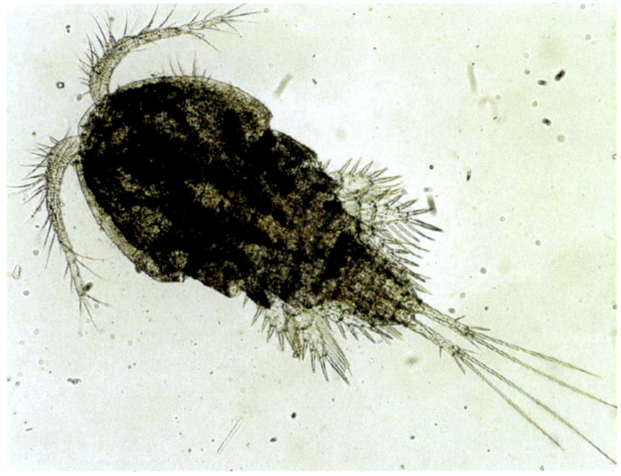

Fig. 7.55 *Cyclops* is the most familiar and abundant copepod. It has a pear-shaped, tapering body divided into a distinct 'cephalothorax' (head and chest) and an abdomen that ends in two finger-like lobes (furcal rami) that bear feathery spines. One of its most conspicuous features is the long first antennae, with which it rows itself along and whose array of sensory hairs gathers information about its surroundings. Another striking feature is the pair of egg sacs the female trails along behind her, one on either side. It has a single ruby-red eye in the middle of its forehead: the feature that drew down upon this tiny marvel the name of the one-eyed monster of mythology who fabricated the armour of Jupiter in its vulcanic workshop on Mount Etna.

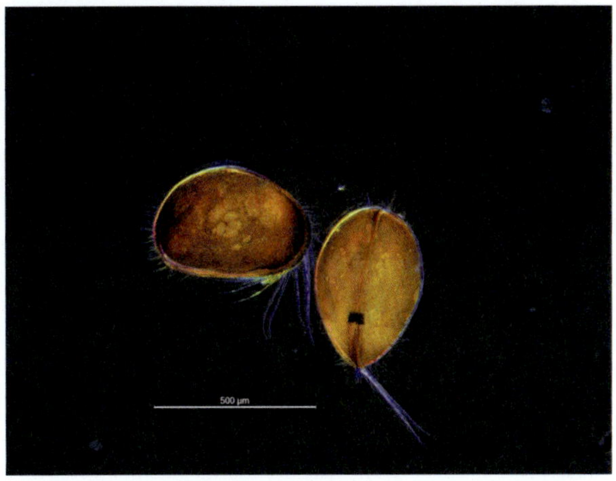

Fig. 7.56 Ostracoda.
Cyclocypris ovum (Markus Lindholm, Anders Hobaek, Norsk institutt for vassforsking (CC BY-SA 3.0).

Lough (17), Lough Avullin (11) and Poirtín Fuinch (15). Larger water bodies tend to have higher numbers of species because of the wider range of habitats they present; the slightly depauperate nature of the island's cladoceran fauna (61% of the Irish total) is a reflection of the absence of larger lakes. *Chydorus sphaericus* remains the most common cladoceran on the island—as it was in Scourfield's day (and as it is everywhere in the world); it even occurs on the summit of Knockmore (Figure 7.54).

Three species that are common on the mainland but not recorded in 1912 are now present: *Diaphanosoma brachyurum*, *Bosmina longirostris* and *Alonopsis elongata*. Surprising absentees include the widespread predatory species *Polyphemus pediculus*, and *Bosmina longispina* (= *B. obtusirostris*), and several typical pond species of *Daphnia*.

Copepods were collected in only a small number of water bodies over the course of the New Survey, whereas Scourfield sampled a wide variety of habitats, his records including some dozen *Cyclops* species, of which the most widely distributed was *Cyclops serrulatus* (Figure 7.455). Nevertheless, the modern authors were struck by the continued absence of *Cyclops abyssorum*, which is such a common and widespread species elsewhere. Its absence (and that of the cladoceran absentees just mentioned) is all the more puzzling in view of the belief that other introductions have taken place over the past century.

Ostracods are more restricted in distribution. The acidic Sphagnum pools that are such prolific habitats for other microarthropods have few; they are altogether absent on Knockmore, and Scourfield never recorded any species at more than 3 of the 21 stations he sampled on the island. The commonest species is *Cyclocypris ovum* (Figure 7.56).

For the New Survey Gillie McCall studied the ostracods over a four-year period (2002–05), recording seventeen species, an increase of seven on Scourfield's list of ten.[235] Only five species are reliably common to both surveys, and three of the newly-recorded species are actually ones that Scourfield had found on the mainland during his survey. At least two of the species recorded by both Scourfield and McCall (*Bradleycypris obliqua* and *Bradleystrandesia fuscata*) have as yet been found nowhere else in Ireland, and *Cypridopsis lusatica* was recorded from Ireland for the first time. Another two species (both of them species of *Cryptocandona*) are thought to be recent arrivals. In spite of the difference between the two lists, McCall sees little significant difference between today's ostracod fauna and that of Scourfield's day, attributing what variations there are to their different sampling strategies, greater sampling effort in the New Survey, possible seasonal variation, and taxonomic revision. Moreover, it appears probable that some of Scourfield's species were misidentified.[236]

Class Arachnida
Subclass Acari (Mites and Ticks)

Family Hydracarina (Water Mites)

Water mites are among the most gorgeously attired of freshwater creatures, looking quite metallic and coloured in reds, blues, greens and yellows of every hue, colours that warn would-be predators of their distasteful nature. More primitive forms tend to favour red colours and have ovate bodies; more highly evolved species favour other colours and a greater variety of body shape, sometimes with pronounced sexual dimorphism. They are all committed carnivores, with beak-like mouth parts adapted for sucking body fluids from their prey, such as crustacea, insect larvae, and other animals they are big enough to tackle. Their juvenile stages are often spent as parasites attached to water beetles and other aquatic insects, but they break away to live independent lives as adults. Attachment to winged insects also enables them to disperse beyond the confines of their narrow home world.

Their arachnid affinities are much more apparent in water mites than in their terrestrial cousins, because the eight legs are long and spider-like. In more actively-swimming species these are elaborately fringed with hairs. Most are 1–2mm in size, and they occur in freshwater bodies of every sort.[237] These hairs are absent from 'crawlers', species that spend most of their time slowly moving about on plants or on the bottom in search of prey.

Water mites were the subject of a separate report by Halbert in the Original Survey, written several years earlier than that on terrestrial mites, based on work mainly carried out in the summer of 1909, to which were added a few scattered earlier records.[238] As in the case of terrestrial mites, this was a regional study, without any particular focus on Clare Island itself. As with many

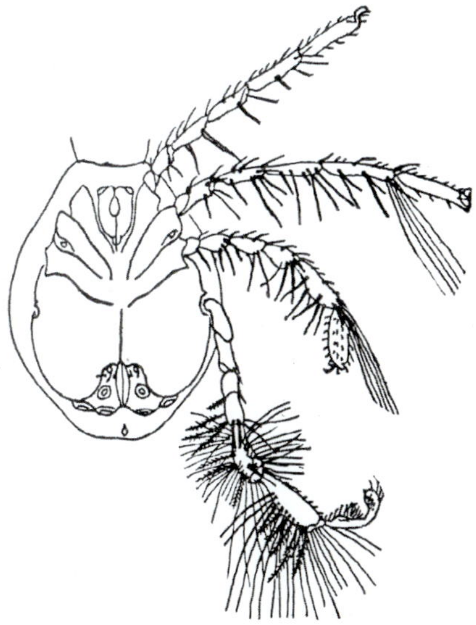

Fig. 7.57
(a) Water mite (Anatoly Mikhaltsov (CC BY-SA 4.0). The body length of this species is 1.1mm.

(b) *Pionacercus leuckarti*, showing the extreme modification of the last pair of legs in the male (shown on one side only, omitting the palps) (Ward and Whipple, *Freshwater Biology*, 853).

freshwater groups, water mites are 'remarkably well-spread animals', and there were few novelties even in this remote location. Eighty species of water mites in 31 genera were collected, including four new species. Twelve of the eighty species were new to Britain and Ireland, and an additional ten were new to Ireland. Sixteen species were recorded for Clare Island itself, half of them in or around Lough Avullin. It is to be hoped that they have been able to find alternative accommodation elsewhere.

Halbert divided the water mites into two 'great faunistic groups': a widely-distributed 'cosmopolitan' group and a more restricted 'northern or alpine' group, both well-represented in this area. He quotes Dr Zschokke's opinion that the latter are glacial relict species. In addition, however, he recognised 'a distinct northern element' represented by such Clare Island species as *Lebertia fimbriata*, hitherto recorded only from Norway. The study also provided evidence of a distinct 'moor-fauna', comprising species characteristic of peatland habitats, represented by such Clare Island species as *Thyas vigilans*, *Lebertia fimbriata* and *Pionacercus leuckarti*.

Halbert believed the hydrarachnid fauna of Clare Island to be a survival from a time when it formed part of the mainland. It has nothing distinctive that could be attributed to its present insular character.

Some notable species on Clare Island

Eylais similis is an egg-shaped red mite in the more primitive family of the Limnocharidae. It is often very abundant, and typically 2–5mm in size. It has feathery palps, and all except the fourth pair of legs has swimming hairs. *Thyas*, on the other hand—another genus of ovoid red mites of moderate size (1-2mm), represented on the island by *Thyas vigilans*—only has short spines on its legs, because it is not a swimmer, but lives on the bottom in swampy situations. The larvae of this species are believed to be parasitic on gnats. *Hydrarachna* is a large genus of globular red or brown mites in the same family, found in swamps, lakes and ponds everywhere and usually common, some up to 8mm in size. Halbert recorded two species on the island. *Sperchon setiger* is an oval mite that lacks well-developed swimming hairs, and is characteristic of mountain streams and lakes. The closely-related *Pseudosperchon verrucosus* lives on the island among mosses in small streams flowing from the hills.

Lebertia fimbriata is a member of a genus of medium-sized mites (0.8-1.5mm), whose most eye-catching feature is a highly-arched body. Most of the species are associated with colder waters. *Unionicola* is a genus that includes both parasitic and non-parasitic species. Although the generic name suggests it should be a parasite of freshwater mussels (*Unio*), *Unionicola crassipes* (recorded in Lough Avullin) is a free-swimming mite—

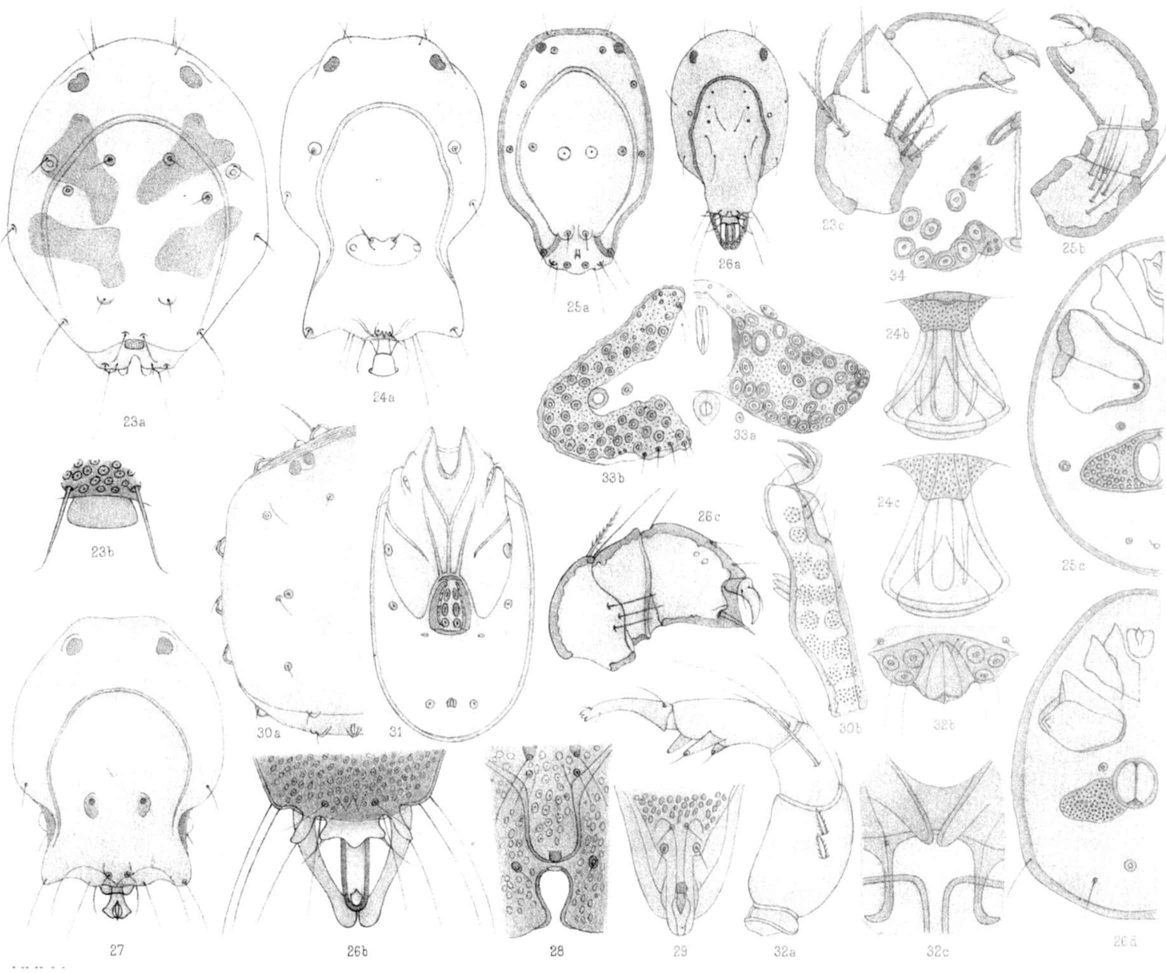

Fig. 7.58 Plate III from Halbert's fascicle on Water Mites (OS 39i).

23. *Arrhenurus bipapillosus* nov.sp.; (a) dorsal view of male, (b) petiolus organ, (c) palp of male x100.
24. *Arrhenurus battilifer* var. *dilatatus*; (a) dorsal view of male, (b) petiolus x74, (c) petiolus of typical form (Germany) for comparison x74.
25. *Arrhenurus sculptus*; (a) dorsal view of male, (b) palp of male x74, (c) ventral view of female.
26. *Arrhenurus Kanei*; (a) dorsal view of male, (b) armature of male, (c) palp of male x100, (d) ventral view of female.
27. *Arrhenurus ornatus*; dorsal view of male.
28. *Arrhenurus forpicatus* var. *perforatus*; Petiolus region x51.
29. *Arrhenurus forpicatus*; petiolus region x51.
30. *Feltria Rouxi*; (a) dorsal view of male, (b) end segment of third leg x170.
31. *Lebertia fimbriata*; ventral view x20.
32. *Unionicola rivularis* nov. sp.; (a) palp of female x100, (b) genital area of male, (c) central epimeral region of female.
33. *Piona rotundoides*; (a) genital area of male, (b) genital area of female x38.
34. *Piona paucipora*; genital area of female.

sometimes as much as 2mm in size—with long legs abundantly equipped with swimming hairs.

Piona are hardy, active mites of middling size (0.5–3mm), often brightly coloured and with well-developed swimming hair, represented by eleven species in the wider area and three on the island. The male has extraordinary modifications on the two hind pairs of legs: on the fourth segment of the fourth pair to assist the male in grasping the female during mating, and on the sixth segment of the third pair to carry the semen to the female's genital opening. In no other species however are these modifications so spectacularly seen as in *Pionacercus leuckarti*, which lives in bog pools on the island (Figure 7.57b).

Arrhenurus is another genus in which there is highly-developed sexual dimorphism. It is one of the most abundant and widely distributed genera (well over 200 species, 11 of which were recorded by Halbert, including two on Clare Island), found in clear, shallow, hard waters where there is plenty of vegetation. The females are oval in form, but the males are very different, with complicated accessory sexual structures (Figure 7.58).

Fig. 7.59 Imago of *Baetis rhodani* (Bryan Kennedy).

Class Insecta (insects)

In the Original Survey a mere four pages were allocated to the several major orders of aquatic insects grouped together as the 'Neuroptera'at that time: 'such insects as the dragonflies, mayflies, caddisflies, and other insects comprising the Neuroptera of the older classifications'.[239] This was partly because the organising committee failed to enlist the help of the acknowledged experts of the day (J.J.F.X. King and K.J. Morton, who were too busy), and the task accordingly fell to J.N. Halbert, whose main interests lay elsewhere (see bibliography). From existing records, and by collecting during the course of the survey, he managed to compile a rather impressive list of 124 species for Clare Island and the adjacent mainland, which was almost half the number known in Ireland as a whole at that time. However, there is little out of the ordinary in these findings—just 24 species were recorded on the island: 12 caddisflies, 8 dragonflies, 3 mayflies and a single stonefly.

Halbert had little doubt that most of his species must have arrived on Clare Island by air, 'in former times while it was still joined to the mainland ... the comparatively weak and fitful flight of the majority of these insects [precluding] the possibility of their having flown across the three miles of ocean, which, at its narrowest part, separates the island from the mainland'.[240] He did except the dragonflies however, which are very strong and agile fliers.

Order Ephemeroptera (mayflies)

Mayflies are one of four very important insect orders that spend the first part of their lives as fully aquatic larvae, which are referred to as naiads or nymphs.[241] These naiads moult several times, before emerging from the water to become winged adults. All mayflies are sensitive to pollution to varying extents, which makes them a particularly valuable group for assessing water quality. Given that they typically dominate invertebrate communities in clean rivers they are a core indicator group in assessing ecological conditions. Of other 3,300 species worldwide 48 occur in Great Britain and 33 in Ireland.

Fig. 7.60 Nymph of *Baetis rhodani* (Bryan Kennedy).

Only three species were recorded during Halbert's Neuroptera review for the Original Survey: two species of *Baetis* (Baetis [*Alainites*] *muticus* and the very rare *Baetis fuscatus* – whose presence in Ireland still requires clarification) along with *Electrogena lateralis*. Recording by Mary Kelly-Quinn in recent years has added *Baetis rhodani* and *Ephemerella ignita*.[242] However, *Rhitrogena semicolorata* was the most abundant invertebrate encountered in both the Owenmore River and the River of Rouing during the 2019 Survey (see below, pages 215–7). A further cosmospolitan species, *Cloeon simile*, was also taken on the same occasion in a plankton haul of Creggan Lough. It is difficult to imagine that further sampling would not lead to the discovery of additional ubiquitous species (such as Caenidae). The absence – so far – of *Ecdyonurus*, a third widely distributed genus in the Heptageniidae, is intriguing and worthy of further investigation, as is the absence of certain stoneflies.

Order Odonata (dragonflies and damselflies)

There are 5,500 species of dragonflies and damselflies worldwide, some of them with wingspans of many centimetres. Adult damselflies are readily distinguished from dragonflies by their more slender bodies and the way the wings are held upright above the body, whereas dragonflies are more robustly built and hold their wings sideways. Both groups are mainly confined to still or gently-flowing waters. Like the adults, the nymphs are predatory, but the two groups lead quite different underwater lives. Dragonfly nymphs are rather stocky-looking and crawl slowly about on the bottom in search of prey, relying on their mud-coloured bodies for camouflage; damselfly nymphs by contrast move

Fig. 7.61 Beautiful demoiselle (*Calopteryx virgo*) (Juri Kowski (CC BY-SA 3.0).

slowly about in bottom vegetation, where the green, yellow and brown tones of their narrow, cylindrical bodies make them very hard to spot.

Nine species of Odonata are now known from Clare Island, just one more than the number recorded by Halbert. All except one of these are dragonflies, the exception being the Beautiful Demoiselle (*Calopteryx virgo*), which is unusual in its preference for fast-flowing waters, which the metallic-bodied adults can be seen patrolling on their colour-panelled wings. Among the more striking species are the two Darters (*Sympetrum striolatum* and *S. danae*), which breed in bogs, wet heaths and similar waterlogged places, and so are very much at home on Clare Island.

The dragonflies of Clare Island	
Sympetrum striolatum	Common Darter
Sympetrum danae	Black Darter
Libellula quadrimaculata	Four-spotted Chaser
Aeschna juncea	Moorland Hawker
Calopteryx virgo	Beautiful Demoiselle
Pyrrhosoma nymphula	Spring Redtail
Ischnura elegans	Common Bluetip
Enallagma cyathigerum	Common Bluet
Lestes sponsa	Common Spreadwing

Order Plecoptera (stoneflies)

Stoneflies inhabit well-oxygenated rivers and streams. Most are very sensitive to pollution, which makes them valuable bioindicators of water quality. Like mayflies, caddisflies, dragonflies and damselflies, they spend their early days as fully aquatic larvae and are important in the diet of fish, emerging as winged adults after moulting perhaps ten times.

The group is very poorly represented on Clare Island, a single species only—*Siphonoperla torrentium*—being recorded in Halbert's Neuroptera survey.[243] Nymphs of this widely-distributed species were easily found again, though not abundantly so, in the 2019 survey of the Dorreen and Owenmore Rivers. A single nymph of *Leuctra inermis* was found on the same occasion in the River of Rouing. The low species diversity surprising in view of the pronounced preference of most for well-oxygenated streams, of which there are plenty on the Island.

Order Trichoptera (caddisflies)

Caddisflies look rather like drab-coloured moths as adults, and generally go unnoticed on that account, especially as they are mainly nocturnal (and, like moths, attracted to light, often in great numbers). Young caddisflies (caddis-worms) are unfamiliar for a different reason: they live in water. There are two main groups, those that construct cases of stones, sand or plant material in which to live – although some only make these in the later larval stages – and the free-living caseless forms. Variously-shaped nets act as snares to catch swimming, drifting or walking live prey, while they function as sieves to trap small food-sized particles in others, such as *Hydropsyche*. From an ecological viewpoint web-spinners like *Polycentropus* can be thought of as underwater spider-substitutes (there are almost no aquatic spiders) while active predators like *Rhyacophila* feed on a wide variety of other invertebrate prey. The case-makers include different species of *Glossosoma* and *Agapetus* and others

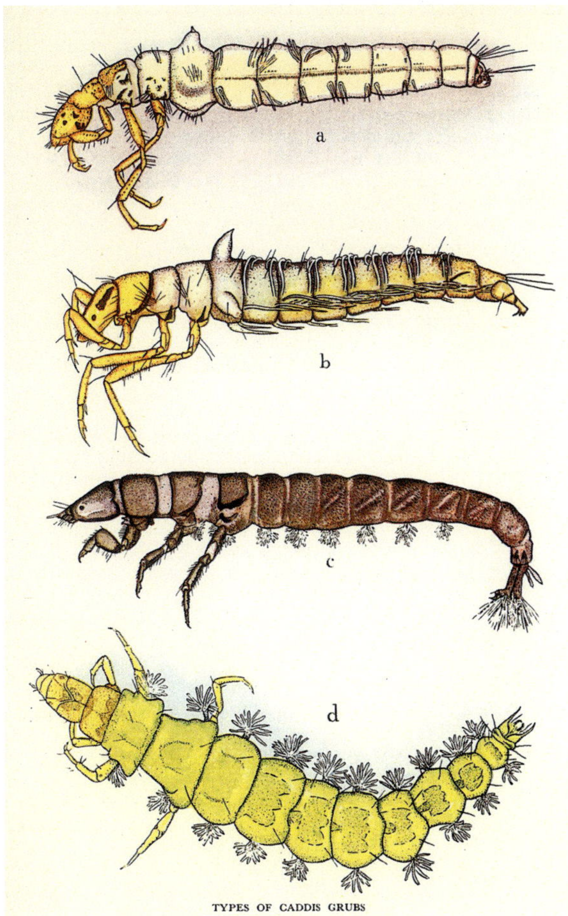

Fig. 7.62 Trichoptera.
Caddis larvae: (a) *Anabolia*, (b) *Phryganea*, (c) *Hydropsyche*, (d) *Rhyacophila* (Norman Hickin, Caddis (1952)

Fig. 7.63 Larva and adult of *Philopotamus* (Bryan Kennedy).

from the families Hydrophilidae, Limnephilidae, Lepidostomatidae, Phryganaeidae and Sericostomatidae.

Just twelve species of caddiflies were recorded on the island during the Old Survey. The group was studied by Sharon Fahy of NUIG as part of her work on the island's aquatic insects in the early 2000s, and by Ken Bond of UCC some years later. Between them they have recorded 38 species of caddisflies. Of the 20 species identified by Bond, only 9 were on Fahy's list of 27. This shows how incomplete all lists are likely to be – even now! Four of the species names in Halbert's list are absent from the modern total.

Thirty-eight species is 25.5% of the Irish total of 149, which is relatively high for an island of Clare Island's size, in contrast to the diversity of mayflies and stoneflies. Although Bond remarks that most of these species are generally widespread, it is intriguing that Claude Morley in his account of the Hymenoptera for the Old Survey, having rather dismissively stated that 'the fauna [in general] bore none of that individual stamp one so confidently anticipated in so distinct a locality,' singled out caddisflies as the exception: 'The only novelties I noticed were among the Trichoptera of the mountain streams.'[244] Unfortunately, he added nothing further by way of explanation. However, it is worth remembering the extent to which our knowledge of caddisflies is based on adults drawn to light and captured, but that the brief adult moment tells us little about the underwater world from which it has come, and in which it has spent most of its life. One wonders whether Morley's remark is based on observation of 'the mountain streams' to which he refers.[245]

Order Hemiptera (bugs)

Many of the less common species of Hemiptera found on the island (see above) are species characteristic of bog pools: *Gerris costai* (a pond skater) *Nononecta furcata* (a water-boatman), *Corixa scotti* and *Corixa Bonsdorffi* (lesser water boatmen).[246]

Order Coleoptera (water beetles)

In preparing his report for the Original Survey Frank Balfour-Browne paid two visits to the Clare Island district, but several others also collected water beetles for him. Between them they identified 86 species, 'a number which compares not unfavourably with that for other Irish counties which have been worked'.[247] Overall there was nothing peculiar about this fauna, certainly nothing that set it apart from the west of Ireland as a whole.[248] Forty-seven species were found on Clare Island itself, which he explored with particular care. He divided the island into three or four 'main classes of habitat', each characterised by a distinctive assemblage of species. By far the most important in terms of the ground it occupied were the peaty pools and drains on and around Knockmore and Knocknaveen; then there were open lakes and ponds such as Creggan and Loughanaphuca, the small area of freshwater marsh in the area of farmland, and finally the several small streams on the island: this last a habitat which unfortunately he entirely neglected. The cultivated area had 34 of the 47 species found on the island, including 11 that occurred only here and a further four that occurred mainly here.

Aquatic beetles were surveyed for the New Survey by Stephen McCormack and Kieran McCarthy, who recorded 76 species in 2002, bringing the total number of aquatic species to 87. Twenty-six of their species were new records for the island, but eleven of those on the older list were not seen this time.[249] The changes are attributed by McCormack and McCarthy to a variety of possible causes: taxonomic change (a number of Balfour-Browne's species have been split into several species), differences in sample strategy and timing, habitat change and the relative rarity of many species.

The island's complement of water beetles contains a higher percentage of species with strongly-developed flying ability than that of Ireland as a whole, reflecting the challenge presented by a sea-crossing. A high percentage are predators, scavengers or detrivores, which may be an indication of the difficulty specialist herbivores experience establishing populations where their food plants may be scarce. On the other hand, the presence of several flightless species shows that the ability to fly is not essential to island colonisation. One of these is *Noterus clavicornis*, which was apparently absent at the time of the first survey but is now well established, suggesting that it has arrived (on foot) recently.[250]

The absence from Balfour-Browne's list of several species that are widely distributed today—and which as experienced a collector as he is unlikely to have missed—does suggest that they are recent arrivals. Many other species however occupy very specific ecological niches, and there is an unavoidable element of luck involved in their capture. A good example is *Octhebius lejolisii*, which is only found along a narrow belt close enough to the shore to receive intermittent but frequent seawater in the form of spray, but far enough above high tide to be sure of occasional flushings of its pool with freshwater. Balfour-Browne described it as common, 'on both sides of Clare Island, east and west', but it was not found during the recent survey. A more puzzling absentee however is *Hydroporus palustris*, which was widespread in 1909–10 but not seen at all in 2002.

Stephen McCormack also looked at Poirtín Fuinch in greater detail in an attempt to correlate differences in water beetle assemblages with different vegetation types: a pioneering attempt to give more quantitative expression to Balfour Browne's observation that habitat names such as 'marsh' or 'peat-moss' really cover a number of different habitats that grade into one another in such a way that only a very critical examination would reveal the differences.[251] With 27 species (including two recorded only here) Poirtín Fuinch is relatively rich in water beetles, a particularly good example of the peaty pools that are the main reservoir for water beetles on the island. In this regard it is interesting to note that the lakelet is a relatively new feature, absent from the Fair Plan and first edition of the Ordnance Survey six-inch map of 1838.

Fig. 7.64 Riffle beetles. *Elmis* (now *Esolus*) *paralellipedes* was common in moss in a rapid stream flowing from Knockmore. This tiny (1mm) riffle beetle is shiny black with reddish-brown underside and legs (Udo Schmidt (CC BY-SA 2.0)).

The vegetation was divided into four broad categories: 1. shallow *Sphagnum*-dominated vegetation; 2. vegetation rich in forbs; 3. floating vegetation in deeper water; and 4. exposed substrates with sparse vegetation. The first category had a distinctive fauna dominated by hydrophilid beetles, but the highest density and diversity occurred in water a few centimetres deep with abundant macrophytes, where the community included both sphagnum dwellers and species of open water.[252]

Freshwater fishes

With only 25 species, Ireland's freshwater fish fauna is relatively limited by comparison with mainland Britain and Europe (55 species are recorded for Britain and Ireland). Rather surprisingly, no account of freshwater fishes was included in the Original Survey, apart from George Farran's observation that small brown trout (*Salmo trutta*) and eels (*Anguilla anguilla*) occurred in some of the streams.[253] Sampling for the New Survey confirmed the depauperate nature of the freshwater fish fauna. Only four species were found, and of these one (three-spined

Fig. 7.65 European eel (*Anguilla anguilla*) (© David Pérez (DPC), Wikimedia Commons, License CC BY-SA 4.0).

stickleback, *Gasterosteus aculeatus*) occurs only in inter-tidal rock pools. The fourth species is flounder (*Platichthys flesus*), which has a restricted distribution and reaches a relatively small size. A total of 210 fish were studied at three locations: 30 brown trout, 30 flounder and 150 eels.[254]

Eels are widespread and native to the island, occurring in all the streams sampled. Brown trout are thought to have been introduced in the nineteenth century in the hope of establishing an angling facility in Lough Avullin, which at that time had a large area of open water but is now almost entirely marsh. Brown trout were abundant in the deep pools of the Mill River when Jackie O'Grady was growing up in the 1940s: 'I would rarely go home with less than a dozen trout, ranging from a quarter pound to three quarters pound in my bag. In recent years the small lake [Lough Avullin] into which the river flows has become overgrown with reeds and is no longer fishable, but the numerous streams that feed the river still hold a large number of small trout.'[255]

Careful attention was paid to the fish parasite fauna as part of the New Survey, which established its depauperate character, only thirteen species being recorded. All three host populations are characterised by low parasite species richness and the dominance of a single parasite species in each host species. The island's trout only carry 5 species of parasites, whereas mainland populations may carry as many as 20 or 30. This is due in large measure to the fact that Clare Island trout share the rivers with so few other fish species.[256] The fact that Clare island trout are resident, non-migratory forms has meant that they are not affected by the sea lice infestations that trouble sea trout in the vicinity of salmon farms.

All of the eels studied were in the yellow-phase foraging stage of their life history. They too carry exceedingly low parasite burdens, again due in large measure to the unusual circumstance that they share the rivers with so few other fish species; indeed, eel is the only fish in the Owenmore catchment. The flounder also has few parasite species and low parasite burdens, more than half of the fish sampled being entirely parasite-free, again possibly due to the insularity of the population.

Fig. 7.66 The lower course of the River of Rouing in the southwest corner of Clare Island (Bryan Kennedy).

Stream ecology

The rocky streams that cut deeply into the rocks of Knockmore and Knocknaveen flow through some of the island's loveliest locations, and many groups of plants and animals that live here were explored by workers on both the Original and New Surveys. At the conclusion of the New Survey some of these streams were assessed by Bryan Kennedy of the Environmental Protection Agency in order to arrive at a preliminary characterisation of their overall ecological condition.[257]

Two reaches of the Owenmore River were examined, along with stretches of the River of Rouing and the Dorreen River. The algae and invertebrates recorded in the Owenmore River and River of Rouing during the course of this survey are indicative of water of the highest quality. The long-term persistence of conspicuous cyanobacteria like *Dichothrix orsiniana*, *Rivularia beccariana* and occasional *Nostoc verrucosum* – documented by Brian Whitton – is evidence of this.[258] The morphology and physiology of these heterocystous forms (characterised by having nitrogen-fixing cells), with their well-developed tapering multicellular hairs, are typical of watercourses deficient in nitrate (NO_3-N) for part of the year, and in phosphorus (PO_4-P) for the remainder of the time. Epilithic diatom assemblages in the streams were found to be dominated by nutrient sensitive taxa— *Achnanthidium minutissimum*, *Platessa oblongella* and *Diatoma moniliformis*. Tolerant diatoms were absent while those with an intermediate sensitivity (*Gomphonema parvulum* and *Cocconeis euglypta*) were of secondary importance. Comparable assemblages were found in the lower reaches of the Owenmore, although specialist cyanobacteria were less well established in the area examined here. While opportunistic green algae (*Microspora* and *Oedogonium*) and *Phormidium autumnale* suggest some differences in the nutrient balance, their biomass remained low. Sensitive invertebrate indicator taxa like *Rhithrogena semicolorata* dominated the aquatic insect community at all locations, and *Electrogena lateralis* was also present, and especially prevalent in headwater streams.

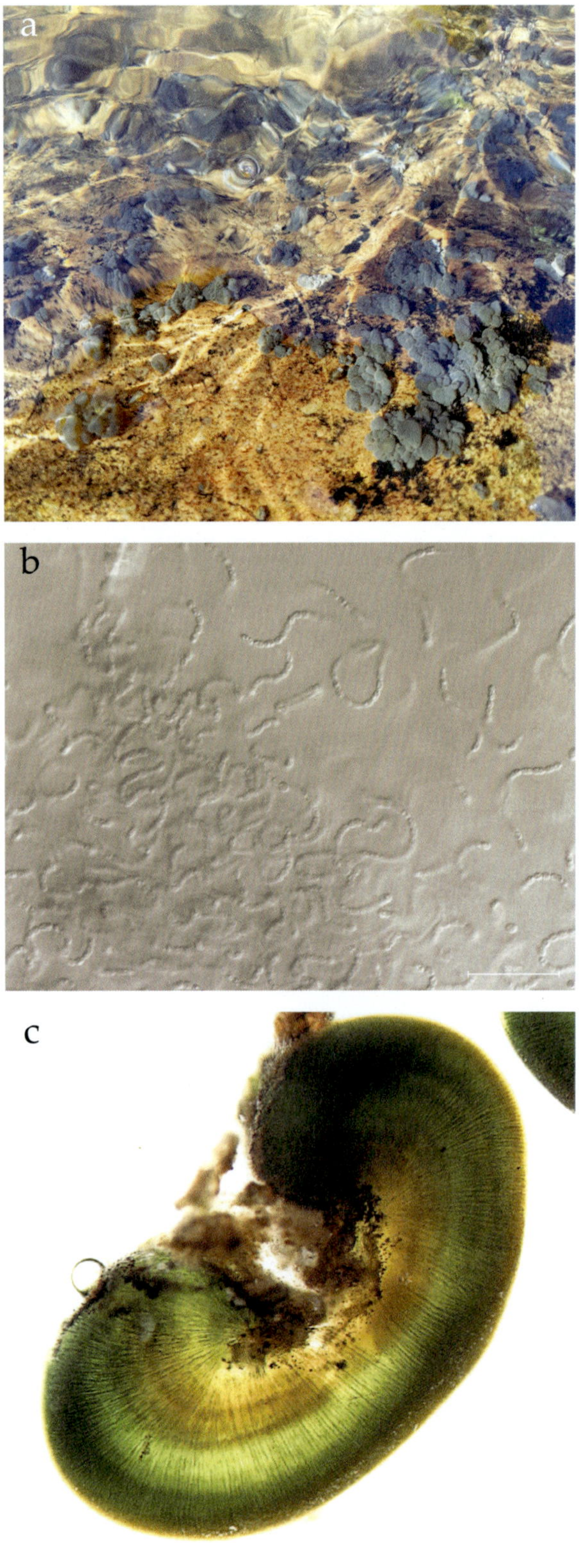

Fig. 7.67

(a) Macroscopic colonies of *Nostoc verrucosum* on bedrock in the Owenmore River.

(b) Microscopic view of a thin section of the same material.

(c) Section through one of the spherical colonies of cyanobacteria shows abundant filaments with small rounded cells at their base (these are the heterocysts), ending in long, tapering hairs at the colony edge (Bryan Kennedy).

Fig. 7.68 Characteristic flowing tress-like tufts of *Lemanea* in the Dorreen River (Bryan Kennedy).

Cased and caseless caddisflies like *Philopotamus* were common while invertebrates with a tolerance to pollution were almost absent.

Extensive ferric-hydroxide deposits and the iron bacteria *Lepthothrix ochracea* were conspicuous along the lower course of the Dorreen River, which cuts its way through the very different rocks of the Ballytuohy Formation. Red algae replace cyanobacteria as the most conspicuous macroalgae: *Batrachospermum*-like algae in the headwaters and *Lemanea fucina* (easily recognised by its resemblance to turfs of coarse hair attached to rocks) succeeding it further downstream. The green alga *Microspora* had an intermittently high cover in sections of the river, but benthic assemblages continued to be dominated by relatively sensitive diatoms. The flattened mayflies (family Heptageniidae) characteristic of fast-flowing streams – and which are so abundant in the Knockmore streams – were uncommon here, where the fauna was dominated by the swimming mayflies *Baetis rhodani*, *Baetis rhodani* and *Alaintes* (*Baetis*) *muticus*. More trophic-tolerant invertebrates were not well represented in the fauna.

Among the outstanding character of the western streams on Clare Island is the extraordinary profusion of colonial cyanobacteria: at a level of diversity and abundance that has few parallels anywhere else in Ireland, the maintenance of which will be an important consideration in the management of the island's biodiversity. Further investigation in this area is planned. Future surveys to establish the distribution of stoneflies and *Ecdyonurus* in the island's streams will enable their reference assemblages to be characterised in greater detail.

Notes

1 Atherton *et al.* (eds.) 2010.

2 Lett 1911, OS 11–12, 1–18.

3 Synnott 2013.

4 Synnott 2013, 212.

5 *Ibid.* An annotated check-list is included in Donal Synnott's account for the New Survey (214–231).

6 Synnott 2013, 214.

7 Smith 1911, OS 11–12, 3. *Mnium hornum* is common in some places on the island, but so are many other species.

8 Smith 1911, OS 11–12, 3–4.

9 Smith 1911, OS 11-12, 5. Lough Curraun is in the centre of the Curraun peninsula on the opposite side of Clew Bay.

10 Blockeel 1995.

11 Praeger 1903; Praeger 1911, OS 10, 2.

12 Lysaght 1998, 68–75.

13 Praeger 1911, OS 10, 3.

14 Jebb 2013, 179–208; Ryle 2013, 27–177.

15 Ryle 2013, 165.

16 Forbes 1914, OS 9, 2.

17 Mary Ann Gallagher in Burns 2001, 42.

18 Ordnance Survey Name Books, 30b.

19 *Doirln, Leathchoill, Foithir na Craoibhe.* Possibly also *Derraghyemon, Derraghgarriff.*

20 Praeger 1950, 114.

21 Praeger 1911, OS 10, 20.

22 Ryle 2013, 163.

23 Coxon and O'Connell 1994.

24 Ryle 2013, 83.

25 Cf White and Doyle 1982, 338.

26 Praeger 1913.

27 Praeger 1903, 284.

28 Ryle 2013, 65–66.

29 Praeger 1911, OS 10, 30.

30 Doyle and Foss 1986; Jebb 2013, 181.

31 Jebb 2013, 196.

32 Praeger 1911, OS 10, 30.

33 Praeger 1911, OS 31.

34 Westropp 1911, OS 2, 68.

35 OS 13, 12. *Candelospora ilicicola* is now considered a synonym of *Calonectria ilicicola* Boedjin and Reitsma 1950.

36 Feehan and McHugh 1992.

37 Oster 2008.

38 However, two species—one of the waxcaps and an earth tongue—are now considered *nomina confusa.*

39 Mitchel 2013, 253–327. He was accompanied by Hubert Fuller, Jolanda Mitchel-Smit and Kieran Connolly.

40 Mitchel 2013, 258.

41 McHugh *et al.* 2002.

42 Vesterholt *et al.* 1999.

43 Dahlberg and Croneborg 2006.

44 www.britmycolsoc.org.uk.

45 Protoctists are nucleated micro-organisms (i.e. excluding plants, animals and fungi) that evolved by symbiotic integration of at least two different kinds of former free-living bacteria. Some 1,000 species of slime moulds have been described.

46 Lister 1912, OS 63, 2.

47 Lister 1912, OS 63, 3. *Margarita metallica* is now called *Calomyxa metallica*, and *Physarum nutans* is now *Physarum album.*

48 'It may be permissible, since the group is but little known, in the first place to rehearse briefly the chief facts of their life-history.' (OS 63, 1).

49 Lister 1912, OS 63, 3–4.

50 Lister 1912, OS 63, 6.

51 *Proc. R. Ir. Academy* 112B, 133–155.

52 *Bull. Ir. biogeog. Soc.* 1995 (No. 18), 132–5.

53 Seaward and Richardson 2013, 233.

54 Smith 1911, OS 14, 2.

55 Several other lichenologists were also involved in early survey work between 1909 and 1911: see Seaward and Richardson 2013, 234–235.

56 Eighteen species recorded during earlier surveys were not re-found during the visits in the 1990s.

57 Seaward and Richardson 2013, 233–251.

58 Smith 1911, OS 14, 3–4; modern names added in square brackets.

[59] Among the groups of insects that slipped through the review net of both the Original and New Surveys are the fleas (Siphonaptera), of which several dozen species undoubtedly occur. The only ones that have been formally recorded appear to be two common wood mouse flea species: *Typhloceras poppei poppei* and *Ctenophthalamus nobilis*.

[60] Southern 1913a, OS 48. 4.

[61] Southern 1913a, OS 48. 1.

[62] One of Southern's 14 lumbricids was not recorded on the Island itself.

[63] Now *Marionina cliarensis*.

[64] Southern 1913b, OS 48 13.

[65] Olaf Schmidt, *pers. comm.*

[66] Welch's contribution to the Survey was his superb collection of photographs, but he was also a leading authority on land and freshwater molluscs.

[67] Stelfox 1912, OS 23, 1.

[68] Colgan 1911, OS 22, 2.

[69] Colgan 1911, OS 22, 7.

[70] Stelfox 1912, OS 23, 2.

[71] Stelfox 1912, OS 23, 12.

[72] Stelfox 1912, OS 23, 9.

[73] When freshwater species are added the total for the island is 58.

[74] Stelfox 1912, OS 23, 13–14.

[75] Stelfox 1912, OS 23, 4.

[76] Stelfox 1912. OS 23, 15.

[77] Anderson (forthcoming). New Survey 9, *Zoology* (in preparation).

[78] *Arion hortensis, Tandonia budapensis, Limacus maculatus, Deroceras invadens.*

[79] Similarly, it is not known which of its segregate species the *Columella edentula* of 1911 refers to. There are several other examples.

[80] Anderson (forthcoming). New Survey 9, *Zoology* (in preparation).

[81] Pack-Beresford 1911, OS 35, 1.

[82] *Ibid.*

[83] Pack-Beresford 1911, OS 35, 2.

[84] Pack-Beresford collected only one specimen of *Argyroneta*; several were collected during the resurvey of 2002, but many more were seen in the pools on the Island.

[85] The woodlouse spider is one of the handful of species that escaped detection during the New Survey!

[86] Nolan (forthcoming). New Survey 9, *Terrestrial Zoology* (in preparation).

[87] *Ibid. Hypomma bituberculatum* is the most abundant spider on the island (accounting for nearly 11% of the total catch).

[88] Nolan (forthcoming). New Survey 9, *Zoology* (in preparation).

[89] Pack-Beresford 1911, OS 35, 5.

[90] *Opilio* in Latin is a sheep master. The name comes from a time when European shepherds used stilts to keep an eye of their scattered flocks.

[91] Wallis Kew 1911, OS 38.

[92] Wallis Kew 1911, OS 38, 2.

[93] Cawley (forthcoming), New Survey 10, *Birds* (in preparation).

[94] Halbert 1915, OS 39 ii, 48.

[95] Order Mesostigmata in modern taxonomic terminology.

[96] Ticks are a family of mites (Ixodidae) in the suborder Ixodoidea.

[97] Order Astigmata.

[98] Halbert 1915, OS 39 ii, 108. Several species of *Rhombognathus* occur in these pools.

[99] Order Cryptostigmata. Sclerotin is an insoluble protein that hardens the chitin of which the skeleton of arthropods is made.

[100] Order Prostigmata.

[101] Halbert 1915, OS 39 ii, 47.

[102] Halbert 1915, OS 39 ii, 48.

[103] A recent list of the acarine fauna of sphagnum in Britain and Ireland has 65 species (Hingley 1993).

[104] Bolger (forthcoming). New Survey 9, *Zoology* (in preparation).

[105] Foster 1912, OS 44, 3.

[106] Foster 1912, OS 44, 2.

[107] Johnson 1912, OS 33, 1.

[108] Cawley (forthcoming). New Survey 9, *Zoology* (in preparation).

[109] *Lamcytes emarginatus* is Johnson's *Lamcytes fulvicornis*, and *Hydroschendyla submarina* his *Schendyla marina*.

[110] Johnson 1912, OS 33, 3.

[111] *Stigmogaster subterranea* and *Geophilus insculptus*.

[112] *'Natura in minimis maxime miranda.'*

[113] Most are between 100 and 500 microns.

[114] Tudge 2000.

[115] Murray 1911, OS 37. For a brief account of Murray's extraordinary life see DeMilio *et al.*, (forthcoming), New Survey 9, *Terrestrial Zoology* (in preparation).

[116] In fact, it belongs to a complex of cryptic species,

all very similar. *Macrobiotus richtersii sensu stricto* is not known for Ireland.

117 Murray 1911, OS 37, 13.

118 De Milio *et al.* (forthcoming). New Survey 9, *Zoology* (in preparation).

119 *Ibid.*

120 Carpenter 1913, OS 32, 11.

121 Bolger (forthcoming). New Survey 9, *Zoology* (in preparation).

122 Carpenter OS 31, 1–4.

123 Carpenter OS 31, 1.

124 Halbert 1912, OS 30, 1–2.

125 Halbert 1912, OS 30, 6-7. Although it was not found on Clare Island it may well occur here.

126 Johnson and Halbert 1912, OS 28, 1.

127 Johnson and Halbert 1912, OS 28, 3.

128 Should translation be necessary: *in stercore bovino* = in cow dung; *in stercore ovino* = in sheep dung.

129 Johnson and Halbert 1912, OS 28, 2. There is every likelihood that the taxonomy has changed. It would be interesting to see how the situation regarding dung beetles has changed with the change in stocking regime. Twelve *Aphodius* species were recorded on Clare Island.

130 Anderson and McCormack (forthcoming). New Survey 9, *Zoology* (in preparation).

131 Greer collected in June 1910, Johnson in August 1910 and May 1911.

132 De Vismes Kane, W.F. 1912, OS 26, 1–2.

133 Bond (forthcoming). New Survey 9, *Zoology* (in preparation).

134 Noctuids (Noctuidae) and geometrids (Geometridae) are the two largest families of moths. The overwhelming majority of noctuids fly at night and are almost invariably strongly attracted to light. Some of the family are preyed upon by bats, but many species have tiny organs in their ears that respond to bat echolocation calls, sending their wing muscles into spasm and causing the moths to dart erratically.

135 Very few micromoths have common names in English.

136 Bond (forthcoming). New Survey 9, *Zoology* (in preparation).

137 Morley 1911, OS 24, 1.

138 Morley 1911, OS 24, 2. Many of Morley's species names are now obsolete. *Mesoleius maculicollis* is now *Campodorus maculicollis*, *Cremastus albipennis* is *Temelucha albipennis* and *Rhogas bicolor* is *Eucystomastax bicolor*.

139 Morley 1911, OS 24, 4.

140 Breen, Paxton and O'Grady (forthcoming). New Survey 9, *Zoology* (in preparation).

141 *Ibid.*

142 *Ibid.*

143 Grimshaw 1912, OS 25, 2.

144 Recorded at two mainland stations, and probably generally distributed.

145 Grimshaw 1912. He included them in his 'Tendipedidae'.

146 Murray (forthcoming). New Survey 9, *Zoology* (in preparation).

147 Murray (2004), 1—2.

148 Murray (forthcoming). New Survey 9, *Zoology* (in preparation); Murray 2004; Murray and Murray 2003.

149 Grimshaw 1912, OS 25, 1.

150 Ashe, O'Connor and Chandler 2012.

151 *Serromyia femorata*, *Culicoides pulicaris*, *Palpomyia flavipes* and *Atrichopogon rostratus* (as *Helea rostrata*); the one most commonly encountered is probably *C. pulicaris*, which 'has elegantly spotted wings, and is easily recognized' (Grimshaw, 9).

152 Grimshaw 1912, OS 25, 29–30; McDonnell *et al.* 2018.

153 Ussher 1912, OS 20, 1.

154 Ussher 1912, OS 20, 5.

155 Lloyd 1982.

156 The various expeditions are discussed in detail by Thomas Kelly in Volume 10 of the New Survey (in preparation), where the difficulties of comparison between them are also analysed.

157 Kelly and Keating (forthcoming). New Survey 10, *Birds* (in preparation). The study emphasises that the results should be regarded as preliminary rather than definitive because of the various factors whose influence cannot be estimated with certainty.

158 Ussher 1912, OS 20, 19.

159 Kelly and Keating (forthcoming). New Survey 10, *Birds* (in preparation).

160 Kelly, Keating, Coombes and Smiddy (forthcoming). New Survey 10, *Birds* (in preparation).

161 Such as such as Long-tailed Skua (*Stercorarius longicaudus*), Sabines (*Larus sabini*) and Little Gull (*Larus minutus*), Whinchat (*Saxicola rupestra*), Barred Warbler (*Sylvia nisoria*) and Wood Warbler (*Phylloscopus sibilatrix*).

162 Kelly, Keating, Coombes and Smiddy (forthcoming). New Survey 10, *Birds* (in preparation).

[163] *Ibid*. Gulls, crows and rats are probably important predators however; exceptionally inclement weather during the nesting season, and occasional trampling by livestock, may also be significant factors.

[164] Praeger 1930, 20.

[165] Ussher 1912, OS 20, 41–43.

[166] Liam Winters, in Crosby 2004, 22–23.

[167] Cussen, McGreal, Warner and Kelly (forthcoming). New Survey 10, *Birds* (in preparation). This figure of 9 excludes the Arctic Tern, which was never seen on the island again. Since the Original Survey 6 new species have arrived on the island, of which 1 is no longer present. This gives a total of 15 breeding seabirds in the period 1909-1917. The addition of 6 new species since the Original Survey is an increase in species richness of 66.7%, but if the calculation is based on a figure of 8 new species (i.e. if we assume that the Arctic Tern disappeared before the arrival of the Fulmar) then the increase is in the order of 75%.

[168] *Ibid*.

[169] *Ibid*.

[170] Newton, Harris and Murray 2015.

[171] 510 adults at almost 200 discernible nests, with 142 nearly or fully-fledged young.

[172] Cussen, McGreal, Warner and Kelly (forthcoming). New Survey 10, *Birds* (in preparation).

[173] *Ibid*.

[174] The last formal record (in the *Irish Naturalists' Journal*) seems to have been on Knockmore in 1891.

[175] Quoted in D'Arcy 1999, 64.

[176] This was probably a young bird of Scottish origin (Praeger 1939, 197).

[177] O'Grady 2008, 101. The skin is preserved in the Natural History Museum (NMINH: 1936.2.2).

[178] Ussher 1912, OS 20, 6–7.

[179] Corncrake not heard since 1988; a single calling bird was heard in 2002.

[180] Ussher 1912, OS 20, 15.

[181] Ussher 1912, OS 20.

[182] O'Grady 2008, 101.

[183] Kelly and Smiddy (forthcoming). New Survey 10, *Birds* (in preparation).

[184] *Ibid*.

[185] Kelly, Smiddy and Keating (forthcoming). New Survey 10, *Birds* (in preparation).

[186] Burns 2001; within living memory seal oil was treasured for its medicinal value.

[187] Praeger 1930, 29.

[188] Barrett-Hamilton 1912, OS 17, 7.

[189] Lawton and McAney (forthcoming). New Survey 9, *Zoology* (in preparation).

[190] *Ibid*.

[191] *Ibid*.

[192] Scharff 1912, OS 18.

[193] Guiry *et al*. 2007.

[194] Lough Merrignagh, a pond on Toormore and the ponds that formed Loughanaphuca are either much reduced or are just wet hollows.

[195] John 2007b, 202.

[196] Lett 1912, OS 11–12, 3.

[197] Loughanaphuca was further west of this pond, and had filled in by the 1830s.

[198] John 2007c, 21.

[199] John 2007c, 22.

[200] Williamson 2007, 27–65.

[201] The Wests wrote five monographs on British desmids that were standard reference works worldwide.

[202] Sims 2007, 115.

[203] Sims 2007, 139.

[204] Bryant 2007.

[205] There are photosynthetic bacteria that do have chlorophyll-like pigments; they usually utilise hydrogen sulphide and produce sulphur.

[206] Rindi and Guiry 2007, 179–192.

[207] West 1912, OS 16, 46–49.

[208] Dunkerly 1913, OS 61–62, 20.

[209] Wailes and Penard 1911, OS 65. This is not a joint report; they kept their contributions separate.

[210] Wailes and Penard 1911, OS 65, 3.

[211] Wailes and Penard 1911, OS 65, 1.

[212] Wailes and Penard 1911, OS 65, 3, 10.

[213] Wailes and Penard 1911, OS 65, 9.

[214] Stephens 1912b, OS 60, 11.

[215] Southern 1912, OS 56, 13.

[216] Southern 1912, OS 56, 2–3.

[217] In Greek mythology, naiads were lesser deities that presided over rivers, springs, wells and fountains; they were generally represented as beautiful young virgins, often depicted leaning on an urn from which a stream of water flowed.

[218] Southern 1913b, OS 50, 5.

[219] Southern 1913b, 0S 50, 5–6.

[220] See the account of marine roundworms on page 110.

[221] Rousellet 1911, OS 51, 8.

[222] Rousellet 1911, OS 51, 2.

[223] Rousellet 1911, OS 51, 3.

[224] Including mosses collected by schoolchildren around Westport, in which he found 26 species.

[225] Murray 1911, OS 52, 7.

[226] Entomostraca is a somewhat old-fashioned name used to collectively describe the smaller crustaceans (as distinct from the Malacostraca). Cirripedia are the barnacles, which we meet on page 107; and the Malacostraca are the amphipods, isopods and decapods met on page 105. What distinguishes crustaceans from other arthropods is that they have *two* pairs of antennae.

[227] Ward and Whipple 1918, 683.

[228] Scourfield 1912, OS 46.

[229] Among the new Irish species recorded on the island were *Cyclops bisetosus* and *Canthocamptus* (*Attheyella*) *hirticornis* in Kinnacorra marsh, and *Cyclops rubellus* in several of the loughs; *Canthocamptus* (*Attheyella*) *crassus* was found in Creggan Lough and in a little lough close to Craigmore; *Nitocra spinipes* was living in little pools not far above high-water mark on the north-east coast.

[230] Scourfield 1912, OS 46, 14. Scourfield's name for the species was *Canthocamptus praegeri*. The accepted name is now *Bryocamptus* (*Echinocamptus*) *praegeri*.

[231] Scourfield 1912, OS 46, 3.

[232] Scourfield 1912, OS 46, 10.

[233] De Eyto *et al.* (forthcoming). New Survey 9, *Zoology* (in preparation).

[234] *Ibid.*

[235] Walsh-Kemmis (forthcoming). New Survey 9, *Zoology* (in preparation).

[236] *Ibid.*

[237] The mites that live in the sea belong to a different family, the Halicaridae.

[238] Halbert 1911, OS 39i, 1.

[239] Halbert 1912, OS 27, 1.

[240] Halbert 1912, OS 27, 2.

[241] In Greek mythology the naiads were nymphs of wells, springs, rivers and fountains. We met the nereids—the sea nymphs—in Chapter 6

[242] Kelly-Quinn and Bracken 2000, 17–18. Halbert's three species were *Baetis binoculatus*, *Baetis pumilus* and *Ecdynurus lateralis*. *B. fuscatus* is Halbert's *B. binoculatus*; Halbert's *B. pumulus* is now *B. muticus*; and his *Ecdynurus laeralis* is now *Electrogena* (until recently *Heptagenia*) *lateralis*.

[243] Halbert's *Isopteryx torrentium*.

[244] Morley 1911, OS 24, 1.

[245] O'Connor's Atlas (2015) maps 35 caddisfly species on Clare Island. All but one of these are reported as 'widespread' in Ireland, or 'widespread and common.' Just one species, *Oxyethira sagittifera*, taken as an adult on the island, is classified as a local species.

[246] *Nononecta furcata* is now *N. obliqua*, and *Corixa Bonsdorffi* is *Cymatia Bonsdorffi*.

[247] Balfour Browne 1912, OS 29, 1. He actually recorded 90 species, but taxonomic changes have reduced this to 86.

[248] The same mixture of northern and southern forms evident in other groups can also be discerned in the water beetles, and is discussed in much detail (OS 29, 13-20).

[249] Two of these 76 species have since been split.

[250] McCormack and McCarthy (forthcoming). New Survey 9, *Zoology* (in preparation).

[251] Balfour Browne 1912, OS 29, 6, 12.

[252] McCormack and McCarthy (forthcoming). New Survey 9, *Zoology* (in preparation).

[253] Farran 1912, OS 19, 6.

[254] McCarthy, Creed and Lenihan (forthcoming). New Survey 9, *Zoology* (in preparation).

[255] O'Grady 2008, 102–103.

[256] The nematode *Raphidascaris acus* was the most common parasite of trout.

[257] The section that follows is taken with little change from Kennedy (2019).

[258] Whitton 2002, 2007.

Overleaf

Fig. 8.1 Farming in Toormore during the Bronze Age. The farm is viewed from above, from the edge of a cliff that overlooks the scene from the east. What remains today of the farmstead at the centre is seen in Figure 8.8 (page 235) (Painting Philip Armstrong).

Chapter 8
Archaeology and early history

No aspect of the Clare Island landscape has received greater attention than the archaeology. For the Original Survey Praeger had recruited the services of Thomas Johnson Westropp (1860–1922), one of the leading archaeologists of the day, and he was assisted by George J. Fogerty. Engaged as he was at this time with the spectacular monuments of north Kerry, north-west Clare and the Aran Islands, 'the subject at first seemed to me to be a field likely to yield but little result' he wrote; but 'the task of recording the scanty remains [which at first] seemed a light one, ... became very different in its aspect after three visits'.[1]

Fig. 8.2 Thomas Johnson Westropp.

Between 1992 and 1996 teams of archaeologists carried out a comprehensive study of the island under the direction of Paul Gosling of Galway-Mayo Institute of Technology for the New Survey; members of the island community also played a key role in the identification and recording of monuments. This was complemented by further shorter periods of field work and a programme of small-scale archaeological excavations between 1996 and 2006. All of this work resulted in a seven-fold increase in the number of features of archaeo-

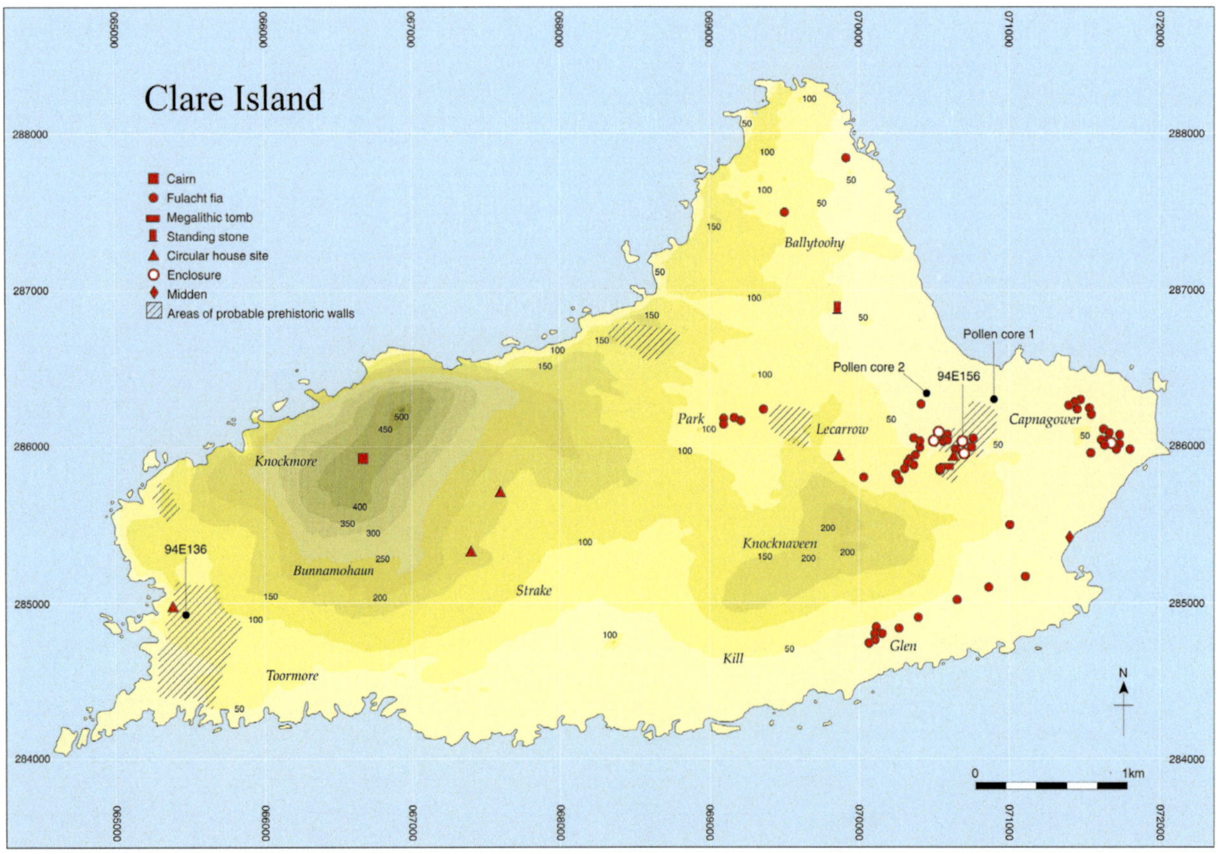

Fig. 8.3 Map showing the distribution of prehistoric settlement (pre-600 BC) on Clare Island. (The inclusion of the houses is speculative). Pollen core 1 marks the location of Pete Coxon's palaeobotanical investigations in Lecarrow (Coxon 1982, 1987); Pollen core 2 marks the location of the palaeobotanical investigations at Lough Avullin (Coxon *et al.* 2013); see page 66. (*New Survey* 5, 32).

Fig. 8.4a Overview of the megalithic court tomb at Lecarrow.

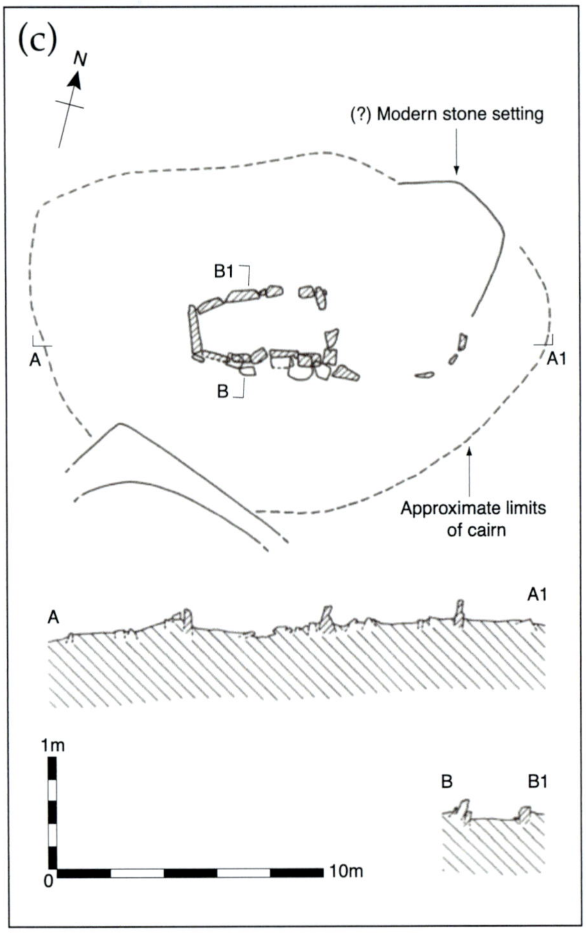

Fig. 8.4 (b) View of the court tomb looking eastwards toward Clew Bay; Poirtín Fuinch is to the right; (c) plan and section of the court tomb (*New Survey* 5, 102).

logical significance recorded (37 to 256).[2] The huge difference in the numbers of monuments recorded by the two surveys is in part due to the much longer time spent by teams of archaeologists for the New Survey, in part to the fact that the great majority of newly-discovered sites have little or no surface expression: moreover, many of the monuments and artefacts recorded during the New Survey were not recognised as antiquities in Westropp's time.[3] Westropp devoted most of his time and attention to the Abbey and the promontory forts.

Its closeness to the mainland and ease of approach on its sheltered eastern side, fertile soils and rich surrounding seas, would have made Clare Island attractive to people from the earliest times; and although we have as yet no artefactual evidence of a Mesolithic presence on the Island (8000 and 4000 BC), the vegetative record does hint at human influence as early as *c.* 5000 BC. In the pollen record preserved in the lake at Poirtín Fuinch to the east of Knocknaveen (at the eastern end of Lecarrow townland), a band of charcoal dated to around 7,500 years ago records clearance of forest by burning, strongly hinting at a Mesolithic presence.[4]

In a landscape studded with great blocks of stone to the extent Clare Island is, it is inevitable that some natural groupings of rocks may be taken to be more than they really are. When Westropp visited the island for the Clare Island Survey he was told about a 'Giant's Grave'—a supposed dolmen—on a slope in Capnagower, but he could find no trace of it. Nearby, however, there was 'a row of large blocks wonderfully like such a monument', but this was entirely natural. At the head of the valley in which Loughanaphuca lies he came on a cist-like cairn of flat stones 'but they formed no definite chamber'. On the shoulder above 'the curious rock-pillar of Budawanna', he described 'an irregular loop of small set stones, the age and purpose alike doubtful'.[5]

Not surprisingly then, it was with some diffidence that Ciara Cullen and Peter Gill drew the attention of archaeologist Paul Gosling to an assemblage of rocks at Poirtín Fuinch to the east of Knocknaveen they thought must be 'something', during the First Clare Island Symposium in 1989. To everyone's astonishment it proved to be a court tomb, establishing beyond all doubt the presence of a settled community on the island some time during the early and/or middle Neolithic (4000–3100 BC).[6] The court tomb was the focal point of this culture, as a church or cathedral might be for ours, with a stunning view across the bay to the north and east, with Croagh Patrick on the horizon. The presence of comparable court tombs on the mainland attests to similar communities here.[7]

These people were farmers, albeit with little in the way of enduring material possessions, their homes and the tools they used all made of stone or wood. Nothing of these has survived on Clare Island, but what has survived are the walls that enclosed the fields in which they grew their corn, now hidden among the field walls of later times. These are especially striking in the eastern part of the island where the court tomb is, and at the extreme western end in Bunnamohaun. In those distant Neolithic days the environment was probably less harsh and exposed than it is today because the island was extensively wooded, with an annual temperature several degrees centigrade higher than it is today.

Although the pollen record suggests a continuous human presence in the later Neolithic, there appears to be nothing in the archaeological record for the interval of time between 3000 and 2100 BC. But there is an abundance of monuments (as well as several artefacts) from the second millennium BC, nearly all of them concentrated in the eastern half of the Island. To date, well over fifty fulachta fiadh have been recorded, many of them in distinct topographical groups.[8] Fulachta fiadh are mounds—typically lunate in plan—of charcoal-blackened soil enclosing fragments of fire-shattered stone: all that remains of the hearths at which a prehistoric community kindled fires and boiled water for their everyday needs. Most of those on Clare Island are between 5 and 11m across and between half and one metre high; the largest measures 12m across and 1.7m high. Among the most accessible well-preserved examples are several not far from the court tomb. Dating evidence suggests they were in use for almost a millennium (*c.* 2000–1100 BC): the lives of 30 human generations. The burnt stone in the mounds is nearly always sandstone, because it is less likely to splinter than most other kinds of rock. The only artefacts that have been recorded to date are a stone maul found near the largest of the fulachta fiadh; as well as a barbed and tanged chert arrowhead from a midden at the harbour. Westropp also recorded a a bronze spearhead (now lost) from the western foot of Knocknaveen.

In many places the fulachta fiadh are associated with settlement features located on drier ground close by, and in some cases the traces of old field walls survive. Over sixty prehistoric homes and hut sites have been located, at least some of which were circular houses around 8m in diameter and enclosed within a ring of stone. Although we have no real evidence to support such a figure, 'a speculative mind might risk a population estimate no greater than *c.* 100 to 150 souls' over the many centuries of the Bronze Age.[9] The distribution of the

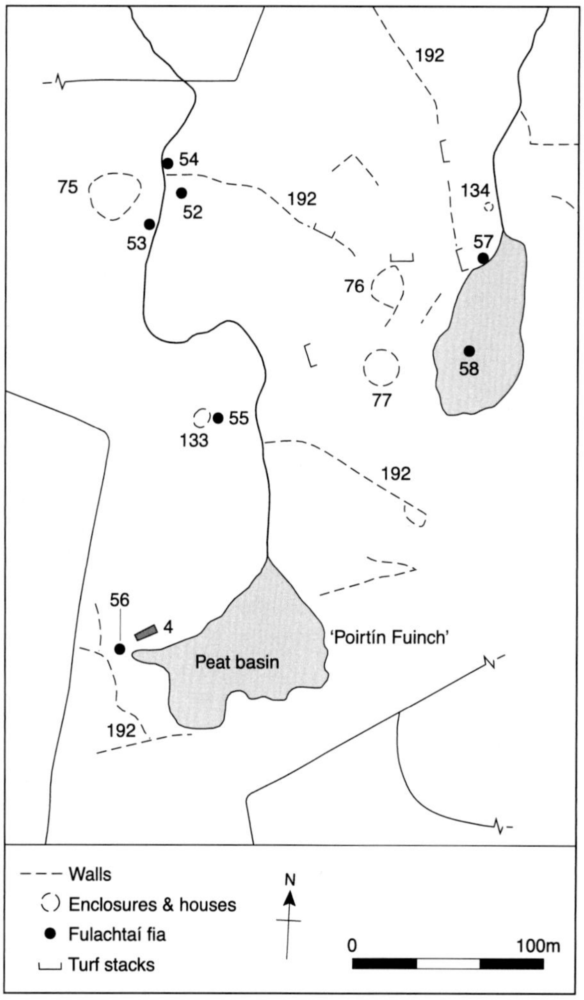

Fig. 8.5 Map showing the locations and spatial relationships between prehistoric monuments in Lecarrow townland on the eastern commonages. These include a megalithic tomb (Cat. No. 4), a number of fulachta fiadh (Cat. Nos 52–58), three enclosures (Cat. Nos 75–77), two houses (Cat. Nos 133–134) and a field system (Cat. No. 192) (*New Survey* 5, page 73).

fulachta fiadh is generally thought to reflect the pattern of prehistoric settlement in Clare Island, although it is possible that they are absent from other parts of the Island because of later intensive land use, especially in the areas of spade cultivation. As it is, only half of the monuments are still relatively intact, the remainder having no surface expression. The availability of sandstone is not likely to have been a factor, but proximity to suitable water bodies may well have been. The clustering of fulachta fiadh along suitable streams is striking even today; close to half are by the banks of streams (for example, twelve of the nineteen monuments in the Lecarrow Group are situated along the upper reaches of what the Ordnance Survey six-inch map calls the Pollbrandy stream. It is less obvious that fulachta fiadh later covered by peat are close to

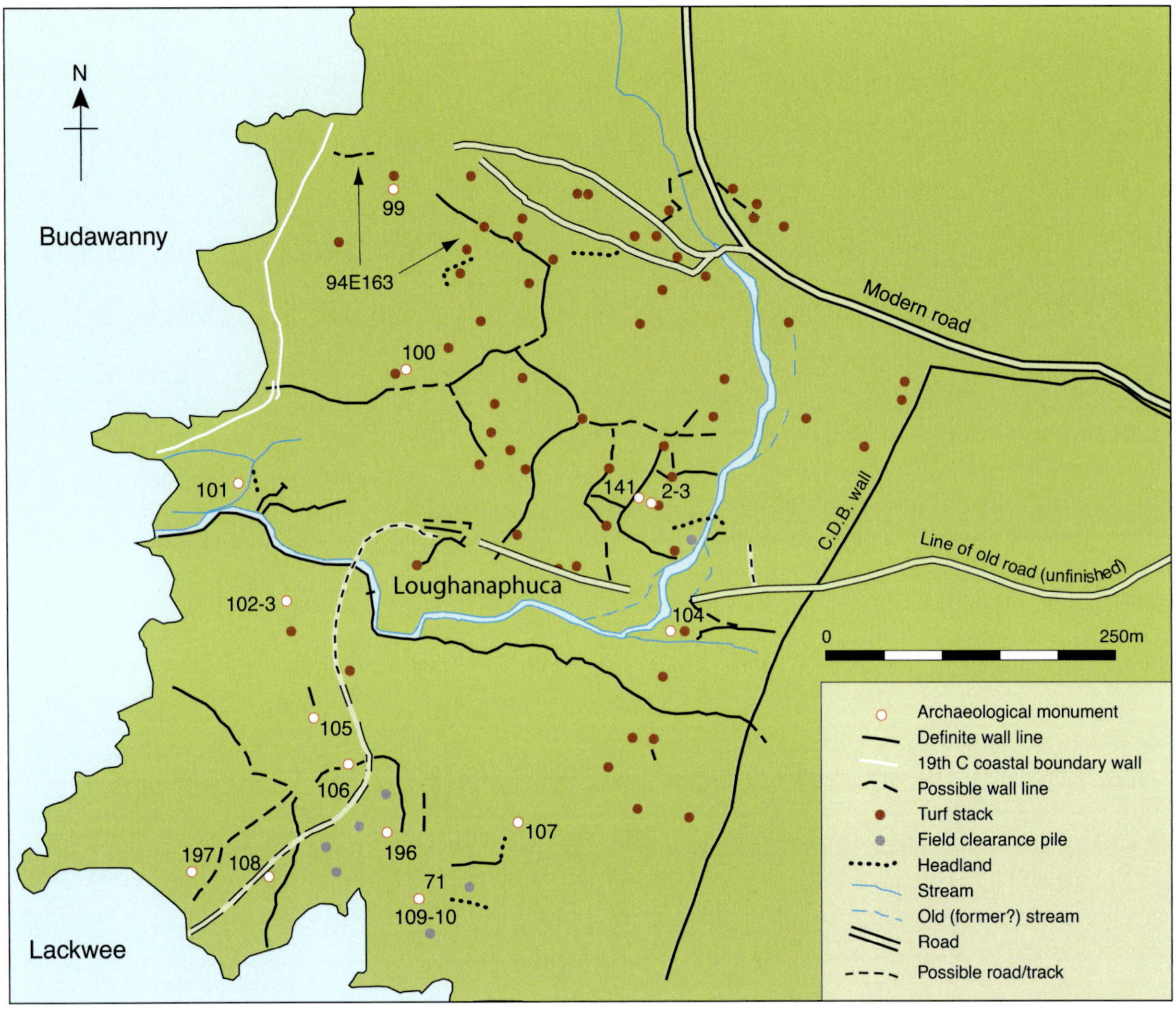

Fig. 8.6 Sketch map of the boundaries (Cat. No. 183) and associated features on the commonage at Bunnamohaun, to the west of Toormore. The detail was plotted manually from the 1:5400 scale vertical aerial photographs (Department of Defence: Air Corp Photos, ref. V171/118–119) and verified in the field. As a result, the scale is approximate. In addition, certain aspects of the topographical detail have been exaggerated for clarity (*New Survey* 5, Page 196).

Fig. 8.7 Fulacht fiadh along the Pollbrandy.

Fig. 8.8 Circular enclosure containing two prehistoric huts on the commonage at the extreme south-western end of the island in Bunnamohaun (Cat. No. 71). It is situated in a sheltered hollow close to the cliffs, its south-eastern end opening out onto an artificial terrace that terminates right on the edge of a deep sea cove.

Fig. 8.9 Prehistoric walls in Bunnamohaun.

suitable streams, but we must remember that there has been extensive peat growth since prehistoric times, and the pattern of surface streams is likely to have been quite different in the Bronze Age. It would be wonderful if we could see the land as these early people did, if we could understand how they assessed and utilised its resources: but this must of necessity remain hidden from us, at least for now.

But some things we can know. The horizon encompassed the entire world of their experience in a way impossible for us to appreciate. Although their direct experience hardly extended beyond the seen horizon that was the edge of their little world, they knew this world in the way the animal knows its niche. They knew every bush and every stone, every slope and shadow, the variation of its soils and what would grow in them: as know it they must because upon this knowledge their lives depended, and their future. For in those distant days each community had to be sufficient to itself in almost all things: in food and shelter and raiment, and for most of the tools and weapons with which they slowly changed their world and responded to its challenges. Those things shaped by art and craft, honed by wisdom tried and tested, and adapted to the peculiarity of this place were a cultural efflorescence born of its timber and its stone: trackway and fireplace, weapon and wall. And yet theirs was, so to speak, an 'open intimacy', reaching through imagination and intellect to frame their intimate territory against the movement of the great lights of the sky and the seasons they controlled, and raising their stone monuments to anchor that precious all-determining relationship.

Archaeological evidence of human activity is quite sparse for the last millennium BC, but it is of interest that most of it comes from the mountain bog in Bunnamohaun, at the extreme western end of the island. This was one of the important turbary areas on Clare Island, with peat up to 1.5m deep. A meandering pre-bog wall here—part of an extensive network of ancient walls indicative of enclosed farmland—was excavated by Paula King in 1994.[10] This wall is thought to have been built in the late Bronze Age or early Iron Age – some time before 500 BC – not long before significant peat accumulation began in this area.[11] Palaeobotanical evidence from the investigation allows us to see what the landscape was like before the bog took over. It was an open, dry-heath type of place, dominated by grasses and abundant ribwort plantain, with many of the plants you see in similar coastal grasslands today: buttercup and dock well represented, dandelion-type flowers, lots of bird's-foot trefoil (*Lotus corniculatus*), tormentil (*Potentilla erecta*), sheep's-bit (*Jasione montana*), meadowsweet (*Filipendula*) and devil's-bit scabious (*Succisa pratensis*), and wild carrot (or hogweed). There was a lot of heather and heath, and some mountain ash and willow in the vicinity. The occasional grain of pollen hints at fields of cereal elsewhere on the Island.[12]

Promontory forts, of which at least eleven have been noted, are the only other evidence for settlement in the last millennium BC and the early centuries AD—at least if we subscribe to the view that they can be assigned to this period: many archaeologists believe they are medieval in age. Besides which, 'fort' is misleading. All we can be sure of is that they are headlands that were cut off for some reason: exclosure or enclosure, defence or refuge. One suggestion is that they may have been refuge or prestige sites, each linked to the land of a particular family grouping.[13] The presence of well-preserved house remains on the promontories at Doon and Doonalia hints at medieval occupation. The most spectacular of these promontory features is Doon, near Toormore (Figure 8.10).

Despite the lack of hard archaeological evidence, the pollen record clearly attests to continued agricultural activity during the early medieval period (400 to 1200 AD). Ringforts are the most ubiquitous type of monument of this period throughout Ireland. Three of the six circular enclosures recorded on the island may be of this type (the most spectacularly sited—though hard to see until you are inside it!—is on the steep lower western slope of Knockmore): but it may be that some of the promontory 'forts' took the place of ringforts along the coast.[14]

The Early Medieval Period

The archaeological evidence for early Christianity centres around the Abbey in Kill. The building we see today only dates to Norman times, but there are traces around it of an earlier Christian presence: what remains of the holy well and altar, an early cross-inscribed pillar, and what are just possibly traces of the early church enclosure. This pillar stone is thought to date to some time between the seventh and twelfth centuries, and

(a)

(b)

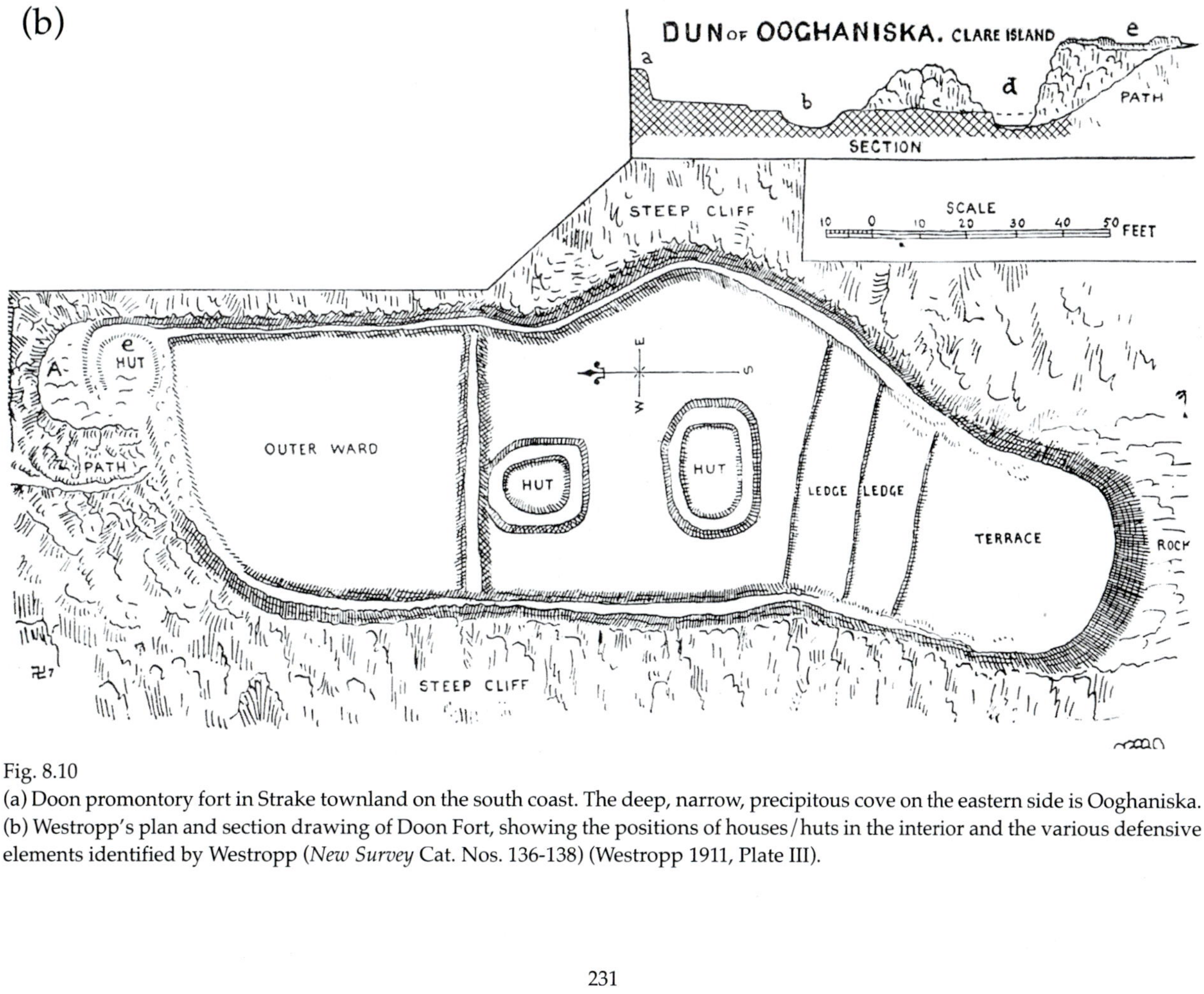

Fig. 8.10
(a) Doon promontory fort in Strake townland on the south coast. The deep, narrow, precipitous cove on the eastern side is Ooghaniska.
(b) Westropp's plan and section drawing of Doon Fort, showing the positions of houses/huts in the interior and the various defensive elements identified by Westropp (*New Survey* Cat. Nos. 136-138) (Westropp 1911, Plate III).

Fig. 8.11 (a) The Pillar Stone at the Abbey; (b) Holy Well at the Abbey (Charles Brown collection, Trinity College).

may possibly have been a termon or boundary marker of some kind (Figure 8.11a).[15] Con Manning has suggested that the first phase in the Abbey's construction conforms to an earlier style of church building, possibly echoing what was on the site before this (see below).

There is a second holy well—Toberfelabreed—'in a bushy hollow' close to a wonderful boulder beach close to the eastern shore in Capnagower. Although the remains we see here today are probably medieval, they may very well mark what was a devotional site from a much earlier time. Many favours were credited to this well (in which lived a mysterious sacred trout, visible to only the most devout pilgrims), and a popular pattern was held here on 15th August.[16] Whatever about the fragility of the threads of historical fact that link her with Clare Island, Granuaile is firmly bound to it by legend and tradition. Apart from giving her name to the tower house on the quay, she is closely associated with this well, which is where island legend claims she was married to a shipwrecked sailor 'by a priest of the station'. Little of the structure as described by Westropp in 1911 remains today (Figure 8.12).[17]

In the Early Medieval period the Ó'Máille emerged as the dominant sept in the Clare Island part of the tuath of Umhall, and it was they who granted the lands associated with the church to the Cistercians of Abbeyknockmoy (probably in the fifteenth century) in order to enhance family prestige. The church continued to serve the local community after this, but when the chancel was added a rood screen was erected to keep the nave—

Fig. 8.12a,b Toberfelabreeda at Kinnacorra, 'traditionally reputed the place of Grace O'Malley's marriage'. (Charles Brown collection, Trinity College)

where the community heard mass—separate from what became in effect a private chapel and a place of burial for the Ó'Máille.[18]

Granuaile's Castle

The absence of solid historical evidence of a connection did not prevent the growth of a rich folk tradition linking Grace O'Malley with Clare Island. According to local lore 'it is said that Gráinne's favourite castle was on Clare Island',[19] but there is no real evidence that links her with the tower house at the harbour. This was built on a notable promontory beside the beach in the early to mid-sixteenth century, but re-modelled extensively in the 1820s, when it was converted to a police barracks (Figure 8.13). The tower house anchors Clare Island in the sixteenth century, but almost nothing remains of the hovels in which the people whose labour on the land and at sea built and supported it.[20] Their imprint on the landscape has been erased by that of later centuries, an impact that reached into every potentially productive corner of the island. Two house clusters and a number of isolated house remains high on the slopes of Knockmore in Bunnamohaun are evidence of the practice of booleying over this period (see page 266–8).[21]

The harbour was probably one of a number of naval bases for the Ó'Máille galleons in Clew Bay, whence they sallied forth to police the bay and on their marauding expeditions further afield. Abbey

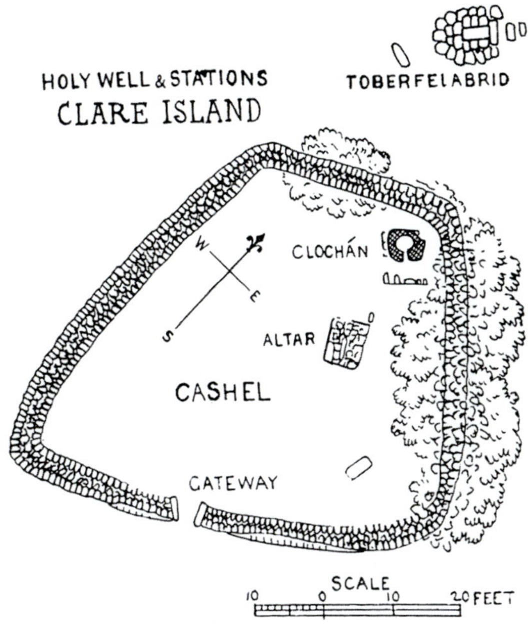

Fig. 8.12c Westropp's plan of Toberfelabreeda.

and Castle bear witness to the wealth and stability of Gaelic culture on the island in the fifteenth and sixteenth centuries, by which time the population may have reached something like 200. This wealth was based on a combination of farming, fishing and commerce. The fertility of the island's lowlands was maintained through an endless supply of seaweed secured from the shore, and there were extensive areas of upland grazing.

Fig. 8.13 The Tower House (Granuaile's castle).

Fig. 8.14a Aerial view of the Abbey at Kill.

The Abbey

There were three main phases of construction in the building of the Abbey (Figure 8.14).[22] The earliest part is the nave, which was built as a simple single-cell church in the early part of the thirteenth century. It had timber rafters, and was probably roofed with stone tiles. The chancel was added two hundred years later in the early to mid-fifththeenth century, and is thought to have been commissioned by Diarmuid Bacach Ó Máille, who probably granted the church and its half quarter of land to the Cistercians of Knockmoy at the same time.[23]

Fig. 8.14b The Abbey at the turn of the nineteenth century (Charles Browne collection, Trinity College).

Different kinds of local rocks were used in the building, all of them abundantly close to hand. These are not easily worked, so the blocks are rather irregular. The stone used to build the chancel 'is generally smaller and has more of a quarried look about it than that used in the nave'.[24] It is perhaps surprising that the much better stone available in the Maum and Capnagower Formations was not used, but these only outcrop on the north-east coast, which may have been considered too distant. This rough masonry was bonded with simple lime mortar using local sand, then pointed, render and plaster applied, and then lime-wash to receive the paintings.

An arcaded and canopied tomb surmounted by an elaborate tracery screen and ogee-headed gable was the central focus of the chancel. A detailed replica of this was painted on the wall to the right of the real one, each taking up exactly one-third of the wall space. The stonework of the tracery is somewhat out of proportion with the tomb chest, suggesting that it may have been prefabricated elsewhere and only added later.[25] We do not know who is or was buried here, but 'tradition states [it] is the final resting place of Grace [O'Malley]'.[26]

Fig. 8.14c The Abbey in 1952, prior to restoration

The interior of the new chancel was elaborately painted with figures. Only two of the earlier motifs survive: a deer flees between the pinnacles of the tomb—the central feature of the chancel—pursued by a hound on the north wall, 'drawn with particular verve and confidence', and a galloping horseman at the west end of the south wall above the door leading to the stairs, which was described as 'a rare and valuable portrait of a Gaelic lord in full cry' (Figure 8.18).[27]

Some decades later it was decided to extend the chancel, so the original roof (probably a flat wooden structure) was replaced by the vaulted roof we see today. This third phase of building provided accommodation above the chancel, and included a small domestic north wing for the priest (or small community of monks) who lived here and looked after the church and its lands. The walls were re-plastered and the original artwork painted over with new designs that were smaller in scale and less impressive than those of the earlier work. Most of the figures to be seen today belong to this second phase of painting, which survive only in the vault. This latest phase of building may have been commissioned by Diarmuid's granddaughter Máire (wife of Ruaidhri Mac Suibhne of Fanad) who was renowned for her piety, and responsible for building many churches in her lifetime.

The walls were plastered inside and outside, and then lime-washed before being painted in the fashion of the day; the only place in Ireland where painting of this sort has survived to any extent is in the Abbey on Clare Island—faint traces have survived in a handful of other medieval Irish churches. All the carved work in the tomb, both pinnacles and tracery, as well as the armorial plaque, is of imported limestone. Several features are made of serpentinite: most notably the inner three orders of the chancel arch, the facings of the original stone altar and the altar top itself. Although serpentinite occurs on the island, it is of lesser quality than that used in the Abbey, which was probably sourced on the mainland.[28]

The link with the Cistercians of Abbeyknock-moy was somewhat tenuous, and it is highly

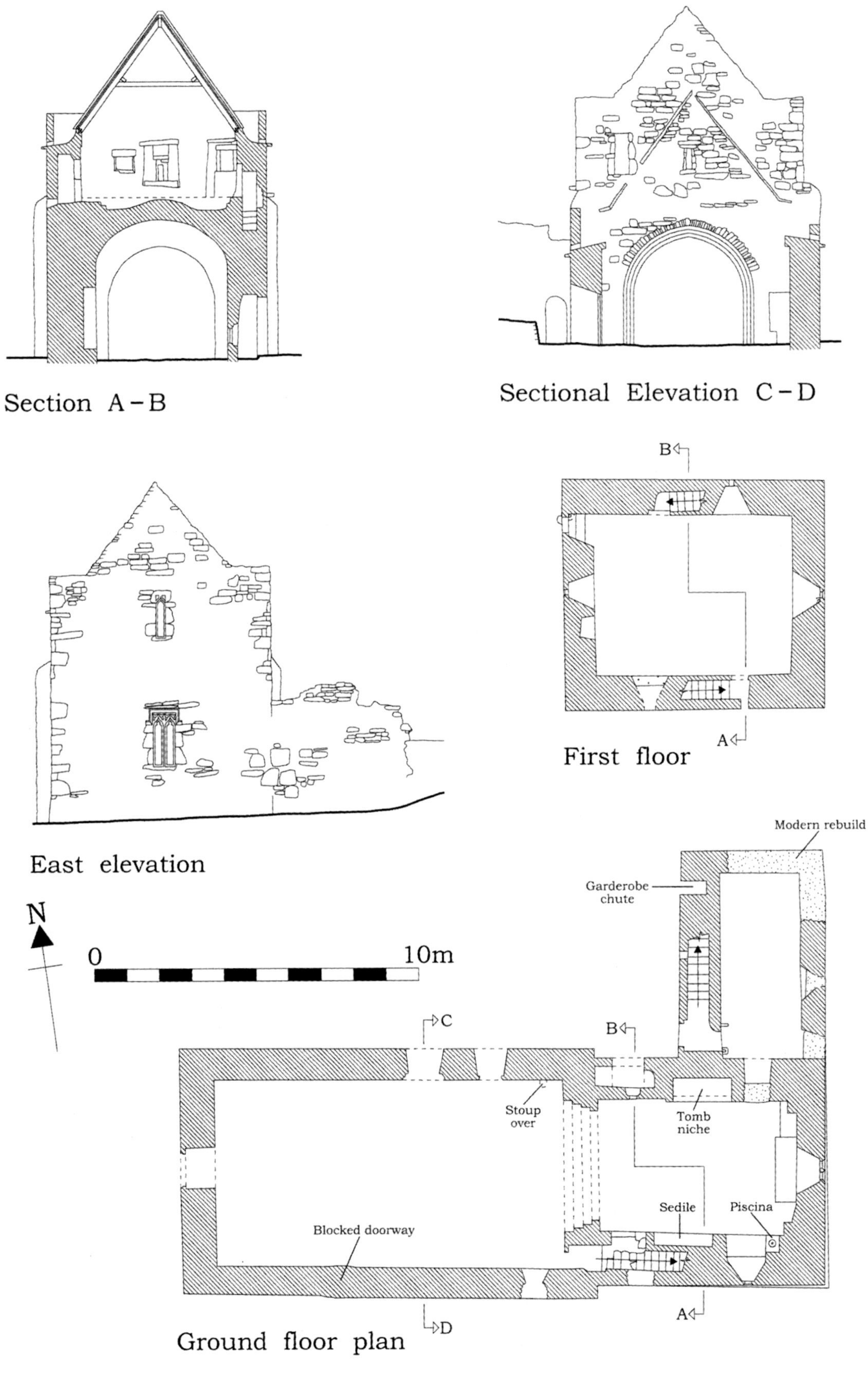

Fig. 8.14d Plans, sections and elevation of the Abbey in 1992 (*New Survey* 4, 14).

unlikely that the Abbey ever harboured a Cistercian community or that monastic hands farmed its 'quarter' of land. The Abbey—and the paintings—are more a celebration of the wealth and status of the ruling family than a monument to the glory of God.[29] However, this is where the people of the island worshipped all through the Middle Ages. Indeed, they appear to have felt a close sense of ownership of the Abbey; in a seventeenth century account we are told that they 'saw themselves as the monks of that place'.[30]

One of the most conspicuous features of the building is a limestone armorial plaque to the west of the tomb. The boar trotting in the centre of this plaque (with three primed longbows aimed at him) is thought to be a figure out of Gaelic mythology: the wild boar that had to die at the hand of the new king.[31] The ship under its front feet represents the maritime prowess of the family, as does the Latin motto, *terra marique potens*: mighty on both land and sea. The plaque is believed to date to the second half of the seventeenth century, and in heraldic terms is considered somewhat idiosyncratic (Figure 8.15).

Fig. 8.15 (a) The canopied tomb; (b) armorial plaque of the Ó Máille.

The paintings on the vault

The most extraordinary aspect of the church is the mural paintings that covered the vault, which are unique in the late medieval Irish context.[32] They are as startling when first you look upon them—so unexpected are they—as the wild animals depicted on the walls of palaeolithic caves. They had deteriorated badly through many centuries of neglect, but that decay was halted by a major conservation project carried out by the National Monuments Service at the same time as the New Survey was in progress.

The paintings appear to have been in good condition as recently as the 1860s, but the elements were given free play when the church was abandoned and allowed to fall into ruin. They had so far deteriorated when Thomas Westropp saw them that he believed they would soon disappear entirely. The account he produced for the Original Survey is a remarkable piece of work, especially considering the limitations under which he worked and the short time he spent at it.[33]

By the time the Abbey was taken into State care in the 1880s it had been a ruin for several centuries. A concrete floor was laid over the vault some time between 1908–1910, but this made things worse because the cement further encouraged the prolif-

Fig. 8.16 Overall view of paintings (*New Survey* 4, Plate v and foldout illustrations 1 and 2).

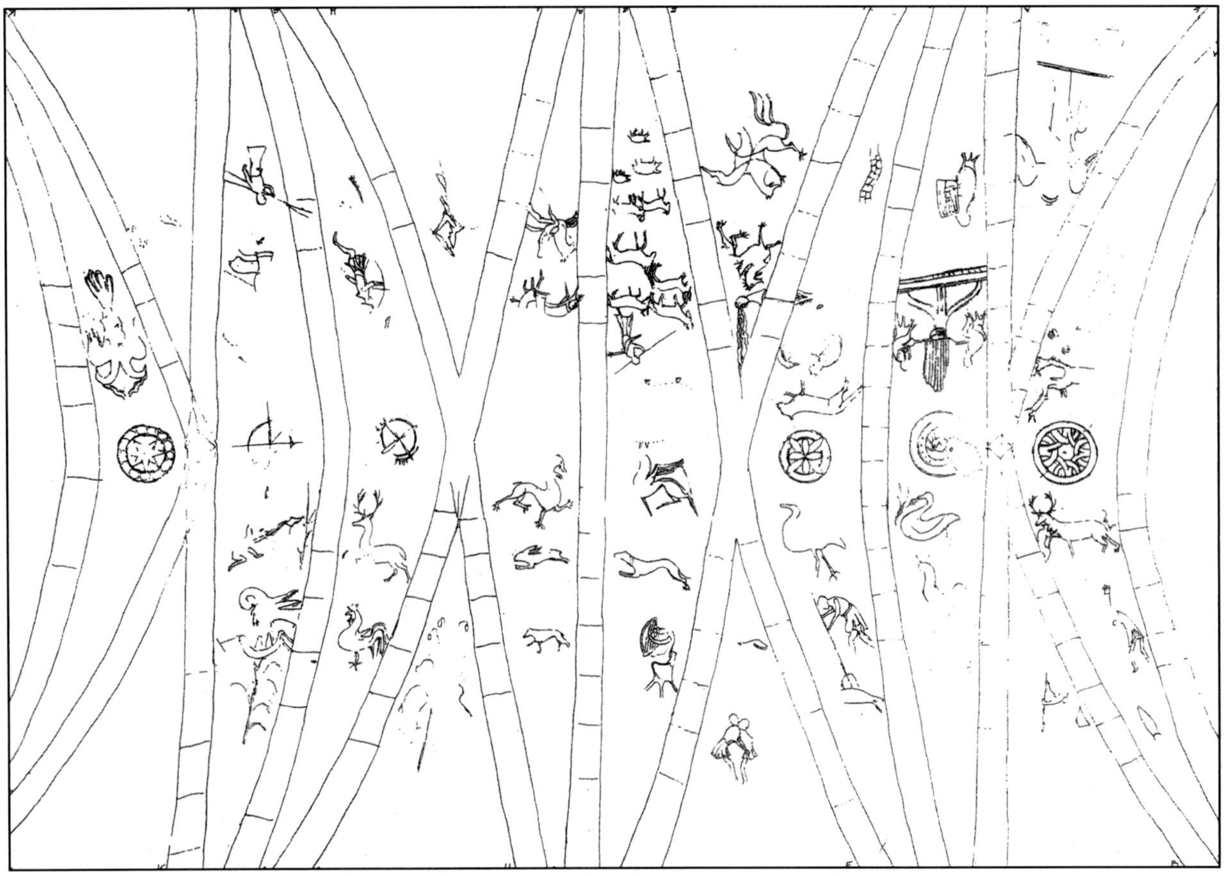

Fig. 8.17 Overall drawing of vault paintings (Christoph Oldenbourg) *New Survey* 4, 102.

eration of a damaging microflora. The chancel was re-roofed in 1952, which only encouraged this growth, allowing moisture to further saturate the fabric. At this time the floor was excavated down to its original levels; over centuries it had risen by an average of 1.2m. Major conservation work was carried out under the direction of Madeleine Karkov and Christoph Oldenbourg over a ten-year period between 1991 and 2001, just in time to rescue the remaining paintings 'from an extreme state of decay and virtual collapse'.[34] The outside of the building has been rendered with a shelter coat of lime render, restoring its original appearance (startling as this appears to our modern eyes!).

The vault is divided into a series of compartments or bays by five deeply-incised false ribs, which are painted black (lamp black rather than charcoal black) to give the impression of stone; and each bay is further divided into four triangular areas by diagonal ribs, the model for which was the real ribbing of the presbytery at Abbeyknockmoy. This was a barrel vault rather than a ribbed vault, but the artificial ribs create the illusion of a more sophisticated architectural construction. The main colours are clear yellow and strong bright red ochres. Lime white was also used. The colours used in Phase 1 were carbon black, high quality red ochre and lime white; black was not used for the actual figures in Phase 2.

The designs were first etched in the still wet plasterwork, then painted in red, yellow, black and white, using high-quality lime-based paints. There are faint hints that there may have been other colours of which no trace now remains. The themes of the paintings are conventional ones widely in vogue throughout Europe at this period, but the 'mode of representation' and the 'individual style' are distinctively idiosyncratic. As the work of painting progressed, slight improvements were made to the outline of most of the figures.

A great variety of subjects are represented, many of which can only be identified by close scrutiny and an informed eye: and of course we must remember we are seeing less than half of what was originally painted (Figure 8.16).[35] Hunting activities and wild animals are the predomi-

Fig. 8.18 Mounted soldier (Christoph Oldenbourg).

Fig. 8.19 The two dragons (Christoph Oldenbourg).

Fig. 8.20 T.J. Westropp's interpretation of some of the motifs.

nant themes. There is a hound chasing hares (or a fox chasing rabbits), a stag being attacked by hounds, a huntsman with a dog on a slip lead, and a variety of other hounds. An armed herdsman pierced with an arrow is depicted with cattle, sheep or goats and a pig in what some people think may be a scene from a cattle raid (Figure 8.20). There are fishermen with spear and trident, an assortment of soldiers, and a pair of wrestlers. Among the wild animals and other creatures depicted are a wildcat, a swan eating a plant, a heron or crane, a hare, a fox, a halting stag, a wildcat, a pelican, a cock, and a cat (or a pine marten?) climbing a tree. There are numerous mythical beasts: a fire-breathing serpent and several dragons, a pair of body-sharing creatures, a double-headed eagle and a griffin.

Religious themes are almost absent. The only unambiguously religious figure is the Archangel Michael, prominently shown with his soul-weighing scales, although the organist and lyre-player in an adjacent panel may be church-based companions. In a distant panel a harper plays on his own. As with so many of the other images, these are themes that were widespread in medieval iconography, although there is a distinctively Irish flavour in the way they are represented.[36] On the whole though, the subject matter has little to do with otherworldly concerns, and indeed at times is 'curiously at variance with monastic ideals'.[37]

Art historians have puzzled over the painting for over a century, examining every detail with forensic precision, conscious all the time that they are only seeing less than half of what was there originally. Some see a pattern, an overall theme binding these varied motifs together, but we should be cautious about reading too much into them.[38] Westropp could see no overarching theme, no coherent narrative. The image of the Archangel Michael, with his scales for weighing the good and bad deeds of each soul on the day of judgement, is really the only convincing religious element, though religious symbolism can be attributed to many of the other figures, such as the pelican feeding her young with her blood, the cock and disembodied torturer's hand from the passion narrative (if that is what it is). However, it is difficult to see how everything can be related to the Archangel in a way that would make this an artistic meditation on the Day of Judgement, as has been suggested.[39]

The range of motifs shows that the artist was familiar with the imagery of his period, but it is debatable as to whether they are combined in a coherent pattern; rather, it appears that this is a selection from a stock of standard motifs, with little intent other than to amuse and entertain.[40] It is difficult to find an overarching theme that links the different panels together.[41] There does however appear to be a peculiarly Irish feel to the manner of execution of many of the figures.[42] Perhaps it is best to see the Clare Island paintings as a modest vernacular imitation, an outlying echo, of more sophisticated and elaborate distant models.[43]

Notes

1 Westropp 1911, OS 2, 1.

2 A detailed catalogue of archaeological sites, monuments and find spots is presented by Paul Gosling in Volume 5 of the New Survey (Gosling 2007c).

3 Gosling, Manning and Waddell 2007, 1–3.

4 Viney 1990.

5 Westropp 1911, OS 2, 17–19.

6 Cullen and Gill 1992c, 5–6; Gosling 2007a.

7 Gosling 2007a, 34–35.

8 Gosling 2007b, 72–4.

9 Gosling 2007a, 39.

10 King 2007, 291–6.

11 Molloy 2007, 305.

12 Molloy 2007, 301–4.

13 Gosling 2007a, 42–43.

14 Gosling 2007a, 43–44.

15 Gosling 2007a, 32–33.

16 Westropp 1911, OS 2, 25–28.

17 Westropp 1911, OS 2, 26.

18 Gosling 2007a, 45–47.

19 Ní Cheanáinn 1988, 39.

20 The tower house also probably dates to the early 16th century; the first floor bartizans and lack of a vault are thought to be late features (Gosling 2007a, 48–50).

21 Gosling 2007a, 58.

22 Manning, Gosling and Waddell 2005; Manning and Morton 2009. Although always referred to as such, the building is not an abbey in the strict sense of the word, housing a monastic community ruled by an abbot.

23 Manning 2007, 15.

24 Manning 2005, 17.

25 Stalley 2005, NS4, 140.

26 Chambers 1983, 35.

27 Stalley 2005, 146. These early figures were only discovered during the recent conservation work. Nothing of them was visible in Westropp's day.

28 Although serpentinite is of frequent occurrence on the island, it is limited both in quality and quantity. It is perhaps more likely that the stone used in the Abbey came from the mainland (perhaps the better-known serpentinite that forms the north side of Croagh Patrick). If it was quarried somewhere such as Portruckagh, where this rock is (just about) exposed, it is impressive evidence of a familiarity with the practical aspects of the island's geology.

29 Stalley 2005, 135, 146.

30 Manning 2007, 18.

31 Gillespie and Ó Comáin 2005, 41–44.

32 Manning 2005, 2.

33 Westropp 1911, OS 2, 29–37.

34 Oldenbourg 2005, 49.

35 Very limited traces of paint from Phase 2 survive on the walls.

36 Buckley 2005, 129.

37 Stalley 2005, 135.

38 Morton 2005, 114.

39 Morton 2005, 118.

40 Stalley 2005, 145.

41 The figures of some panels seem to wander into adjacent panels sometimes.

42 Cf. Stalley 1987, 216. There are Scottish parallels for the imagery that may reflect the close connections between the western seaboard of Ireland and that of Scotland at this time (Manning 2007, 20; Stalley 2005, 144–5).

43 To Paul Gosling the paintings seem to afford a glimpse of a shrinking late medieval Gaelic world view (pers. comm.).

Overleaf

Fig. 9.1 The cluster settlement (clachan) of Glen in 1838 (Painting by Philip Armstrong).

Chapter 9

Landscape and society in the nineteenth and early twentieth century

The clachans and rundale

In the eighteenth and early nineteenth centuries the population of Ireland grew with explosive rapidity, soaring from one million at the start of the seventeenth century to 8.5 million in 1845. Between 1780 and 1845 four million people were added. This increase led to extensive reclamation of increasingly marginal land, resulting in the dense and rapid colonisation of western Ireland in particular as people sought enough land to live off. The unique cultural landscape of the western seaboard, often thought to be the last bastion of an ancient Gaelic way of life which had disappeared elsewhere, is largely the result of this transformation.

On the poorest land, in the areas where population growth was most rapid, a form of communal farming known as *rundale* developed, which later observers thought of as remnants of an ancient and once more widespread tradition, the relics of an ancient settlement pattern, surviving in remote corners where modernising influences had failed to reach. An alternative view has been put forward by Kevin Whelan however, arguing that rundale developed independently in these 'Congested Districts' because it was the most efficient way to utilise a limited area of land with a high and growing population—which was how the medieval openfield system itself had initially developed in the eleventh century.[1] Rundale demanded a new mode of co-operative industriousness: it was inventive and innovative within the strict limits of its possibilities. Rather than a relic from the past, it was a specific adaptation to the peculiar demographic, geographical and agricultural conditions that were developing in the eighteenth century. By the early nineteenth century clachans had become the dominant form of rural settlement in the Congested Districts of Donegal, Mayo and Connemara.[2]

Nineteenth-century relics of joint land-holding are known from most of Ireland, and it is at least possible, perhaps probable, that the rundale economy was shaped or at least influenced in part by an older tradition that had not been entirely abandoned when the need to develop a new communal farming arose along the western seaboard.[3]

This form of farming grew up where a lease on marginal land was taken out by a number of people who agreed to farm the land in common. Families built their homes in the centre of the farm, an apparently random cluster of houses known as a *clachan*. It was not a village in the ordinary sense. It did not have centuries of history behind it. It had no church, no shops, no school, no public house. There was no castle or great house belonging to lord or priest or wealthy craftsman. There was no plan to it. It just grew as the population grew, as best it could. Yet this was an institution ideally adapted to permit the steady rise in population, largely dependent on a unique crop grown on marginal land suited to no other: the potato.

People lived in two-roomed *byre dwellings* (kitchen and room); at night animals (cattle, pigs and fowl) were housed at the end of the kitchen furthest from the fire. The floor was paved and had a drainage channel for animal urine.Ventilation was more or less non-existent, chimneys being a late nineteenth-century innovation; up to the middle of the century there was just a smoke hole in the thatch, and the interior was often so smoky you could hardly see.[4] The climate on Clare Island is mild enough for animals to be out in the open for most of the year, but if the weather was very cold they were taken into the house at night.[5] When the ethnographer Charles Browne visited Clare Island in 1896, people and animals were housed under the one roof in all but five or six households.[6]

Surrounding the clachan was an aureole of unenclosed cultivated *infield*, in which each family had a number of strips of land, often widely scattered to ensure a fair allocation of land of different quality. These strips were periodically redistributed as long-term insurance of fair distribution. The infield was permanently cultivated with oats and potatoes, 'a see-saw between potatoes and oats until it was thought that the land required a rest, when it was left to Nature, generally without any laying down in grass'.[7] On Clare Island in the late nineteenth century a four-year potato-oats rotation was generally followed, with the land being laid down to grass after the arable rotation.[8] Hay only became important in the twentieth century.

On the edge of the permanent cultivated infield was an area of *outfield*, parts of which might on occasion be taken into cultivation. Beyond this again was the common grazing land, where each family had its allocated number of *collops*, the standard measure for the maintenance of sustainable grazing. There was obviously a limit to expansion of the clachan because there was a limit to the land available (it was often co-extensive with the townland), and that limit was reached within a few generations.

Each little community was alert to the danger of overstepping the limits this commonage could support. The carrying capacity was carefully assessed, and each family allocated a share of the grazing rights proportional to their share of the arable land. A careful balance had to be maintained between the number of animals kept on the commonage, not only because this meant more for their owners, but because after the harvest they were turned into the fields and given the run of the infield through the winter, enriching them with their dung. Indeed, this winter dunging of the infield was its most important nutrient input.

Before the Famine there was no restriction of grazing rights on the commonage; the entire island was 'undivided and held in common by the inhabitants', all the rent on the island being paid 'by the bulk'. Once the crops were harvested the unfenced arable field became a common. While they were growing, a constant watch had to be kept to keep cattle and sheep out of the tillage land.

The increase in population in the congested districts was partly due to encouragement by landlords who wanted to increase the number of voters under their control.[9] However, the communal farming system went against everything the capitalist agrarian reformers of the time believed in, and several improving landlords made attempts to eradicate rundale in the decades immediately before and after the Famine. It was effectively brought to an end on Clare Island by the Congested Districts Board (CDB) in the final decade of the century (see below, page 252). The clachans were broken up, and the tenants re-settled on small, individual enclosed farms.

The simple rotation of potatoes and oats was possible because high nutrient inputs in the form of seaweed, and sand or lime were generally available. Most of these clachan settlements were close to the sea, where an almost limitless supply of the seaweed and sand needed to boost the fertility of the infield could be harvested. Rights of seaweed were accordingly carefully demarcated and guarded.[10] Great quantities of seaweed and sand were hauled from the shore up to the fields, and spread there every year, transforming rocky, acidic land to a complex of productive soils in which the potato could thrive.

Fig. 9.2 One of several lime kilns on top of the eastern till cliffs in Ballytuohy More.

What little farmyard manure was available was carefully saved for the potatoes. Manure was vitally important, but in very short supply, with the amount available determined by the number of cattle that could be grazed. The dungstead outside the door, of which outside commentators had such an adverse opinion, had therefore a very important role to play. An extensive network of artificial new soils (*plaggen* soils) resulted, developed with vast amounts of nutrient-enriching materials brought from the seashore. The livestock that grazed on the outfield and common beyond the arable land were allowed to graze the infield stubble once the crops were harvested, providing a valuable nutrient supplement in the form of dung as they did so. However, this grazing of the infield in winter made it impossible to grow winter root crops or introduce improved rotations.

As the aureole of cultivated land expanded with the growth of the clachan the area of grazing available was reduced, sometimes resulting in overstocking. Also, continuous cultivation inevitably led to soil impoverishment, which the meagre return from dung was not able to ameliorate in inland clachans.[11]

Clare Island also had numerous lime kilns; over thirty have been recorded, some very well preserved (Figure 9.2).[12]

As the nineteenth century advanced, oats became increasingly a cash crop, and a cow was beyond the economic reach of most in the poorest areas. For those who could afford to keep one, the pig saw many families through the season before the end of last year's harvest and the arrival of the new potatoes. The diet was accordingly stripped of oatmeal and milk, down to the barest essentials. In

Fig. 9.3 House on Clare Island said to be the oldest on the island at the time the photograph was taken (1896) (Charles Browne collection, Trinity College).

the late eighteenth century oaten bread was an important part of people's diet eaten on the island, but by the 1830s 'large quantities of grain' were being exported from Clare Island.[13] By the 1830s one third of the people of Ireland had come to rely on one species of plant—and one variety for the most part—for 90% of their calorie requirements.

The diet of potatoes, dairy products, eggs and mutton was supplemented by an abundance of fish: fresh in summer and autumn, pickled or dried over the winter.[14] At the time of Charles Browne's visit people had three meals a day, many taking strong tea at each. Everybody dressed in bulky homespun.

The pattern of land-holding in those parts of the country where clachan settlement had developed in the 18th century had become extremely complicated by the time of the Famine. Increasing subdivisionfrom one generation to the next had led to a situation where the holding consisted of several tiny parcels of land often widely separated from each other. For the outsider it was well-nigh impossible to make sense of this system before the large-scale maps of the Ordnance Survey became available.

The system approached breaking point as the burgeoning population broke the bounds of possible containment. The pressure was accentuated by a series of downturns in the world beyond the clachan, which cut off the vital arteries that had brought money into the system from the outside world.[15] The prices of agricultural goods halved in the depression that followed the ending of the Napoleonic wars. The cottage linen industry struggled as factory-based spinning and weaving took hold in Ulster, displacing the farmer-weaver. And nature itself lent a hand: a series of wet summers led to bad harvests, and the herring—so important and hitherto a reliable part of the summer diet in the west during the late eighteenth and early nineteenth centuries—deserted the west coast. The annual migrations of the herring were notoriously unpredictable. Great shoals might turn up at the same place every year for a generation, and then suddenly stop coming. Within two decades of the catastrophic collapse, which the Great Famine precipitated, one million were dead and two million had emigrated, the great majority of whom came from the areas of clachan settlement.

The cultivation of the potato

The potato was cultivated with the spade in the long ridges that outside observers called 'lazy beds'. The ancient implement with which this transformation was effected was the one-sided spade or *loy*—the most distinctively Irish of agricultural implements. Without the potato and the lazy bed the extraordinary rise in population that took place in the least fertile parts of Ireland during the eighteenth and early nineteenth centuries could never have occurred.[16]

Although they were used primarily for potatoes, oats were planted in lazy beds later in the rotation, at which stage the ridges tended to be lower and the furrows more narrow. On poorly-drained land ridges tended to be higher and narrower.[17] Lazy beds were a special variant of the ridges traditionally used to sow corn, adapted for the spade cultivation of potatoes on previously untilled land. It was a system that (in the words of the Devon Commission) 'supplied the most minute system of artificial drainage that can be imagined for that one crop', a method that could be sensitively adapted to the idiosyncracies of local topography. It was in particular a system perfectly adapted to the physical difficulty of the irregular terrain in which potatoes were often grown, especially in the west, where a plough could not be used. On the wetter soils of Clare Island in particular the furrow or 'sheugh' between the ridges was vital for drainage. On the island the ridges often had a pronounced bow-shaped profile, which increased the surface area, and tended where possible to be oriented north-south, with curved ridges on sloping ground to reduce erosion.[18]

In the breaking of untilled ground the loy was a very efficient and sensitive tool, especially in heavier soils. It was not a spade in the modern sense; it was essentially a hand plough, designed to undercut and turn the sod, rather than a tool for lifting like the modern spade, but it was twice as efficient as the two-sided spade at its proper task. A straight young sapling or branch, pointed and shod with iron, was generally used for planting the potatoes.

The lazy bed method kept the roots aerated, well-drained, and the foliage well-lit. It allowed the cultivation of steep slopes and rocky places no plough could go, allowing every inch of ground to be utilised. At its most efficient it could be up to three times as productive as drills ploughed with a horse, but it was immensely labour-intensive, absorbing the one resource that was not in short supply at this time of rapid population expansion made possible by the produce of those same ridges.

The cultivation of potatoes in lazy beds was five times as labour-intensive as cultivation in drills with horse and plough, but the yield was three times higher. It was 20 times more labour-intensive than keeping stock. The energy devoted to ridge management, it was felt by some, would be better spent on drainage. The Devon Commission in 1847 argued that the labour saved by cultivating potatoes in drills during the preceding 20 years would have been sufficient to drain all the productive land in need of drainage in the whole of Ireland.[19]

Agriculturalists pointed out that it was more efficient to sow potatoes in drills—it saved labour, time, seed and manure, and they could be weeded and tended more easily—although they admitted the value of lazy beds for breaking in new ground.[20] Drill cultivation was adopted over a wide area in the early nineteenth century; but on the small farms on the edges of cultivation, where time and labour had a different meaning and value, the lazy bed continued to reign supreme. The spade lost its advantage in large-scale tillage when labour ceased to be cheap and abundant.

On Clare Island

As population grew on the mainland in the early years of the nineteenth century, large numbers of people migrated to Clare Island and fitted into its tenancy structure in whatever way they could.[21] There is evidence that newcomers from the congested mainland (or evicted tenants) not infrequently reclaimed plots for themselves in the outfield commonage.[22] By the time of the 1841 Census the population of the island had risen to 1,615: a density of one person per hectare.

The old townlands acquired a new human focus of development and exploitation with the establishment of the clachans, each of which was located in a uniquely pivotal position to exploit to the full the resources of the land available to it. At the high point of this development each house cluster was surrounded by a large aureole of unenclosed, permanently cultivated land, in which every family had its allotted share of stripes, each share a mixture of good and less productive land, an

Fig. 9.4 The Geological Survey field sheet for the area east of the cluster settlement of Glen (inset) noted the wide extent of cultivation ('Low moory ground well cultivated' (see also Figure 5.5).

attempt to ensure equitable distribution. Every few years these stripes were re-allocated. Around the grassland of the infield was an aureole of more marginal land, the outfield, which might be cultivated from time to time, but in any case formed a barrier or transitional zone between the cultivated land and the often wilder and always more barren land beyond, generally above the level of the glacial till: the bounds between these categories often strikingly coincident with the geological boundaries. Here the animals lived during the growing season, supporting themselves on what nature provided.

The villages were generally sited on the lee slopes of higher ground, the buildings always oriented as much as possible at right angles to the slope, and usually fronting south or east.[23] There was generally a small garden next to the house in which vegetables were grown. The site for the settlement was carefully chosen so as to facilitate the transport of seaweed and manure to the arable ground (which was usually between the village and the coast), the movement of the animals onto the mountain commonage during the growing season, and onto the cultivated land once the crops were saved.

Although the periodic redistribution of tillage plots had ceased at an early stage, and certainly by 1890, subdivision of land over the generations had become so complex that when the CDB bought the island only the tenacious memories of the inhabitants could decide the boundaries of each man's scattered holdings. A man who paid rent for ten acres might have it in twenty or thirty scattered patches. Many single acre fields were tilled in tiny plots by six or eight different tenants.[24] Here is an extreme example of how complicated and unmanageable this system had become: three tenants shared two small fields totalling about 8 acres (3.2ha) in the north of the island. One tenant had 22 separate plots, another had 16 and the third had 14. Many of these 'plots' were single tillage ridges, an indication that by this time the ridge had become the tillage unit in practice.[25]

There were no wheeled vehicles at this time, which meant the roads were in good repair. Turf, seaweed, fish or potatoes were carried in baskets on peoples' backs, or in panniers slung over the back of a donkey or horse (there were 80 horses) (Figure 9.6).[26] Cattle remained more important than sheep until after the Famine.

The oft-quoted description of Clare Island immediately before the Famine years (from the *Parliamentary Gazeteer of Ireland* published in 1846) painted an idyllic picture of a rustic Arcadia more suited perhaps to a sales prospectus. Knockmore, we are told, 'might be termed a mountain ... but for its being capable of cultivation to the summit'.

> The island appears to be one of the most fertile of the many which surround the mainland of the kingdom. It contains a valley which produces as luxuriant crops of corn and potatoes as are anywhere to be seen; it is nearly all in a tolerably good state of cultivation; and it is susceptible, under the management of skill and taste, of being made, in a sublunary sense, a miniature paradise ... Much grain is shipped for Westport; large quantities of sea-manure are landed, and about 340 men and boys divide their cares between farming and the fisheries.[27]

The recently completed Valuation Survey estimated the value of individual lots at between 1/2d and 15/= an acre where the national range was from under a penny (1d) to 30/= an acre.

Clare Island after the Famine

When in 1845 the potato blight with which the air above Ireland was charged that black summer crossed the sea to Clare Island the bottom fell out of this small world that had grown so rapidly in the short space of a century and a half. The Great Famine solved the problem of landless congestion, the greatest obstacle to agricultural improvement in Ireland. It precipitated the remodelling of Irish agriculture, though it was decades before there was any real improvement in the efficiency of agriculture over much of the country. Its most devastating effects were felt by the poorest, especially in western districts, where depopulation enabled many landlords to rationalise their minutely subdivided holdings.

The Famine brought utter devastation to the life on the island. Before it the population density had been 1 to every 1.2ha (1,615 in 1841). Within a decade less than half this number lived on the island (745 in 1851[28]), partly because of the direct effects of crop failure, in part due to the abandonment of cabins and to emigration. It continued to decline thereafter: to 716 by 1861, 494 in 1871, rising briefly to 621 in 1881, and down again to 490 in 1901.[29]

Large areas of cultivated land were abandoned, and the settlement pattern altered dramatically. The Land Judges Court Map prepared for the sale of the island in 1895 shows that the large rundale villages in Glen, Lecarrow and Ballytoohy More had disintegrated (Figure 9.7). The roads and paths that had permeated the island like arteries decayed as the life blood of the human footfall drained away. But even with this stemming of the tide of depopulation there was little improvement in the life of the people. There was now no shortage of land, but the system of working it had broken down and the wherewithal to adapt to the changed circumstances was not there.

The famine crisis did not end with the 1840s. It flared up intermittently for another half century. There was recurring blight in 1859–64, 1879–84 (the potato crop was a total failure in 1879 and the poor were once more faced with famine), and in 1889–90 the west experienced famine conditions once more. The power of the Famine to sweep people off the land was augmented by new legislation that made it in the landlord's interest to get rid of the poorest tenants.

In 1886 notices of eviction were served to 61 destitute farmers—on top of 22 earlier evictions (the owners at this time were 'ladies residing near Dublin'). 'Famine and fever claim hundreds of victims on these islands, where the distress is now such as was never experienced since 1847. There is neither food, fuel nor clothing left to the miserable people of Clare Island, Inishturk and Inishbofin' wrote a correspondent for the *Illustrated London News* in 1886 (Figure 9.5). In that year, the poor yield of potatoes, the low price of cattle and the failure of many seasonal labourers to find work in England or Scotland had 'reduced them to absolute destitution'.[30]

The account of Clare Island that appeared in the *National Gazeteer of Great Britain and Ireland* in 1868, twenty years after the *Parliamentary Gazeteer* painted its idyllic picture before the Famine, was condensed to a few lines: 'This spot is one of extreme beauty and fertility.'

Fig. 9.5 The *Illustrated London News* ran several articles on conditions on Clare Island during the 1880s. These two sketches appeared in a report published on 10 April, 1886, one depicting an eviction scene, the other a visit to a cabin on the island.

The Congested Districts Board

The continued poverty and 'congestion' of the western counties in particular finally precipitated large-scale state intervention, following a series of highly publicised bad years, with near famine conditions prevailing in the western counties again in 1889–90. In 1891 the CDB was established, initially for the western seaboard, but later extending its territorial remit eastwards, so that it was responsible for some 1.4 million hectares—nearly a third of the country—by 1909: 'the most far-reaching state intervention in European rural society prior to the Russian Revolution'.

The CDB promoted widespread infrastructural development and the modernisation of agriculture; proper rotations with good seeds, and new and better breeds of animals were introduced. New harbours, piers, roads and railways were built. It initiated a plethora of projects to extend education, industry and co-operation, as well as better standards of health, diet and hygiene. In the poorest areas especially, the improvement in agriculture even in the first ten years was remarkable. Where in 1891 the cattle were 'thin, light, raw-boned, degenerate, long-horned specimens of the old worn-out breed, good for neither beef nor milk', by 1901 the majority were of the polled Galloway breed, which did much better under the conditions prevailing in Connemara, or Shorthorn and Aberdeen Angus where conditions were suitable for them.[31] An entire new stock of well-bred pigs, based on the Large White Yorkshire, was established. From these beginnings the CDB moved to a second phase of compulsory purchase and redistribution of land in order to create viable holdings, remodelling extensive areas of the west in a very short time.

This quiet revolution was administered on the ground by a small group of intelligent and committed field personnel. It was a model in social reform—inclusive, democratic and progressive—and its agenda was determined by local and regional needs.[32]

In 1892 Major Robert Ruttledge-Fair visited Clare Island and prepared a report on conditions there for the CDB.[33] At this time the island's population consisted of 125 families (692 people in all), 25 of them 'in very poor circumstances'.

The lease was held by 76 tenants, but things were 'in such a state of chaos' that it was impossible for most of the people to pay their way; no rent, rates or taxes had been paid for years—there was nothing to pay it with.[34] When the island was put up for sale it remained on the market for several years.[35] Most people lived in thatched two-roomed houses, into which cattle, pigs and poultry were brought during the winter months. On their scattered holdings families grew an acre or two of potatoes, the same of oats, a little hay and perhaps a few perches of cabbage. All had an equal right to graze the 2,000 acres of 'superior mountain' (two-thirds of the total area was commonage). There was no fencing on the island, so that when the crops were in the ground 'the people would have to continue to herd the sheep as they used to do from daylight until dark ... to save the crops'.[36]

Fig. 9.6 Stone cleaves for carrying building stones. Wheeled vehicles were almost unknown on Clare Island in the nineteenth century (Charles Browne collection, Trinity College).

Potatoes were grown on lea land at the start of a rotation, followed by oats, then potatoes followed by oats again until falling yield made it necessary to break up another piece of lea. The potatoes were manured with abundant seaweed and shellfish, and what little farmyard manure could be saved. The practice of paring and burning was carried on extensively.

Those who were managing reasonably well generally had tea and bread for breakfast, potatoes and fish or eggs for dinner, and potatoes with milk for supper. Maize was often eaten when the potatoes ran out. Although there were 27 boats fishing part-time, including five pookauns, two hookers and twelve curraghs, little effort was made to develop what was one of the richest fisheries in the west, partly because the people of the island saw themselves as farmers, and partly because they lacked suitable boats and a proper harbour.[37] In his account of agriculture at the time of the original Clare Island Survey, James Wilson reported that most people added to their income by taking part in the autumn herring and mackerel fishing, and by lobster fishing. There were also still a few weavers and stocking makers.[38]

Ruttledge-Fair's report echoed earlier favourable evaluations of the island's agricultural potential. The land was 'of good quality, a superior quality of both cattle and sheep are reared than on any of the other islands in the immediate vicinity'. Galloway cattle and both black-faced and Cheviot sheep had been introduced by this time. There were fifty pony mares, but the two stallions on the island were 'miserable specimens of horse flesh'.

To an outside observer, even one as sympathetic as Major Ruttledge-Fair, there was a slovenliness about this way of life, one at variance with the diligence required of a yeoman farmer:

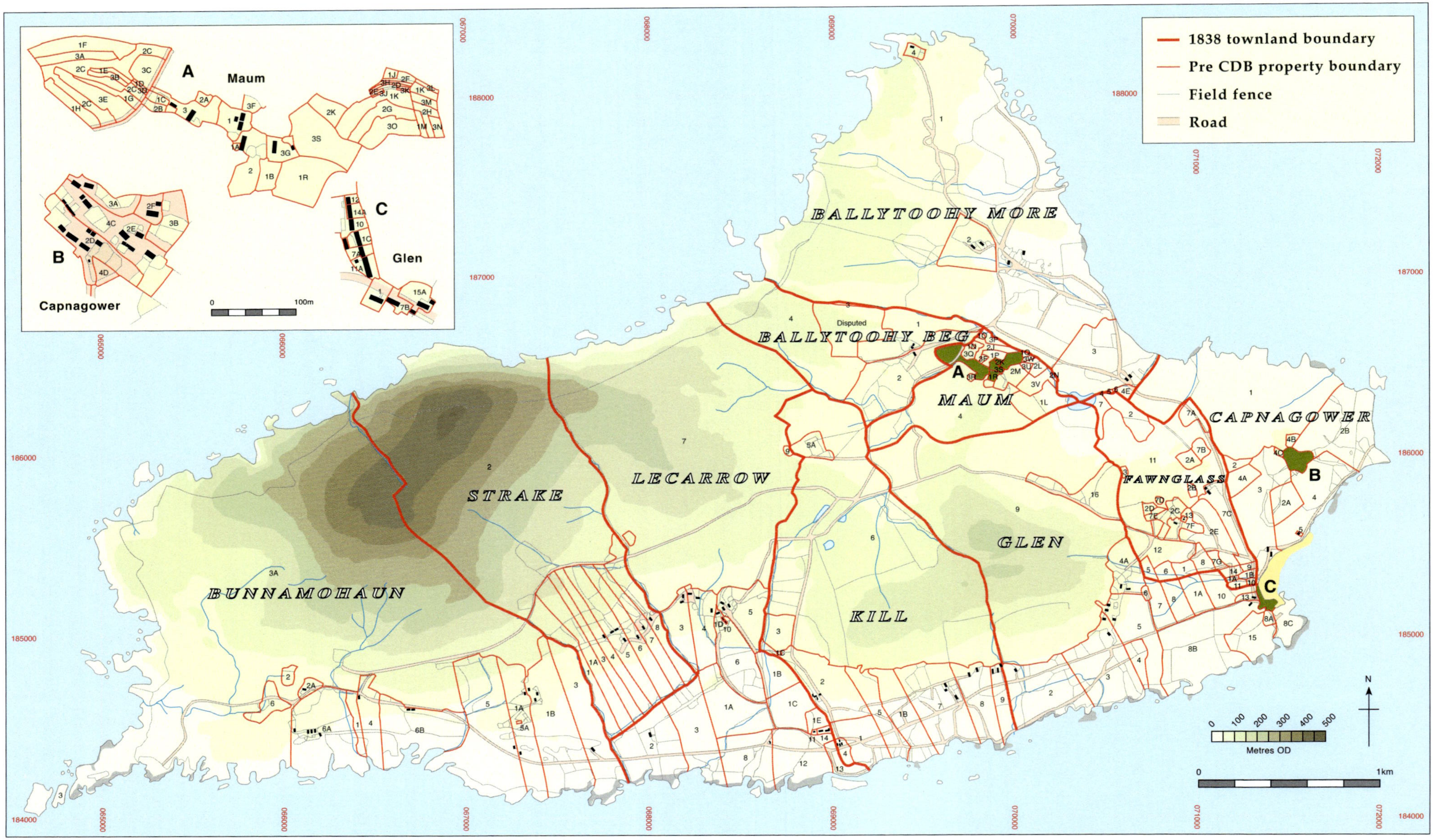

Fig. 9.7 This map was prepared around 1895 to illustrate the landholding situation on Clare Island when the property was bought by the Congested District Board (through the Land Judges Court) from the estate of Ismena McDonnell. It shows the highly fragmented pattern characteristic of rundale settlement (*New Survey*, Volume 1; the original Land Judges map is held in the National Archives of Ireland).

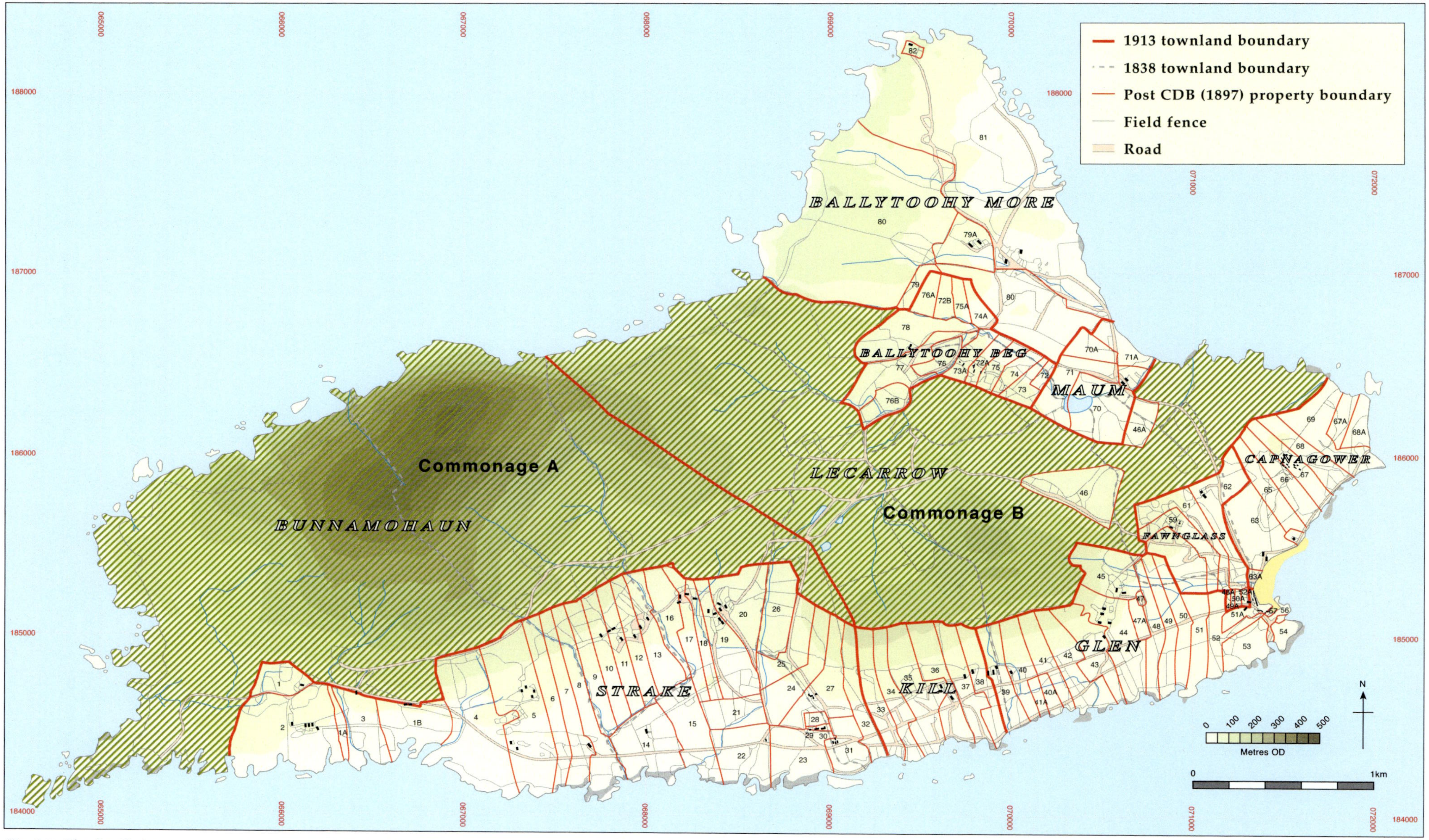

Fig. 9.8 The detailed map prepared by Congested District Board inspector Henry Doran in 1897 to depict the landholding arrangements agreed between the Board and the Islanders. The original of this map (which is held in the National Library of Ireland) was individually signed by the tenants and the arbitrators (Henry Doran and J.J. Burke) and was a legally-binding document. It shows the two major commonages in Bunnamohaun and Lecarrow, and the Great Wall dividing them from the enclosed land and the 76 consolidated 'ladder farms' created out of the old infields (*New Survey*, Volume 1).

Fig. 9.9 C. Kelly, C.R. Browne and H. Doran, examining a bird Browne has just shot (1896). Doran was the Congested District Board's inspector on Clare Island between 1895 and 1907 (Charles Browne collection, Trinity College).

Fig. 9.10 A typical late nineteenth-century home on Clare Island. Note fish spread on the thatch to dry (Charles Browne collection, Trinity College).

> The inhabitants of both islands [Clare Island and Inisturk] have lost almost all habits of industry and self-reliance. They have good holdings of land as a rule, and the mountains adjacent, on which their stock graze, are celebrated for their feeding qualities; but they live very extravagantly, and in good years make no effort to lay by anything to meet adverse circumstances.

A second report was prepared for the Board two years later by Henry Doran.[39] 'From an agricultural point of view Clare Island is not an inviting place', he wrote:

> Nature did little for it, and mankind has robbed it of all it could ... The physical conformation of the island is very rugged and irregular, offering serious obstacles to improvement. ... The methods of husbandry adopted are of the most careless and primitive kind and there is not a man on the island who makes, or has made, the slightest improvement of his land ... Much of what was then [1840s] reclaimed has since returned to its original state.[40]

Nevertheless he saw it as capable of 'much improvement'. He echoed Ruttledge-Fair's condemnation of the practice of using peaty scraws from the commonage as fuel.

On the other hand, Charles Browne found the islanders kind, obliging and communicative, if shy and reserved at first with strangers. They had a keen sense of humour and an excitable nature. Drunkenness was unknown.

> They are very kindly to one another in times of trouble or distress. The charge of laziness has been brought against them, and with some degree of justification; but the manner in which they worked when organised by the Congested Districts Board, and when they had some real inducement to do so, leads one to think that they did not work on account of having no real interest in doing so.[41]

This last point is very relevant. There was no need or scope under the old clachan system for the kind of diligence appropriate to the independent yeoman farmer, but when work needed to be done it was done by all.[42]

It is hard for us today to imagine the cultural disorientation that came in the wake of the Famine. Before this everything had been done as a community activity, and there were no territorial walls in the space between people.

They found themselves unaccustomed to the kind of industriousness needed by the new breed of farmer they now had to become, a kind of industriousness that had neither been needed nor possible under the moribund way of life they now had to move away from and for which they had little preparation.

The remaking of Clare Island

The Congested District Board bought Clare Island for £5,000 on 15 March 1895, and proceeded to reorganise the chaotic pattern of land-holding into one of viable farms, where each farmer had all his land together. The 76 new farms were divided from each other by sod-banked fences. They often took the form of narrow 'stripes' of land extending from the mountains right down to the sea, so that everybody still had a fair share of land of varying

Fig. 9.11 Men of Clare Island photographed by Charles Browne in 1896 (Charles Browne collection, Trinity College).

Fig. 9.12 Women of Clare Island photographed by Charles Browne in 1896 (Charles Browne collection, Trinity College).

quality, the stripes being assigned to each tenant on the basis of the rent of the old holding.

The new farmhouses that replaced the pre-CDB homes in the clachans were little more than an enlarged and more refined version of the existing dwelling type. The homes were built by the islanders themselves, using local materials—often from demolition of the older buildings; although two-roomed houses were still in the majority by 1901, most were three-roomed by 1911.[43] Sheds were built for the animals, and main drains were opened. The mountain commonage was fenced off from the newly enclosed ladder farms by an immense stone wall nearly seven miles long and six feet high, built by the tenants as a contribution towards the arrears on the purchase price (at a cost of £1,600). The original recommendation was for a barbed wire fence, but nothing less than a six-foot wall would have confined the mountain sheep to the commonage. Thirty-five miles of mearing fences (sod fences for the most part) were constructed from the wall to the sea.

The island's townlands were completely re-modelled. The striped land was confined to the new seven townlands of Strake, Kill, Glen, Fawnglass, Capnagower, Maum and Ballytoohey Beg. Two new commonages were defined: Bunnamohaun (A) and Lecarrow (B). The valuations assigned to the land element of holdings were also substantially altered.

The work took about five years (1895–1901) at a cost of £8,589, the working day on the island during those years governed by a bell erected near the quay. When the rent collected was deducted the net cost to the Board was £10,547, twice the purchase price of £5,000 paid for the island. On 25 October 1897 Clare Island was re-sold to the tenants for £10,000, to be repaid in the form of annuities of 3.5% (£325 a year) over a period of about 68 years. The tenants undertook to pay a portion of the arrears (an average of two years rent) within six months, which every one of them duly did. People who did not have enough money to buy stock were given loans of up to £15, all of which was repaid in full. A condition of the sale (imposed by the Land Purchase Acts) was that the holding could not be subdivided until the purchase price had been paid back in full.

New grazing, turbary and seaweed rights were defined and upheld by arbitration and careful

Fig. 9.13 Erection of the first 'congested house' on Clare Island in 1896 (Charles Browne collection, Trinity College).

supervision. Even geese (of which Clare Island had a large number) were tagged and monitored. Each tenant was allowed the grazing of a specified number of 'sums' or 'collops' of cattle, sheep or horses (a 'sum' was defined as a cow or other beast over three years old or an equivalent number of sheep—equivalent being eight) 'but the number of sheep was stinted'.

The number of grazing animals allowed to any tenant was calculated on the basis of his tillage area, which was measured not in acres but in *collops* or *sums*, the actual extent of which varied with the quality of the soil (on Clare Island it probably averaged around .3-.4ha). There were many attempts by individual tenants to exceed the number of animals permitted on the commonage, but this was strictly monitored and enforced.[44]

Stock on the island in 1905

145 cows	2,009 sheep
60 horses	110 calves
6 foals	72 yearling cattle
9 pigs	44 two-year old cattle
39 asses	1 three-year old cattle

Each tenant was allocated an area of turbary on one or other of the two commonages, 'where the bog is of such depth that turf can be procured and leave at least six inches of peaty surface over the subsoil or underlying rock'. Lack of proper access to the good bogs higher up led people to cut peaty scraws from grazeable land closer to home, leaving nothing but bare rock.[45] Major Ruttledge-Fair had found that people were often not prepared to make the effort to get the good turf that was to be had on higher ground, preferring instead to 'cut scraws, half peat and half clay, in the vicinity of their houses. In this way acres upon acres of very fairly good grazing land are being annually destroyed, and rendered useless for any purpose.'[46] Each tenant was allowed an equal share in all seaweed along the shore.

Each farmer ended up with 15 to 20 acres of potentially arable land on the lower ground and a right of grazing for so many sheep or cattle on the mountains. Three or four acres of this might be tilled at any one time, half under oats and half under potatoes, alternating year about for 8–10 years, after which the land was allowed to revert

Fig. 9.14 The Great CDB Wall running across the south-eastern slopes of Knocknaveen above Glen.

to grass and a new patch broken up. There might be a little rye, barley and turnips or mangels. The remaining acres of the farm south of the wall were grazed by the stock.[47] The new arrangement gave sufficient incentive for some farmers to begin the practice of re-seeding around the turn of the century, but many continued to let nature look after the fallow.

Fig. 9.15 **Last of their race**
Modern representatives of the greyhound pig, photographed on Clare Island by Sir Arthur Ball in the early twentieth century (Scharff, 1917).

The old Irish farm pig was a very different animal from today's pig: an independent, tall, long-tailed hairy animal, so fast on its feet that it earned for itself the name of the greyhound pig. It was a life-saver for the small cottier, and often had the run of the miserable hovel in which he lived, to the horror of many travellers in 19th century Ireland. Its downfall was its failure to fatten as readily as more modern breeds, which drew upon its head the condemnation of the Congested Districts Board (Feehan, 2003, 178–9).

In areas close to the sea the potato land was heavily manured with seaweed in November, and again in spring, after which 6–10′ ridges were marked out, separated by a 2–3′ trench. The potatoes were planted along the ridges, and covered with soil from the trenches. On a holding of this size the average family now kept a cow or two, at least a score of sheep, a couple of pigs (fattened with Indian meal), hens, ducks or geese, a donkey and a brood mare (there were 80 horses on the island); the calves would be sold at 15–18 months.[48] The Board had to work hard to prevent overstocking of the commonages, which Doran calculated could only carry a maximum of 800 sums of mixed stock, and he recommended that sheep should account for no more than 500 of these. In 1897 however, the total number of mixed sums was 953.[49]

Poultry were the responsibility of the women, and there was a considerable trade in the export of poultry and eggs to Westport. There was further improvement in the breeds of stock on the island: although Clare Island was one of the last strongholds of the old Irish farm pig, that extraordinary animal survived long enough to be photographed here in the early years of the twentieth century.[50]

Into this newly-made landscape the naturalists and antiquarians of the Praeger Survey came, no more than a few years after its making: and it is important to remember that all they recorded had survived the enormous human impact of the previous century on the landscape: just as everything that the New Clare Island Survey recorded has survived all that happened in the hundred years since.

The remodeled landscape of the CDB is essentially what we see today. Nothing remains of the clachan settlements but scattered banks and wall footings, house platforms and grassy

Fig. 9.16a The purchase and reorganisation of Clare Island was one of the first and most successful of the Congested District Board's projects. When it was bought from the landlord there was not a single fence on the island: a constant guard had to be mounted when the crops were growing to keep the animals out of the tillage land, where individual holdings consisted of several scattered patches.

Fig. 9.16b The new face of Clare Island after its reorganisation by the Congested Districts Board: new farmhouse and stripes of land running from the shore to the mountain commonage. The photograph shows the newly-built wall separating the farmland from mountain commonage (Coyne (1902), *Ireland Industrial and Agricultural*).

footpaths. Scarcely a stone remains upon a stone, the walls having been dismantled and incorporated in the walls of the new farmscape. They have melted back into the ground to become another layer in the archaeological palimpsest, less conspicuous by far than the ridged fields of which they were once the very heart.

Trí nithe is sia 'na bhian ann a rian: rian guail i gcoill, rian siséil i lig, agus rian suic i gcrich.
Three things which leave the longest traces: charcoal in a wood, a chisel on a stone, and a spade on a furrow.

Notes

1 Whelan 1994, 1997.

2 Although the term 'clachan' is generally used by historical geographers for these cluster settlements, it has a different meaning in Scotland, where it originated and where its primary meaning is a village with a church. In Ireland such a meaning is secondary (Críostóir MacCárthaigh, pers. comm.).

3 The rundale and clachan landscape may have originated on Clare Island in the seventeenth century, as the island passed to absentee landlords.

4 Micks 1925, 10.

5 MacCárthaigh 1999, 65.

6 Browne 1898–1900, 62.

7 Micks 1925, 25.

8 MacCárthaigh 1999, 49.

9 Micks 1925, 11.

10 Ní Chinnéide 1976-77; Knight 1836.

11 Feehan 2003, 115–120

12 Vullings *et al.* 2013, 155.

13 Lewis 1837, 326.

14 MacCárthaigh 1999, 49.

15 Much of this summary is based on Kevin Whelan's superb précis in *The Atlas of the Irish Rural Landscape* (1997).

16 On Clare island (according to Críostóir MacCárthaigh) 'the older loy, forged by island blacksmiths', was replaced by an Ulster type spade in the late 19th century (MacCárthaigh 1999, 53).

17 Bell 1984.

18 MacCárthaigh 1999, 51.

19 Devon Commission 1847, 84–85.

20 Bell and Watson 1986, 112–37; *The Munster Farmer's Magazine* (1812), Vol. 1, 52–53.

21 MacCárthaigh 1999, 41.

22 MacCárthaigh 1999, 43.

23 MacCárthaigh 1999, 42.

24 Sutherland 1909, 133–34.

25 MacCárthaigh 1999, 43–44.

26 Browne 1898-1900, 63.

27 Quoted in Vullings, Collins and Smyllie 2013, 4–5.

28 'In 1846 there were 1616 people on the island, which was at that time the property of Sir Samuel O'Malley of which name there were 67 families resident in 1821' (Browne 1898–1900).

29 Although the General Valuation listed 61 joint tenancies in the mid-nineteenth century, the 1851 Census has 175 households, the apparent discrepancy being the result of the subdivision of holdings (MacCárthaigh 1999, 43).

30 *Illustrated London News*, 24 April 1886.

31 Micks 1925, 31–32.

32 Micks 1925, *passim*.

33 Ruttledge-Fair 1892.

34 *Ireland Industrial and Agricultural* gives different figures: 95 tenants, a rental of £495.16.6d, one-fifth held in commonage.

35 Appendices to the *Third Report: Minutes of Evidence. Royal Commission on Congestion in Ireland* (1906), 106. It was sold by the mortgagees, an insurance company.

36 Appendices to the *Third Report: Minutes of Evidence. Royal Commission on Congestion in Ireland* (1906), 101.

37 Although the Confidential Report of 1892 tells us that there was hardly any fishing, and in the *Minutes of Evidence in the First Report*, we read that 'there is no fishing worth talking about, except that they get coarse fish for food round the rocks.'

38 Whelan 1999, 90.

39 Henry Doran was the Board's inspector between 1895 and 1907.

40 Doran 1908, *Tenth Report: Minutes of Evidence. Royal Commission on Congestion in Ireland*.

41 Browne 1898–1900, 58.

42 *Ibid*.

43 MacCárthaigh 1999, 62, 65.

44 MacCárthaigh 1999, 45–46.

45 *Tenth Report: Minutes of Evidence. Royal Commission on Congestion in Ireland* (1908), 16.

46 Ruttledge-Fair 1892, 3.

47 Wilson 1911, 44–45.

48 *Ibid*.

49 MacCárthaigh 1999, 47.

50 Feehan 2003, 178–79; Scharff 1917.

Opposite

Fig. 10.1 Branding sheep on Clare Island a century ago (Welch photograph) (National Museum of Ireland).

Chapter 10
Living off the land: agriculture past, present and future

The agriculture is improving, and large quantities of grain are shipped here for Westport—but the crops are sometimes seriously injured by storms.

Samuel Lewis (1840)

Introduction

What little we know of the farming practice of the earliest peoples on Clare Island is outlined in earlier chapters. Our evidence is limited to clues provided by the pollen record and the scant remains of field walls and enclosures; artifactual evidence other than this is almost entirely lacking. So what are we to make of the irregular pattern of small fields beneath the bog or smothered by the hill-slope talus of later millennia? What were they intended to enclose or exclude? What crops did they shelter? The collapsed footings we see today must have been substantial drystone walls in their day. We get a glimpse of this where fields at the foot of Knockmore were smothered in talus following their abandonment; elsewhere the walls were exposed to the forces of disintegration for much longer as bog grew up around them (Figure 10.2). Most of the other relics of earlier farming on the island—lime kilns and corn-drying kilns, forges and watermills—date to much later periods.[1]

Ownership and land division

Clare Island is thought to have been divided into four 'quarters' in medieval times. These were quarters of an original larger unit, the 'baile' or 'towne', usually made up of sixteen 'townlands'. Clare Island was probably a single baile. The land that supported the Abbey is recorded in 1584 as being one quarter, but in the late seventeenth century we find it reduced to half this: *Lecarrowna-manestragh*—the half of a quarter belonging to the monastery—whose name survives in the pre-1890s townland of Lecarrow.[2] The townland of Strake, which adjoins Lecarrow on the west, is thought to be more modern, land newly reclaimed and 'striped' (= *stráca*) in the eighteenth century, but part of Lecarrow originally. Lecarrow and Kill therefore, were one of the medieval 'quarters', and Bunnamohaun to the west of it was a second. Five of the remaining townlands meet at Lough Avullin, making it a highly significant topographical 'node'. Maum and the two Balltoohy townlands probably constituted a third quarter, and the remaining townlands of Glen, Fawnglas and Capnagower represented the fourth and richest quarter. By the sixteenth / seventeenth centuries Kill could have been regarded as a full 'quarter' on its own, bringing the number of major land divisions at that time to five.[3] The very different way in which the island is divided up today is a direct legacy of the radical reorganisation of land holding on the island by the Congested Districts Board (CDB) (see Figures 9.7, 9.8).

Having passed through a number of hands in the early part of the seventeenth century (Evelin Moore and her husband Robert in 1625, Sir Richard Blake in 1634) the entire island was granted following the restoration of Charles II in 1660 to the Earl of Clanrickard, from whom George O'Malley of Rossmindle recovered it in the early eighteenth century, but his grandson, Sir Samuel O'Malley sold it in the middle of the nineteenth century. After this it passed through the hands of several absentee owners until it was bought by the CDB in 1895.

The population rise of the nineteenth century transformed the island. The cultivated land was concentrated mainly all along the moraine-covered 'marginal platform' from Toormore eastwards to Capnagower, and in the north-east. This marginal platform occupies more than a third of the island, and from it the hills of Knockmore and Knocknaveen rise 'with remarkable abruptness'. Timothy Hallissy identified this as 'part of the great post-Carboniferous peneplain that extends over the central plain of Ireland, while the hills of Knockmore and Knocknaveen are remnants of a higher and older plateau from which the mountains of the western highlands have been carved'.[4] A striking example of how cultivation followed the glacial till is seen in the col between Knocknaveen and Knockmore, where an impressively preserved system of unusually wide cultivation ridges was developed on the till deposited in the col by stagnating ice at the end of the Ice Age (Figure 10.3).

All of the 'clachan' settlements were located along the boundary between the arable land and the 'rocky mountain pasture', and ever-increasing grazing pressure came to bear on this common pasture.[5] James Wilson (who visited the island in July 1909) was puzzled by this and went to some pains in his attempts to explain it, but as we have seen earlier, this was the usual pattern for cluster settlements of this kind. The lack of settlement near the harbour in 1838 is in striking contrast with today's settlement pattern.

Little trace of an earlier agriculture survived the great changes that accompanied the rise in population of the early nineteenth century and the radical restructuring of the CDB. The pattern of

Fig. 10.2 (a, b) Prehistoric field walls in Bunnamohaun; (c) Walls at the foot of Knockmore were sometimes buried more rapidly under hill-slope talus, allowing the preservation of more substantial remains. These walls outcrop in the banks of the Guaille Lich stream in Bunnamohaun.

Fig. 10.3 Cultivation ridges on the western slopes of Knocknaveen (Glan Hill) viewed from the North Road.

farming on the new compact holdings that followed the restructuring of the island by the CDB was concisely described by James Wilson:

> In some parts, especially along the southern side, a farm strip runs from the mountain down to the sea. Each farmer has from fifteen to twenty acres of arable ground down below and a right of grazing for so many cattle or sheep upon the mountain.
>
> If we take one farm with seventeen acres below, we shall see how the farmer works. Of these seventeen acres he tills about three. This area is divided in two halves; and each half grows oats and potatoes year about for eight or ten years. At the end of that time a new patch is broken up and the patch that has been tilled is left down to pasture. Till recently it was the custom to let the patch sow itself, but now mixtures of grass and clover seeds are sometimes sown down. The remaining fourteen acres are grazed by the farmer's cows and calves and by the pony, if he have one.
>
> Here and there some barley and a small quantity of rye are also grown; there are a few patches of turnips; and recently mangels have been tried. This of course, means drill cultivation.
>
> The land is manured as follows: The stubble or grass intended for potatoes is heavily manured with seaweed in November, and again in spring.[6] Ridges from six to ten feet wide are then marked out and two or three feet of a trench is dug out between them: the dug-out soil being thrown on the ridges to cover the manure. When the potatoes are planted, dung is also added; and when they appear above the ground, a little more soil from the trench is spread over them.
>
> This farmer's stock is usually two cows; two calves which are sold when they are from fifteen to eighteen months old; thirty sheep; a pony mare which rears a foal nearly every year; and one or two pigs. Indian meal is bought to fatten the pigs.[7]

With the abandonment of tillage the arable land was allowed to revert to poor pasture, grazed in the main by sheep.

The booley

In many parts of Ireland it was for a long time the practice to drive cows and other farm animals to hill pastures for the summer months. These summer grazings were known as the *buaile*, and the practice itself as booleying. From early May to the end of October the animals remained at the buaile with their herders, sending milk and butter down to the home farm at intervals. This is the Irish version of a seasonal movement of livestock or transhumance that has been central to the economy of rural regions for millennia, and indeed still survives in some areas.[8] Booleying maximised the use of grazing resources and allowed the grassland around the home farm to be rested for winter grazing or grown on as meadow, and it also protected the often poorly enclosed crops from hungry animals.

The cattle were usually herded and accompanied by the girls and young women of the community. Their spare time was often spent spinning and knitting, but there was abundant leisure for talk and play. The buaile had a deep psychological and social impact on the community; the women who spent their young summers in the buaile carried with them all their lives the memory of a unique freedom and sense of grace. During their stay they lived in small huts, rectan-

Fig. 10.4a The Prawke booley.

gular or circular in shape and often built of unmortared stone, generally thatched with heather and rushes, which also provided bedding. These were generally sited close to mountain streams, because an ample supply of clean water was essential to the butter making.

On Clare Island there were two main clusters of booley huts, which are thought to predate the nineteenth century, when the western part of the island was sparsely populated.[9] One of these consists of a group of six huts and houses situated halfway up the southern slopes of Knockmore between 250 and 300m, on either side of one of the head-streams that merge to form the Owenmore river, along with a handful of others, in the amphitheatre known as Gabhlán. The second group (known as the Prawke booley) was located on a grassy terrace in a sheltered hollow halfway up the south-western flanks of Knockmore at 175m, invisible from below. This comprised eight huts and houses on the right bank of the Guaille Leic river that flows westwards off Knockmore (Figure 10.4). In order to prevent flooding at times of spate, the flow of the stream was altered, and its bed below the hollow deepened by removing boulders; these form a wall-like rampart on the right

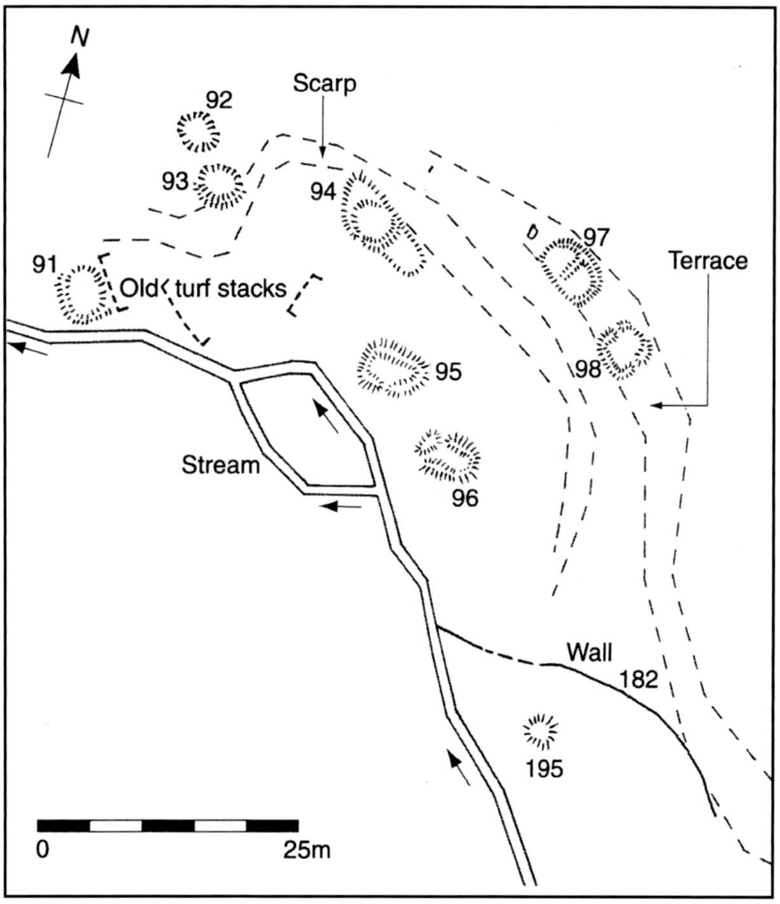

Fig. 10.4b The relative proximity of these groups of huts to permanent settlement suggests that they may have been temporary shelters for herders and milkers in summer rather than true 'booley' settlements. They were oval in plan, 2m x 3m in size, the thick, rounded gable walls providing the main structural support (*New Survey* Volume 5, page 152).

Fig. 10.4c Close-up of one of the huts in the Owenmore group (*New Survey* Volume 5, page ix).

bank. Some of the surviving structures are wall-footings of the huts in which the herdswomen slept and sheltered; others may have had something to do with butter-making and the storage of milk and dairy produce.

Seaweed as a resource

The fertility for which the island was noted in pre-Famine times was to a very real extent a product of the surrounding seas, because it was augmented and sustained by the bounty of the shore, which provided an endless supply of nutrient-rich seaweed.[10] Seaweed was always liberally applied to potato ground before planting, and on oats after harvest and again in spring. A dressing was also applied to meadow land.

The collection and burning of kelp was an important routine in the farming year on Clare Island. Weed was collected in the coves and bays from November to February, and cut from the rocks during March, April and May (Figure 10.5).[11] In May the reddish fronds of the 'May weed' (*Laminaria hyperborea*) and the streaks of white due to patches of bleached *Laminaria digitata* (oarweed) and *Laminaria saccharina* industriously collected during winter and spring (especially after storms) 'give striking colour effects on the green fields'. On Clare Island 'other weeds are despised'.[12] The burning took place in the second half of June and early July, 'when clouds of heavy oily smoke may be seen along the coast-line', but the preparations went on for several months before this.[13]

Seaweed became extremely important for another reason from the late eighteenth century on. A range of new uses was found for the soda, iodine, bromine and other chemical substances in

Fig. 10.4d Close-up of one of the Prawke huts (*New Survey* Volume 1, 63).

Fig. 10.5 Colour print from G.H. Kinahan (1869), showing 'an ideal submarine garden' on the wet coast, depicting *Ascophyllum nodosum* (knotted wrack), *Fucus vesiculosus* (bladder wrack), *Fucus serratus* (serrated wrack), *Laminaria digitata* (oarweed), *Saccharina latissima* (sugar kelp) and *Alaria esculenta* (bladderlocks). From 'The sea-weeds of Yar Connaught, and their uses.'

Fig. 10.6 Edible seaweeds (a) sloke (*Porphyra umbilicalis*); (b) dulse (*Palmaria palmata*) (Mike Guiry).

seaweed, and there was a ready market for kelp (fused seaweed ash), which rose in price from £2 a ton in the 1820s to £15–20 in the early 1840s.[14] It remained profitable for a few more decades, but eventually cheaper sources of iodine, potash and bromine were discovered in great quantities in other places, and the price of kelp fell so low as to be uneconomical. By 1885 kelp-burning was largely a thing of the past, although by the time of the Original Clare Island Survey about thirty tons of burnt kelp were still being exported annually, worth around £4 a ton at this time.[15] The practice survived on a small scale until the end of the Second World War.

> Sea rods were gathered after the winter storms. The blades had to be pulled off by hand and the rods draped over a wall to dry. In March they were tied in bundles and taken across to the mainland. In 1960s you would get £8 or 9 per ton.[16]

Within living memory carrageen moss and sloke were gathered in springtime for local use, and until the 1960s carrageen was picked, dried and bleached for sale.[17] Carrageen (*Chondrus crispus*, with some *Mastocarpus stellatus*[18]) was at an earlier time regularly sold as an article of commerce on the island (destined for Dublin and thence Liverpool, 'where it is said to be chiefly used for fattening calves'). It was boiled to a jelly, and then added to milk. It was also used medicinally, and for 'dressing' linen and rope. Carrageen grows in a dense and tangled turf below the zone of the rockweeds and mussels, the labyrinthine spaces between its branchy fronds and stems a city in which live troglodytes of bewildering variety: brittle stars and sea urchins, baby starfish, tube-dwelling amphipods, nudibranchs and moss animals.

Dilsk (dulse) and crannogh are epiphytic forms of *Palmaria palmata* commonly found growing on wrack and kelp.[19] The fronds were chewed in a fresh state or after drying. Sloke (*Porphyra umbilicalis*) grows abundantly on exposed rocks. It was gathered in spring, stewed at once or boiled into a jelly for future use, when it was re-cooked and fried with butter (Figure 10.6).

Fig. 10.7 (a) Turbary in Ballytoohy; (b) Old turbary pool in Bunnamohaun.

Fig. 10.8 Peat-drying stands (turf-stacks) in Bunnamohaun. Note the shallow depth of the peat, which now remains only under the stand itself.

Fuel: the bogs

The last of the woods disappeared in the eighteenth century, and the islanders turned to the great reserves of peat in the bogs at Park, Ballytoohy and on Knockmore (Figure 10.7). Under the several metres of peat at Park are concealed the remains of an earlier farm landscape; during drainage work around 1980 several fulachta fiadh were uncovered under the bog here, where Bronze Age farmers sat around their cooking fires before a worsening climate precipitated the growth of peat.

Turf was cut wherever there was peat of any depth. There are deep pockets of peat in the hollows of the bouldery moraine at Maum and in the west of the island at Bunnamohaun. Higher up there is deep bog on Knockmore. A more localised area of deep peat on the cliffs east of Doon Fort was extensively exploited and cut away a century or so ago. The island commonages are dotted with the remains of turf stacks, many of recent construction and in a state of good preservation, but most no more than low, irregular cairns of stones. They are particularly numerous on the commonages to the south of Lough Avullin and in the vicinity of Loughanaphuca in Bunnamohaun (Figure 10.8).[20]

Other natural resources

Other than the abundant harvest of the sea, the natural biological resources of the island could contribute little to the human economy once the forests were gone. In addition to fish and seaweed, shellfish were eaten, becoming more important in times of want; periwinkles especially were consumed in vast quantities (mostly plain boiled, with salt and vinegar). The only midden recorded on Clare Island contained shells of oyster, limpet and periwinkle.[21] Thomas Westropp believed the reason middens are scarce on the islands is because prehistoric people tended to avoid them; it seemed to him that 'the early settlers, circle-makers, dolmen- and cairn-builders and sandhill-dwellers did not venture across the fierce currents and waves round Clare Island, Turk, and Caher, to those islands; certainly save for pillars in the first two, I found nothing resembling an early monument'.[22]

As long as the islanders continued to weave their own cloth, wild plants and lichens were gathered to make what dyes were needed. Black dyes were made from bramble roots (*driseóg*), and from the roots of meadowsweet (*airgead luachra*) mixed with copperas (iron sulphate). A superior yellow dye was made from the rock lichen *Parmelia saxatilis* (*scra cloch*), and a saffron dye from *féitóg liath* (*Ramalina siliquosa*) a shrubby lichen that is common on rocks above high water mark. A yellow dye was also made from the flowers of furze (*aiteann*).

A small number of wild herbs were routinely used for medicinal purposes. The leaf rosettes of marsh ragwort (*boglus*) were pounded to make a poultice for sores. Ribwort plantain (*slán lus*) was a well-known remedy for cuts and sores; the leaves were pounded and mixed with those of broad-leaved plantain to make a poultice (*copóg Phádraig*).

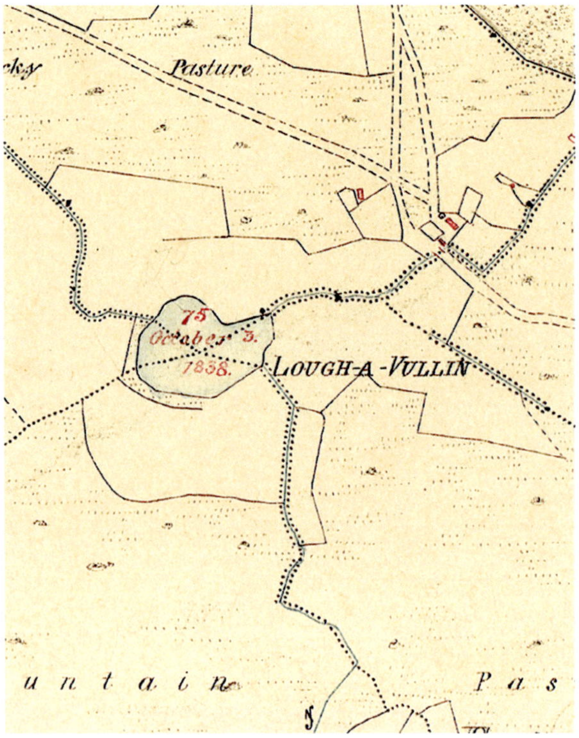

Fig. 10.9 Lough Avullin on the O.S. Fair Plan of 1838.

From self-heal (*dubhán ceann chaorach*) a tisane was made, 'a drink called Cailleach's Tay, that's very good for a weak heart'.[23] The use of orache (*ceathramha caorach*) is also recorded.

Even a few generations ago the island had to be self-sufficient in *materia medica*; there was neither doctor nor vet. In the late nineteenth century, a poultice of mashed ferns was used to treat burns, and for cases of 'gravel' a decoction of wild geranium. For expectorant use a syrup made with sugar and vinegar, with broom-tops and nettle roots was prepared.[24] A standard—and reputedly often effective—prescription for people in 'very delicate health' a century ago was a herbal mixture made from four herbs: plantain (*slánlus*), burdock (*crádán*), dock (*cupóg*, 'without seeds') and '*cróagh paraic*'.[25] The liquorice-flavoured tubers of *Lathyrus macrorrhizus* (*carra milis*) were dug up and eaten by children, as were the roots of silverweed (*blucán*) after roasting.[26]

Querns, watermills and other relics of earlier agriculture

In earlier times grain was ground as required to make meal on a hand-mill or quern. With the growth of population and the advance of technology the work of grinding was mechanised, using the power of water or the wind. There may have been a horizontal mill along the upper reaches of the Toormore river south of Knockmore in Bunnamohaun at an earlier period, but as the population rose in the eighteenth and early nineteenth centuries Sir Samuel O'Malley decided to invest in a watermill.[27] This was an expensive and time-consuming, but potentially profitable, venture. There were few suitable sites, because all the rivers on Clare Island are spate rivers, requiring a mill pond to store and control the flow of water to the wheel. The site chosen was at Maum in the north-east of the island, where Loughavullin (*loch a mhuilinn*) became the millpond (Figure 10.9). At this time however the only stream flowing into Loughavullin was the Pollbrandy, and there was no natural outflow.

Loughavullin is something of an anomaly. It appears on the six-inch map as a substantial shallow lake before the Famine, but was only half its former size by the time Praeger saw it 'owing' (it was thought) 'to the deepening of its outflow',[28] but more likely because ecological succession was even then far advanced. Today it is simply a marsh, with no open water. When the Ordnance Survey was collecting placenames in the late 1830s the only name the people of the island had for this substantial body of water was 'the mill pond', because this is what 'Loughavullin' means. It had no older name. Had there been a lake in this essentially depositional setting in previous centuries, it would within a few hundred years have filled in.

The Mill river starts out as the Doreen stream that flows down the northern slopes of Knocknaveen before curving around to flow through the gap north of the col between it and Knockmore. Then, at Maum Cross it bends sharply to flow eastwards to Lough Avullin. A deep cutting had to be constructed to take the diverted flow, which was augmented with waters gathered from Ballytoohy Beg by constructing a dyke (Figure 10.10). Lough Avullin was named *after* the mill; it did not exist as open water before this. The old course of the river that had carried the stream westward to the sea at Portlea became Lassau (*leas abha*), and an area of alluvial woodland unique on the island grew up in the former floodplain. This had puzzled the early botanists; how had woodland survived here when it had been so thoroughly cleared from the rest of the island? Significantly however, there is no sign of the wood at Lassau on the first edition of the Ordnance Survey six-inch

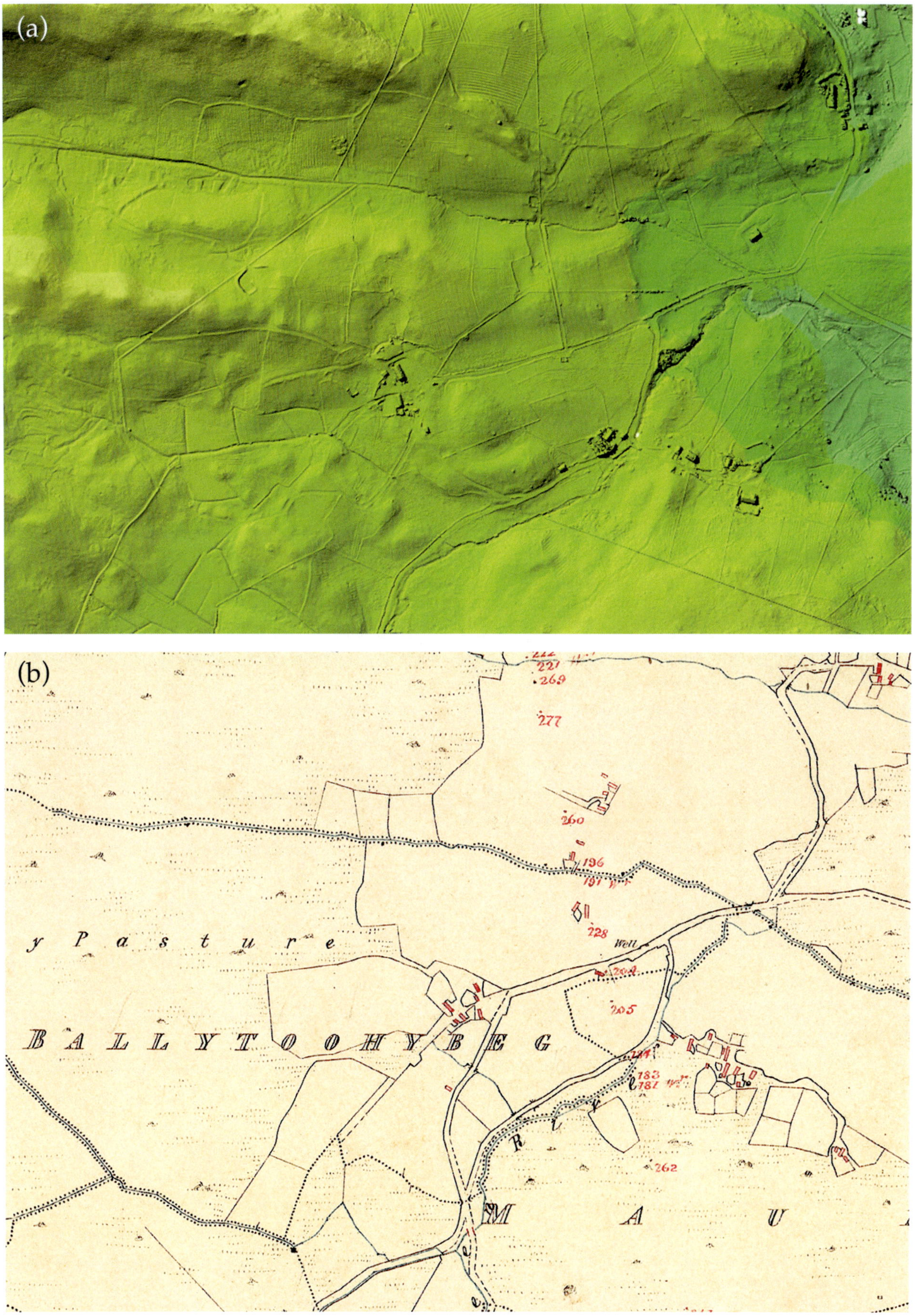

Fig. 10.10 The course of the Dorreen (a) on Lidar and (b) on the Fair Plan for the first edition of the six-inch map (1838).

Fig. 10.11 Lime kilns.

map nor on the Fair Plan (1838); another hint of its recent origin is the presence of faint cultivation ridges in the wood. It seems appropriate that Macalla Farm, through which the vanished river once flowed, should so be called today, for *macalla* means 'echo' in Irish, and surely the river must echo still in some silent way here where its waters ran for millennia.

The mill was not a commercial success; although in the 1830s 'large quantities of corn' were being exported from Clare Island, the mill was in ruins by 1838. Everybody on the island ground their own corn with querns until 1870 (querns frequently turn up in the course of archaeological investigation). The milll's short life may have been because the hydrology had not been sufficiently researched, and the lake did not deliver enough power to operate an undershot waterwheel. Within decades Loughavullin began to revert to its natural state. Had it been a lake for all the centuries before this, it would not have accumulated the peat profile studied by the palynologists.

The mill at Loughavullin was a small rectangular building, about the size of a traditional house of the period. Local tradition maintains that there was a second mill at the other end of the island, beside the Toormore stream in a sheltered oasis of former cultivation near the foot of Knockmore known as Páirc Mhíchíl. Strangely, on the first edition of the six-inch Ordnance Survey map the Green Road crosses the North Road and continues across the lower slopes of Knockmore at the northern end of Lecarrow and Strake to end at this spot. Although the situation is difficult to interpret hydrologically, a horizontal mill at this location is a possibility.[29]

The occurrence of the element '*tuar*' in several placenames is thought to be a relic of a time when the growing and processing of flax was common; three possible bleach greens have been identified on the island.[30] The lake at Toormore may have been used as a retting pond in the days when linen production was one of the island's industries. Four smithies are known from the early modern period, and two corn-drying kilns.

Lime

Calcium is another essential nutrient in which Clare Island's acidic soils are naturally deficient. This deficiency was countered by burning limestone boulders from the Roscahill Till in order to obtain lime for spreading on the land, and to a lesser extent for making lime-wash and mortar for building. Over thirty lime kilns have been recorded on the island, some very well preserved (Figure 10.11). After the CDB reorganisation of farming on the island at the end of the nineteenth century limestone was also brought in from outside. The large kiln close to the pier head at the harbour was built in 1912 to burn limestone imported from the Aran Islands.[31]

Getting around

On Clare Island in earlier times little use was made of wheeled vehicles, and a complex network of paths and trackways developed that were more suited to people travelling on foot or with donkey or pony.

The main roads radiate outwards from the harbour, which has always been the Island's nexus of communication with the wider world across the water. The straight West Road that connects the harbour with the Abbey at Kill was probably laid out by Mayo's Grand Jury (the forerunner of

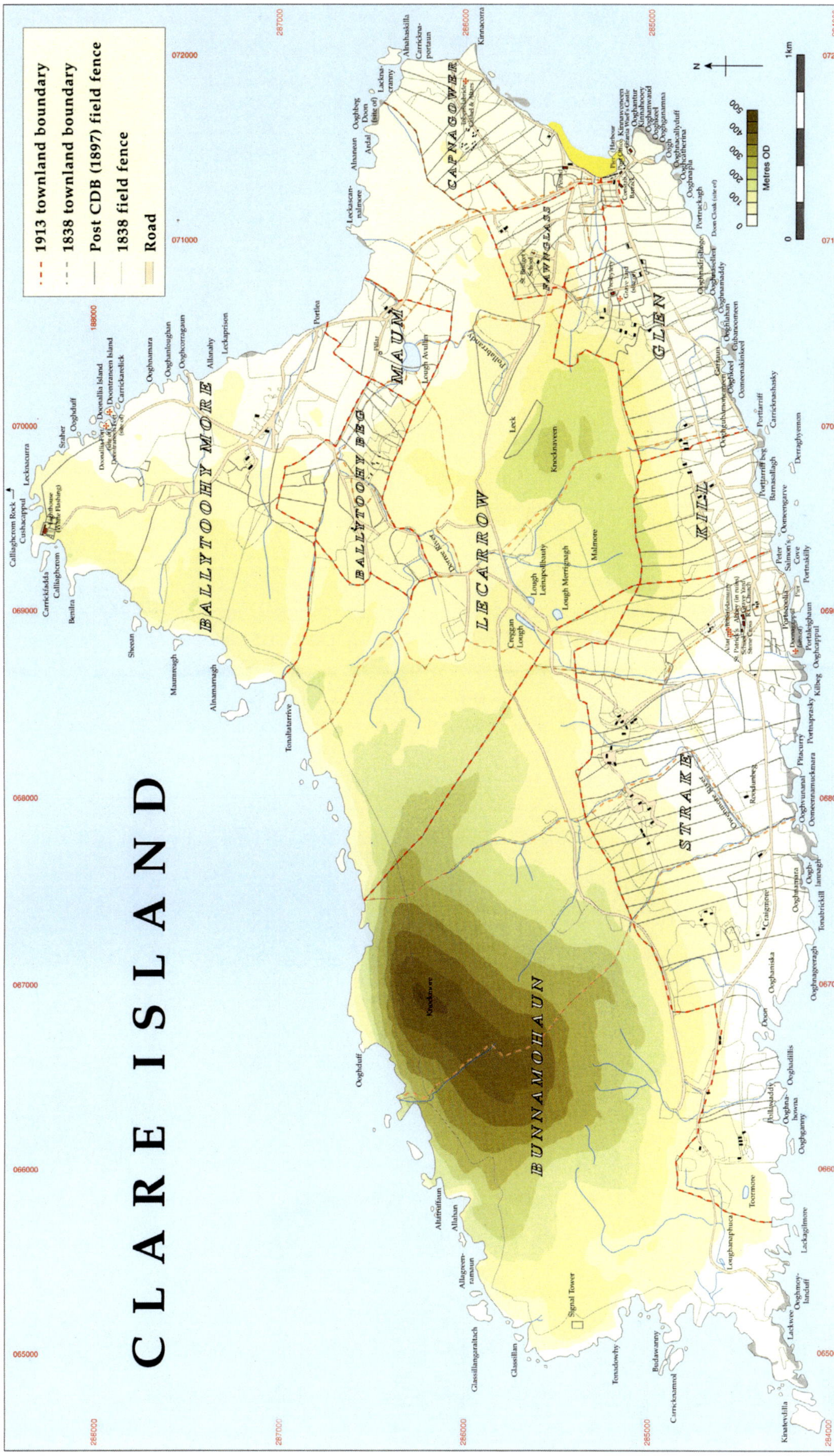

Fig. 10.12 Clare Island at the time of the Original Survey, showing the road network (*New Survey* Volume 1, foldout map at end).

today's County Council) towards the middle of the nineteenth century. The stretch of road west of Kill came later, and originally followed a more southerly route to the house cluster at Toormore and further towards the coast at Leckwee. Before the construction of the present West Road two routes linked the harbour with Kill. One of these ran along the boundary between the cultivated infield and the commonage at the base of Knocknaveen, while the other stayed close to the south coast. These two converged at Kill, and from there followed a looping course to link up the house clusters of Craigmore, Strake and Lecarrow (Figure 10.12).[32]

The present Lighthouse Road, like the West Road, is a product of the nineteenth century. The final stretch was constructed early in that century to link the newly built lighthouse with the rest of the Island. Before this the road into and through Ballytoohy ran closer to the coast via Portlea.

The Green Road that runs along the northern edge of Knocknaveen connected the settlements of the east and south with the deep turf-banks at Park. Even in his later years in the 1980s John Samon of Glen made his way up to the Fawnglass Gap and along the Green Road to the bog every other day, bringing home a supply of turf and stone. The North Road that runs through the gap between Knockmore and Knocknaveen provided access from Kill to the bog at Park. It follows the line of a track that probably goes back to much earlier days, relics of which have come to light in recent years here under the bog at Park.

These roads were the arteries through which the fertility of the surrounding ocean was transferred to the land, and the warmth stored in turf from the cold uplands to the homes and hearths sheltering lower down. All along the south coast in Glen and Kill a series of smaller stretches of road ran between the cultivated infield and the sea, facilitating inland transport of the all-important seaweed.

Architecture

The Church of the Sacred Heart, (erected 1861–62), is one of only three prominent buildings built on the island in the nineteenth century, and a fine example of Catholic rural church architecture of the period (Figure 10.13). In the years leading up to the Famine, worship took place in a narrow rectangular building east of the present graveyard, of which no trace now remains.[33] Before this, as we saw earlier, people worshipped in the ruined abbey.[34]

The Signal Tower above the western cliffs was erected in 1804–06 as part of a network of such towers erected during the Anglo-French wars to give early warning of imminent invasion (there are similar stations on Achill and Inishturk). It was in use for no more than four or five years. A square building of local sandstone, the sides measuring

Fig. 10.13 Inside the Church of the Sacred Heart

Fig. 10.14 The Signal Tower

4.25m on the inside, it originally consisted of two stories and a basement, but the upper part is now collapsed.[35]

The Lighthouse was erected on cliffs close to the northernmost tip of the island around the same time. In fact there were two lighthouses, both of them circular towers, with associated outbuildings. The older of the two was erected in 1806, but this was replaced by a new lighthouse in 1818 after the old one was damaged by fire in 1813. The light was 'extinguished' in 1965; it then passed into private hands and has been extensively refurbished as a hotel in recent years.[36]

Fig. 10.15 The Lighthouse from the air

Farming today

The old marks of cultivation are very curious on the hillsides with curved and even S drills.

Westropp

Clare Island today experiences the same problems as small farms throughout Ireland, but magnified by the extra strains imposed by being an island. By the beginning of the last century the pattern of farming was little different from that practiced all along the western seaboard generally. It remained essentially self-sufficient, but became increasingly dependent on the market as time went by, its course coupled with the fortunes of Irish agriculture as a whole. During the first half of the last century land intended for tillage was first heavily fertilised and sown with oats and potatoes alternately for between four and six years (as a general rule), after which it was laid down to grass with hay seed and again heavily manured. After taking hay for a few years the land was then allowed to revert to pasture for between six and eight years before starting the cycle again.[37]

Down to the middle of last century most farm families remained self-sufficient to a very considerable degree. Everyone had a garden, in which all the family's vegetables were grown: potatoes in first place, along with cabbage, carrots, parsnips and onions. Nearly everyone kept a few cows, poultry, and often a pig.

The increased affluence that followed Ireland's accession to the EEC in 1973 was accompanied by an ever-increasing dependence on external markets for agricultural produce, and an increasing dependence on purchased foodstuffs that in time completely supplanted the traditional self-sufficiency This new dependency was also accompanied by an ever-rising threshold of expectation in relation to comfort, security and choice.

With Ireland's accession, sheep—always important—became more central in the island's economy. The Ewe Premium gave the final push away from cattle. The switch to continental breeds of cattle was not easily made, partly because of the expense of importing the required supplementary feeding and the cost of winter housing. The sheep on the island are mainly mountain blackfaces, bred to a Suffolk and the occasional Texel ram, which gives

Fig. 10.16a Spade ridges in early spring, shortly after planting. Once the first shoots appear above ground more soil will be dug from the sheugh and spread on the ridges. Small heaps of seaweed can be seen on the slope in the background, where ridges are still being made (L.E. Klim, 1934; National Museum of Ireland) (*New Survey* volume 1, 52).

Fig. 10.16b Spade ridges on sloping ground were often curved or even S-shaped to reduce erosion: and also because it was easier to make them this way than straight because the digger worked backwards downhill. (L.E. Klim, 1934; National Museum of Ireland) (*New Survey* 1, page 52).

Fig. 10.16c Fossil ridges in the fields of Gallagher's Farm on the Lecarrow commonage in spring (May 2017) viewed from the eastern edge of Knocknaveen. These are currently overwhelmed by bracken, and almost impossible to see in high summer. Note the cross-digging in the fields at the back. The farmstead is an island surrounded by once-cultivated land now completely smothered in bracken: an example of one of the major problems associated with the abandonment of traditional management, resulting in the loss of good grazing on grassland of high cultural significance and biodiversity value.

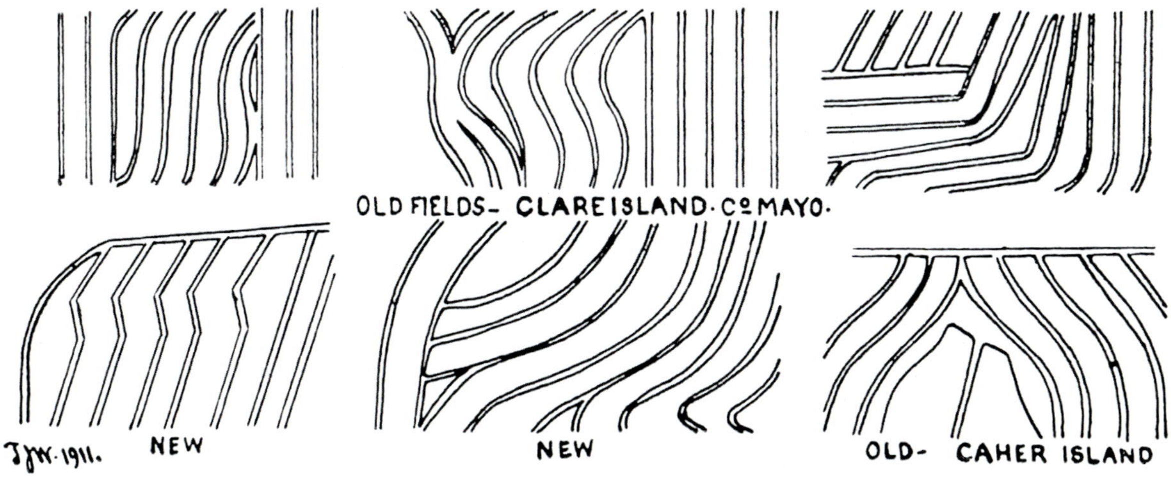

Fig. 10.16d Thomas Westropp commented on the antiquity and curvature of plough ridges in his review of the archaeological monuments. 'On the shoulder above Budawanna', he recorded, 'an irregular loop of small set stones, the age and purpose alike doubtful. The old marks of cultivation are very curious on the hillsides with curved and even S drills.' (*Original Survey* 2, plate vi).

Fig. 10.16e Gallagher's Farm in September, showing the extent of bracken cover on the abandoned cultivation ridges.

small lambs.[38] Most sheep are sold at home as stores to a buyer who visits the island once or twice a year. However, the increased numbers that resulted from sheep headage payments led to severe degradation of areas of the commonage, especially since the mild climate allows the animals to be outwintered.

Arable enterprise has all but disappeared, and no more than a handful of farmers now follow a variant of the traditional oats-potato rotation; these days, rolled oats are the main supplementary feed. Very little nitrogen is applied to land; certainly less than is needed for optimum yield. The only pollution is that which results from inadequate septic tanks. Some 30 acres of silage are made, the rest of the productive grassland being devoted to hay.

The spread of bracken over the last century or so is a major problem, and can be attributed in large part to the overwhelming dominance of sheep and the less-intensive land management of recent decades. One of the main objectives of the Clare Island Research Programme currently in progress is the development of a mixed stocking regime more appropriately tailored to the island's landscape rather than the forced adoption of a farming model considered optimal on the mainland. This new regime is specifically designed to enhance the unique natural and cultural heritage of the farmed landscape at the same time as it boosts farm income and supports rural community.

The high costs associated with intensive agriculture make profit margins tight—and uncertain—so that small farms are very dependent on the Single Farm Payment (and off-farm employment). All farmers in receipt of the Payment are bound by a plethora of regulations governing land use and farm management that are designed to maximise output while protecting the rural environment through codes of good practice in relation to such things as reducing the emission of greenhouse gases, controlling pollution, preserving natural habitats, maintaining water quality and the visual character of the landscape. On earlier maps the farmed landscape is divided into fields, but it is further divided into distinct 'parcels' in Teagasc's Land Parcel Identification System, which identifies each parcel with its own LPIS number. The new digital technology makes it possible to draw up a detailed management plan for every parcel on the farm, and to monitor compliance with all the regulations upon which receipt of the Single Farm Payment depends.

In terms of its farming future, Clare Island is firmly tied in with the fortunes of the country as a whole, and will continue to be dependent on subsidies provided through the agricultural policies of the European Union. The population increased by almost a quarter since 1981 and now stands at 159, with an additional 74 part-time residents. There are twenty children in the two-teacher National School at present.

Population		
1986	140	
1996	136	
2006	136	
2016	159	+19 since 1986

The number of farmers at the time of writing is approximately forty (almost all of them part-time), some thirty of whom have grazing rights on the commonage, the amount depending on the size of their holdings. The number of sheep on the island at present is somewhere between 3,000 and 5,000, but given the small size of farms the market value of the output barely covers costs. Profit margins are so tight that it is impossible to make a comfortable living within the constraints of modern agriculture without the economies of scale possible on larger holdings: especially on poorer land. Some consolidation of holdings is taking place; a farm is sometimes sold to a neighbour when the owner of a holding dies and the next of kin does not want to live on the island.

Improved connectivity with the mainland—due in large part to the vast improvement in ferry transport—and improved connections with the wider world in more recent times (especially with the arrival of the internet and broadband on the island) has broadened the horizon of expectation and ambition by orders of magnitude. The improved ferry service has brought great advancement to the farming sector, as it has to life on the island as a whole. High quality piers and cranes used for lifting make it possible to get stock on or off the island most weeks, which allows for more flexibility in selling times. When stock had to be taken off the island for sale on the mainland there was great pressure on the farmer to sell; it was not easy to take lambs to market, then bring them home again if the price was not high enough. Close to a thousand store lambs are sold directly off the land to buyers who travel to the island in late August and September each year.[39]

The relative isolation—though almost insignificant compared to the past, and compensated to an extent by the communal intimacy of island living—and the patient temperament necessitated by Clare Island's peculiar weather engenders a distinctive outlook better able to appreciate, albeit largely unconsciously, the blessings conferred by island isolation and an intimacy with this place whose roots deepen over time.

The farmer's needs are no longer met by what the land produces. He is entirely dependent on the increased support provided under the EU Common Agricultural Policy, direct payments under which average €120–150 per hectare annually. This is supplemented under a new scheme initiated in August 2015 which compensates farmers in areas of natural constraints (ANC), under which island farmers receive a supplementary payment of €200/ha up to 20ha, €150/ha for the next 15ha and €90/ha up to 40ha to compensate for their insular disadvantage.[40] It is estimated that living costs on the island are over a third higher than for mainland residents: which includes the costs of feed, fertiliser and silage making (which can be as much as €18 a bale higher).

Farmers who participate in the GLAS/AEOS schemes receive in addition a payment of usually between €3,500 and €4,500—much less than they would have received under the Rural Environment Protection Scheme (REPS). At the present time, 60–75% of farmers are in these schemes.[41] The main requirement imposed under the Commonage Framework Plan of 1999 (with which all farmers must comply) was a reduction in the level of stocking by 22.4%. The stocking density today is somewhere in the region of 1.19 ewe equivalents per hectare.

Under current legislation the cliffs and coastline of Clare Island are protected as Proposed Special Areas of Conservation (SACs) under the EU Habitats Directive (Clare Island Cliffs, Site Code 002243; West Connaught Coast, Site Code 002998); most of the cliffs are also designated as a Special Protection Area (SPA) under the Birds Directive. The island in its entirety is a Proposed Natural Heritage Area (Site Code 477) under national legislation, though what obligation this imposes is not spelt out in any detail. To qualify for payments, farmers are required to manage their lands in ways that conform to Good Farm Practice; what this implies is spelt out in the individual plans all farmers are required to prepare.[42]

The future for Clare Island

Clare Island is more than just another small part of rural Ireland. What makes it unique among the islands of Ireland is in the first place its extraordinary transformation under the Congested Districts Board; and secondly the great detail in which it was subsequently studied during the Clare Island Surveys of 1909-11 and 1992-2017. As a result of the attention brought to bear on the island through these surveys, no area of comparable size anywhere else in Ireland – and indeed perhaps anywhere else in the world – can claim to be known in such detail. This unique interest from the natural heritage perspective is coupled with a cultural heritage that is comparably unique. Nowhere else does the landscape so visibly and dramatically evoke the transformation that occurred in the west of Ireland after the Famine.

Over the last century the fortunes of farming have been governed essentially by developments in the wider world outside of Clare Island: but one of the most remarkable things about the island is the completeness with which the pattern of traditional farming a hundred years ago is preserved in its landscape. With the gradual abandonment of arable farming the ridged fields reverted to grassland, which acted as a protective seal on the earlier pattern of ridges. In recent decades farmers have concentrated mainly on sheep, the close grazing of which is optimal for the proliferation of the high nature value waxcap flora this kind of grassland supports, and which does not require the same degree of labour-intensive attention as arable farming. Much of the ridged farmscape has however been lost to encroaching bracken and soft rush, and there has been significant regression to peat formation, often resulting in the suppression of the waxcap flora.

As far as is economically feasible, future land management on Clare Island should seek to maintain all aspects of its natural and cultural diversity; In an earlier chapter we noted that areas with a particularly rich archaeological heritage are sometimes protected and managed *in their entirety* as 'Archaeological Complexes' rather than as a series of isolated features. In a similar way, the 'Landscape Complex' of Clare Island in its entirety might be designated on the twin grounds of unique natural and cultural heritage. The provisions of whatever agri-environment schemes are in effect should be structured with this in view, with individual farm plans tailored specifically to this end, and landowners remunerated accordingly.

The deterioration of this unique landscape is a serious loss; but it is one that can be halted and reversed. The information and insights provided by the two Surveys – summarised in the preceding pages – provide the necessary data for the development and implementation of an agri-environment programme specific to Clare Island that might lead to a new model for farming on the island.

Such a model would be more responsive to the island's unique possibilities and more rewarding in economic terms. The essential first step towards this end would be a comprehensive and detailed study that would tease out in detail the nature and scale of the challenges involved and of the benefits to be delivered. A key element in such a programme would be restoration of the large area of agricultural grassland of an earlier time that has been lost to bracken and rush encroachment, but which can be recovered through carefully targeted changes in the stocking regime and systematic drainage, thereby greatly extending the area of grazing available. This grassland hosts the island's rich waxcap flora and much of its archaeological heritage, as well as the material evidence of the extraordinary arable landscape that was reformed by the CDB's re-direction of the island's agriculture in the decade immediately preceding the arrival of Praeger and his cohort of co-workers.

At the time of this writing the programme of research necessary to the development of such a model for farming on Clare Island is in its early stages. This is in the process of documenting the condition of the farmed landscape in a detailed, quantified way: evaluating the extent of such changes as bracken, soft rush and peat encroachment and the ways in which these changes are grounded in changing agricultural practice. It seeks to explicate the methods to be employed in tackling these challenges under the varying circumstances that prevail in different parts of the island. It will outline in detail the timing and extent of farmer involvement, as well as issues relating to funding and remuneration. Such a study would lay the indispensable foundation for an application for funding to *implement* the proposed Programme. At the same time, it aims to devise effective management strategies, specific to individual land parcels, that can restore these areas to optimal conservation status, while at the same time optimising productivity.

The new research programme is under the direction of the CDB's successors: the School of Agriculture and Food Science at University College Dublin and Teagasc, the State's agricultural advisory body, but driven and managed by the farmers of Clare Island themselves, just as the original reformation of its agriculture was driven a hundred years ago.

Notes

[1] Gosling 2007a, 29–68.

[2] Manning 2007, 22.

[3] Manning 2007, 23.

[4] Hallissy 1914, OS 7, 3.

[5] 'Rocky mountain pasture' is a phrase that appears frequently on the Fair Plan for Clare Island, but not on the printed first edition of the six-inch map that was based on it (National Archives, Ordnance Survey Records, B 248(9)). Some 70% of the island has rockland and moorland ornamentation or shading on the first edition.

[6] In a footnote he noted that the quantity of seaweed thus used was about three times that of farmyard manure.

[7] Half the wool was sold, the other half kept on the island to be woven into cloth.

[8] Feehan 2003, 397–400.

[9] Gosling 2007a, 57, 152–54, 159–60.

[10] It may be possible in the near future to quantify the extent to which the fertility of Clare Island's soils is due to historical inputs from the shore, because the carbon and sulphur present in seaweeds have different stable isotope ratios from those in terrestrial plants.

[11] Wilson 1911, OS 5, 44.

[12] Cotton 1912, OS 15, 152–53.

[13] Cotton 1912, OS 15, 151–52.

[14] Kinahan 1869.

[15] Wilson 1911, OS 5, 44.

[16] Burns 2001, 38.

[17] Burns 2001, 37.

[18] *Gigartina mamillosa* of Cotton.

[19] *Rhodymenia palmata* of Cotton

[20] Gosling 2007a, 57–58. Nearly 60 were recorded in the latter area during the archaeological field work for the New Survey.

[21] A much wider selection occurs in the abundant middens along the mainland coast.

[22] Westropp 1911, OS 2, 6.

[23] Colgan 1911a, OS 4, 10.

[24] Browne 1898–1900, 65.

[25] O'Malley n.d.

[26] Colgan 1911a, OS 4, 7.

[27] Cullen and Gill 1992e, footnote 6.

[28] Lett 1912, OS 11–12, 3.

[29] The remains of the mill in Maum are no more substantial than what little there is to be seen at Páirc Mhíchíl.

[30] Gosling 2007a, 61; see MacCárthaigh 1999, 49.

[31] Vullings, Collins and Smillie 2013.

[32] Gosling 2007a, 53–57.

[33] Gosling 2007c, 188.

[34] Manning 2007, 18.

[35] Gosling 2007c, 181–82.

[36] Gosling 2007c, 178–81; O'Grady 2008.

[37] Dwyer 1963, 266.

[38] Ewes crossed with a mix of Blackface, Cheviot and Texel rams.

[39] Interview with Joe O'Malley: *Farming Independent*, 9 August 2016 (8–9). The number of sheep in May 2019 (including lambs) was around 5,000.

[40] No farmer on the island has a holding greater than 70ha.

[41] REPS payments were worth maybe between €7,000 and €10,000 a year, providing a indispensable economic anchor. The maximum payment under the GLAS/AEOS schemes is €5,000/year.

[42] Good Farm Practice is a set of rules that that all farmers in Ireland have to abide by. Depending on the nature of the individual farm, some have more rules to abide by than others; e.g. sheep-only farmers have less to worry about than farmers who have both cattle and sheep.

Bibliography

Agassiz, L, 1840 *Études sur les Glaciers*.

Anderson, R. (forthcoming) The non-marine Mollusca of Clare Island. In T.K. McCarthy, E. Lenihan and J. Breen (eds), *New Survey of Clare Island. Volume 9: Zoology* (in preparation). Dublin. Royal Irish Academy.

Anderson, R. and McCormack, S. (forthcoming) The Carabidae (Coleoptera) of Clare island. In T.K. McCarthy, E. Lenihan and J. Breen (eds), *New Survey of Clare Island. Volume 9: Zoology* (in preparation). Dublin. Royal Irish Academy.

Andrews, J.H. 1974 *History in the Ordnance Map: an introduction for Irish readers*. Dublin. Ordnance Survey of Ireland.

Ashe, P., O'Connor, J.P. and Chandler, P.J. 2012 A revised checklist of the biting midges (Ceratopogonidae) of Ireland. *Bulletin of the Irish Biogeographical Society* **36**, 190–231.

Atherton, I., Bosanquet, S. and Lawley, M. (eds) 2010 *Mosses and Liverworts of Britain and Ireland; a field guide*. British Bryological Society.

Balfour Browne, F. 1912 *Aquatic Coleoptera*. Clare Island Survey. Part 29. *Proceedings of the Royal Irish Academy* **31**.

Barrett-Hamilton, G.E.H. 1912 *Mammalia*. Clare Island Survey. Part 17. *Proceedings of the Royal Irish Academy* **31**.

Beattie, P. 2001 Whale Watching—by mutual consent. *SubSea* (summer), 12–13.

Bell, J. and Watson, M. 1986 *Irish Farming. Implements and Techniques 1750-1900*. Edinburgh. John Donald.

Bell, J. 1984 A contribution to the study of cultivation ridges in Ireland. *Journal of the Royal Society of Antiquaries in Ireland* 114. Dublin.

Blaxland Benham, W. 1896 Polychaet Worms. *The Cambridge Natural History*, Volume 11.

Blockeel, T.L. 1995 Summer field meeting, 1994, second week, Clifden. *Bulletin of the British Bryological Society* **65**, 29–35.

Bolger, T. (forthcoming) A Resurvey of the Collembola and Acarina (Oribatida and Mesostigmata) of Clare Island. In T.K. McCarthy, E. Lenihan and J. Breen (eds), *New Survey of Clare Island. Volume 9: Zoology* (in preparation). Dublin. Royal Irish Academy.

Bolger, T. 1986 The Collembola of Ireland—A checklist and bibliography. *Proceedings of the Royal Irish Academy* **86B**, 183–218.

Bond, K. (2006) A survey of the Trichoptera (caddis-flies) of Clare Island. Praeger Grant Report for the Royal Irish Academy.

Bond, K.G.M. (forthcoming) The Lepidoptera (Butterflies and Moths) of Clare Island. In T.K. McCarthy, E. Lenihan and J. Breen (eds), *New Survey of Clare Island. Volume 9: Zoology* (in preparation). Dublin. Royal Irish Academy.

Breen, J., Paxton, R. and O'Grady A. (forthcoming) The Social Hymenoptera of Clare Island. In T.K. McCarthy, E. Lenihan and J. Breen (eds), *New Survey of Clare Island. Volume 9: Zoology* (in preparation). Dublin. Royal Irish Academy.

Browne, C.R. 1898–1900 The ethnography of Clare Island and Inisturk, in the county of Mayo. *Proceedings of the Royal Irish Academy* **21**, 40–72.

Browne, J.F. 1991 The glacial geomorphology of Clare Island, Co. Mayo. Unpublished Ph.D. thesis, University of Dublin, Trinity College.

Brück, P.M. and Rushton, W.A. 2011 The Dalradian Supergroup in Scotland and Ireland. Chapter 11 in Rushton, A.W.A., Brück, P.M., Molyneux, S.G., Williams, M. and Woodcock, N.H., *A Revised Correlation of the Cambrian Rocks in the British Isles*. The Geological Society, Special Report Number 25, 40–41.

Bryant, J.A. 2007 The Stoneworts (Chlorophyta, Charales). In M.D. Guiry, D.M. John, F. Rindi and T.K. McCarthy (eds), *New Survey of Clare Island Volume 6: The Freshwater and Terrestrial Algae*, 67–72. Dublin. Royal Irish Academy.

Buckley, A. 2005 The musical instruments depicted at Clare Island. In C. Manning, P. Gosling and J. Waddell (eds), *New Survey of Clare Island, Volume 4: The Abbey*, 123–32. Dublin. Royal Irish Academy.

Burns, A. 2001 *A Time to Remember. Stories, Songs and Recitations from Clare Island*.

Carpenter, G. (1912). *Pycnogonida*. Clare Island Survey. Part 34. *Proceedings of the Royal Irish Academy* **31**.

Carpenter, G. 1913 *Apterygota*. Clare Island Survey. Part 32. *Proceedings of the Royal Irish Academy* **31**.

Carpenter, G.H. 1912 *Orthoptera*. Clare Island Survey. Part 31. *Proceedings of the Royal Irish Academy* **31**.

Carson, R. 1955 *The Edge of the Sea*. London. Staples Press.

Cawley, M. (forthcoming) Chilopoda, Diplopoda, Opiliones, Pseudoscorpiones and Land Isopoda from Clare Island. In T.K. McCarthy, E. Lenihan and J. Breen (eds), *New Survey of Clare Island. Volume 9: Zoology* (in preparation). Dublin. Royal Irish Academy.

Chambers, A. 1983 *Granuaile. The Life and Times of Grace O'Malley*. Dublin. Wolfhound Press.

Chew, D.M., Daly, J.S., Magna, T., Page, L.M., Kirkland, C.L., Whitehouse, M.J. and Lam, R. 2010 Timing of ophiolite obduction in the Grampian orogen. *G.S.A. Bulletin* (November/December 2010), **122**, No. 11/12, 1787–99.

Chew, D.M., Daly, J.S., Page, L.M. and Kennedy, M.J. 2003 Grampian orogenesis and the development of blueschist-facies metamorphism in western Ireland. *Journal of the Geological Society, London* **160**, 911–24.

Cole, G.A.J., Kilroe, J.R., Hallissy, T. and Newell Arber, E.A. 1914 *The Geology of Clare Island, Co. Mayo*. Memoirs of the Geological Survey of Ireland. Dublin. His Majesty's Stationery Office.

Colgan, N. 1911a *Gaelic Plant and Animal Names, and Associated Folk-lore*. Clare Island Survey. Part 4. *Proceedings of the Royal Irish Academy* **31**.

Colgan, N. 1911b *Marine Mollusca*. Clare Island Survey. Part 22. *Proceedings of the Royal Irish Academy* **31**.

Collins, T. 1985 *Floreat Hibernia. A bio-bibliography of Robert Lloyd Praeger 1865–1953*. Royal Dublin Society.

Collins, T. 1994 The Clare Island Survey: an early multidisciplinary success story. In T. Collins (ed.), *Decoding the Landscape. Papers Read at the Inaugural Conference of the Centre for Landscape Studies*, 114–32. Centre for Landscape Studies, University College Galway.

Collins, T. 1999 The Clare Island Survey of 1909–11: participants, papers and progress. In C. Mac Cárthaigh and K. Whelan (eds), *New Survey of Clare Island. Volume 1: History and Cultural Landscape*, 1–40. Dublin. Royal Irish Academy.

Collins, T. 2009 'Robert Lloyd Praeger', in Jones and Steer (eds), 5–9.

Cotton, A.D. 1912 *Marine Algae*. Clare Island Survey. Part 15. *Proceedings of the Royal Irish Academy* **31**.

Coxon, P. 1987 A post-glacial diagram from Clare Island, Co. Mayo. *Irish Naturalists' Journal* **22**, 219–23.

Coxon, P. 2001 The Quaternary History of Clare Island. In J. Graham (ed.), *New Survey of Clare Island. Volume 2: Geology*, 87–112. Dublin. Royal Irish Academy.

Coxon, P. and McCarron, S.G. 2009 Cenozoic: Tertiary and Quaternary (until 11,700 years before 2000). In Holland, C.H. and Sanders, I.S. (eds), *The Geology of Ireland*, Second edition, 355–96.

Coxon, P. and O'Connell, M. (eds) 1994 Clare Island and Inishbofin. *IQUA Field Guide No. 17*. Dublin, Irish Association for Quaternary Studies.

Coxon, P., Corcoran, R., Gibson, P. and McCarron, S. 2013 A Holocene pollen diagram from Lough Avullin, Clare Island, Western Ireland. In D. Synnott (ed.), *New Survey of Clare Island. Volume 7: Plants and Fungi*, 1–26. Dublin. Royal Irish Academy.

Coyne, W.P. (ed.) 1902 *Irish Industrial and Agricultural*. Dublin. Browne and Nolan.

Cronin, M., Myers, A. and O'Riordan, R. 2002 The reproductive cycle of the intertidal gastropod *Melarhaphe neritoides* (Linnaeus) on Clare Island. In A. Myers (ed.), *New Survey of Clare Island. Volume 3: Marine Intertidal Ecology*, 133–43. Dublin. Royal Irish Academy.

Crosby, N. 2004 *Glimpses of Clare Island*.

Cullen, C. and Gill, P. 1992a *Studying an island: Clare Island, Co. Mayo*. Clare Island Series 1. Clare Island Centre for Island Studies.

Cullen, C. and Gill, P. 1992b *Rocks, trees and bogs of Clare Island, Co. Mayo*. Clare Island Series 2. Clare Island Centre for Island Studies.

Cullen, C. and Gill, P. 1992c *Neolithic, Bronze and Iron Age man on Clare Island, Co. Mayo*. Clare Island Series 3. Clare Island Centre for Island Studies.

Cullen, C. and Gill, P. 1992d *Holy Wells and Christian settlement on Clare Island, Co. Mayo*. Clare Island Series 4. Clare Island Centre for Island Studies.

Cullen, C. and Gill, P. 1992e *recorded history from Grace O'Malley to the present day on Clare Island, Co. Mayo*. Clare Island Series 5. Clare Island Centre for Island Studies.

Curran-Cournane, F. 2007 Diet of marine-feeding European otters, *Lutra lutra*, in Clare Island, Co. Mayo. B.Sc. (hons.) thesis, National University of Ireland Galway.

Cussen, C., Myers, A., McGrath, D., Delany, J. and O'Riordan, R. 2002 Turnover in a mussel population on an exposed rocky shore in the west of Ireland. In A. Myers (ed.), *New Survey of Clare Island. Volume 3: Marine Intertidal Ecology*, 121–27. Dublin. Royal Irish Academy.

Cussen, R. and Myers, A. 2002 Dog whelk *Nucella lapillus* (Linnaeus), a significant prey item in the diet of herring gulls on Clare Island. *New Survey of Clare Island. Volume 3: Marine Intertidal Ecology*, 151–57. Dublin. Royal Irish Academy.

Cussen, R.E., McGreal, E., Warner, S. and Kelly, T.C. (forthcoming) The Seabirds of Clare Island. In T. Kelly (ed.), *New Survey of Clare Island. Volume 10: Birds* (in preparation). Dublin. Royal Irish Academy.

Dahlberg, Anders and Croneborg, Hjalmar 2006 The 33 threatened fungi in Europe. *Nature and Environment* **136**.

D'Arcy, G. 1999. *Ireland's Lost Birds*. Dublin. Four Courts Press.

De Eyto, E., Trojanowska-Olichwer, A., Nilssen, Jens Petter and McCarthy, K. (forthcoming) Cladocera and Copepoda of the inland waters of Clare Island. In T.K. McCarthy, E. Lenihan and J. Breen (eds), *New Survey of Clare Island. Volume 9: Zoology* (in preparation). Dublin. Royal Irish Academy.

Delany, J., Myers, A. and McGrath, D. 2002a The tide pools at Portnakilly: experimental disturbance. In A. Myers (ed.), *New Survey of Clare Island. Volume 3: Marine Intertidal Ecology*, 29–39. Dublin. Royal Irish Academy.

Delany, J., Myers, A. and McGrath, D. 2002b A comparison of the interactions of the limpets *Patella vulgata* and *Patella ulyssiponensis* with crustose coralline algae. In A. Myers (ed.), *New Survey of Clare Island. Volume 3: Marine Intertidal Ecology*, 79–90. Dublin. Royal Irish Academy.

DeMilio, E., Marley, N.J., Lawton, C. and Kristensen, R.M. (forthcoming) Tardigrada of Clare Island. In T.K. McCarthy, E. Lenihan and J. Breen (eds), *New Survey of Clare Island. Volume 9: Zoology* (in preparation). Dublin. Royal Irish Academy.

Dewey, J.F. and McKerrow, W.S. 1963 An outline of the Geomorphology of Murrisk and northwest Galway. *Geological Magazine* **100**, 260–75.

Doyle, G.J. and Foss, P.J. 1986 A resurvey of the Clare Island flora. *Irish Naturalists' Journal* **22**, 85–89.

Doyle, T.K. 2007 Leatherback Sea turtles (*Dermochelys coriacea*) in Irish waters. *Irish Wildlife Manuals No. 32*. National Parks and Wildlife Service.

Drummond, J.L. 1832. *Letters to a Young Naturalist on the Study of Nature and Natural Theology*.

Dunkerly, J.S. 1913 *Flagellata and Ciliata*. Clare Island Survey. Parts 61–62. *Proceedings of the Royal Irish Academy* **31**.

Dwyer, D.J. 1963 Farming an Atlantic Outpost: Clare Island, County Mayo. *Irish Geography* **48**, 255–67.

EastWest Mapping, Enniscorthy 2016 *Achill and Corraun, and Clare Island. Detailed Maps 1:25,000 and 1:7,500*.

Evans, D.H. and Graham, J.R. 2001 The Geology of Clare Island: Perspectives and Problems. In J. Graham (ed.), *New Survey of Clare Island. Volume 2: Geology*, 49–62. Dublin. Royal Irish Academy.

Farran, G.P. 1912 *Decapoda*. Clare Island Survey. Part 40. *Proceedings of the Royal Irish Academy* **31**.

Farran, G.P. 1912 *Pisces* Clare Island Survey. Part 19. *Proceedings of the Royal Irish Academy* **31**.

Farran, G.P. 1913 *Marine Entomostraca*. Clare Island Survey. Part 45. *Proceedings of the Royal Irish Academy* **31**.

Farran, G.P. 1914 *Notes on Marine Plankton*. Clare Island Survey. Part 66. *Proceedings of the Royal Irish Academy* **31**.

Farran, G.P. and Southern, R. 1914 *Tunicata and Hemichordata*. Clare Island Survey. Part 21. *Proceedings of the Royal Irish Academy* **31**.

Feehan, J. 2003 *Farming in Ireland: History, Heritage and Environment*. Faculty of Agriculture, University College Dublin.

Feehan, J. and McHugh, R. 1992 The Curragh of Kildare as a *Hygrocybe* grassland. *Irish Naturalists' Journal* **24**, 13–17.

Feeley, H.B., Baars, J.R. and Kelly-Quinn, M. 2016 *The Stonefly (Plecoptera) of Ireland—Distribution, Life Histories and Ecology*. National Biodiversity Data Centre, Waterford. Ireland.

Forbes, A.C. 1914 *Tree Growth*. Clare Island Survey. Part 9. *Proceedings of the Royal Irish Academy* **31**.

Foster, K.H. 1912 *Land and Fresh-water Isopoda*. Clare Island Survey. Part 44. *Proceedings of the Royal Irish Academy* **31**.

Gardiner, M. and Radford, T. 1980 *Soil Associations of Ireland and their land use potential*. Soil Survey Bulletin No. 36. Dublin. An Foras Talúntais.

Gillespie, F. and Ó Comáin, M. 2005 The Ó Máille memorial plaque and its heraldic achievement. In C. Manning, P. Gosling and J. Waddell (eds), *New Survey of Clare Island. Volume 4: The Abbey*, 41–4. Dublin. Royal Irish Academy.

Gleeson, E., McGrath, R. and Treanor, M. (eds) 2013 *Ireland's climate: the road ahead*. Dublin. Met Éireann.

Gleeson, E., Drijfhout, S., Dijkstra, H. and Livina, V. 2013 EC-Earth's "Little Ice Age". In Gleeson *et al*. (eds), 91–4.

Gosling, P. 2007a The human settlement history of Clare Island. *New Clare Island Survey. Volume 5: Archaeology*, 29–68. Dublin. Royal Irish Academy.

Gosling, P. 2007b A distributional and morphological analysis of fulachtaí fiadh on Clare Island. *New Clare Island Survey. Volume 5: Archaeology*, 69–80. Dublin. Royal Irish Academy.

Gosling, P. 2007c Catalogue of archaeological and architectural monuments, sites and stray finds on Clare Island. In C. Manning, P. Gosling and J. Waddell (eds), *New Clare Island Survey. Volume 5: Archaeology*, 83–212. Dublin. Royal Irish Academy.

Gosling, P., Manning, C. and Waddell, J. (eds) 2007 *New Survey of Clare Island. Volume 5: Archaeology*. Dublin. Royal Irish Academy.

Gosling, P., Manning, C. and Waddell, J. 2007 Introduction. In C. Manning, P. Gosling and J. Waddell (eds), *New Clare Island Survey. Volume 5: Archaeology*, 1–5. Dublin. Royal Irish Academy.

Graham, J. 2002 Enigmatic fossils from Knockmore. *New Survey of Clare Island, Newsletter* No. 5 (March), 9.

Graham, J.R. 1981 The Old Red Sandstone of County Mayo, northwest Ireland. *Geological Journal* **22**, 347–70.

Graham, J.R. (ed.) 2001 *New Survey of Clare Island. Volume 2: Geology*. Dublin. Royal Irish Academy.

Graham, J.R. 2001a The Geology of Clare Island: Perspectives and Problems. In J. Graham (ed.), *New Survey of Clare Island. Volume 2: Geology*, 1–18. Dublin. Royal Irish Academy.

Graham, J.R. 2001b The Silurian Rocks of Southern Clare Island. In J. Graham (ed.), *New Survey of Clare Island. Volume 2: Geology*, 34–48. Dublin. Royal Irish Academy.

Graham, J.R. 2001c The Carboniferous Rocks of Clare Island. *New Survey of Clare Island. Volume 2: Geology*, 75–86. Dublin. Royal Irish Academy.

Grimshaw, P.H. 1912 *Diptera*. Clare Island Survey. Part 25. *Proceedings of the Royal Irish Academy* **31**.

Guiry, M.D. 1997 No Stone Unturned: Robert Lloyd Praeger and the Major Surveys. In J.W. Foster (ed.), *Nature in Ireland. A Scientific and Cultural History*, 299–307. Dublin. The Lilliput Press.

Guiry, M.D., John, D.M., Rindi, F. and McCarthy, T.K. (eds) 2007 *New Survey of Clare Island. Volume 6: The Freshwater and Terrestrial Algae*. Dublin. Royal Irish Academy.

Guiry, M.D., Maggs, C.A. and Adair, C. 1991 *Antithamnion densum* (Suhr) Howe from Clare Island, Ireland: a marine red alga new to the British Isles. *Cryptogamie: Algologie* 189–94.

Halbert, J.N. 1911 *Acarinida: i – Hydracarina*. Clare Island Survey. Part 39 ii. *Proceedings of the Royal Irish Academy* **31**.

Halbert, J.N. 1912a *Hemiptera*. Clare Island Survey. Part 30. *Proceedings of the Royal Irish Academy* **31**.

Halbert, J.N. 1912b *Neuroptera*. Clare Island Survey. Part 27. *Proceedings of the Royal Irish Academy* **31**.

Halbert, J.N. 1915 *Acarinida: ii—Terrestrial and Marine Acarina*. Clare Island Survey. Part 39 ii. *Proceedings of the Royal Irish Academy* **31**.

Hallissy, T. 1914 *Geology*. Clare Island Survey. Part 7. *Proceedings of the Royal Irish Academy* **31**.

Heron-Allen, E. and Earland, A. 1913 *Foraminifera*. Clare Island Survey. Part 64. *Proceedings of the Royal Irish Academy* **31**.

Herries-Davies, G.L. 1983 *Sheets of many colours: the mapping of Ireland's rocks 1750–1890*. Dublin. Royal Dublin Society.

Hickin, N. 1952 *Caddis*. London. Methuen.

Higgs, K.T. and Williams, D.M. 2001 A Silurian microflora from the Ballytoohy Formation (Upper Clew Bay Complex) of Clare Island. In J. Graham (ed.), *New Survey of Clare Island. Volume 2: Geology*, 19–34.

Hingley, M. 1993 *Microscopic Life in Sphagnum*. Naturalists' Handbooks 20, Richmond Publishing.

Hopkin, S.P. 2007 *A key to the Collembola (Springtails) of Britain and Ireland*. Field Studies Council Publications.

Jebb, M. 2013 An annotated checklist of the vascular flora of Clare Island. In D. Synnott (ed.), *New Survey of Clare Island. Volume 7: Plants and Fungi*, 179–208. Dublin. Royal Irish Academy.

John, D.M. 2007a The green algae (Chlorophyta, in part) and most other phyla. In M.D. Guiry, D.M. John, F. Rindi and T.K. McCarthy (eds), *New Survey of Clare Island. Volume 6: The Freshwater and Terrestrial Algae*, 73–113. Dublin. Royal Irish Academy.

John, D.M. 2007b Summary of Results and Conclusions. In M.D. Guiry, D.M. John, F. Rindi and T.K. McCarthy (eds), *New Survey of Clare Island Volume 6: The Freshwater and Terrestrial Algae*, 193–204. Dublin. Royal Irish Academy.

John, D.M. 2007c Distribution and Ecology. In M.D. Guiry, D.M. John, F. Rindi and T.K. McCarthy (eds), *New Survey of Clare Island Volume 6: The Freshwater and Terrestrial Algae*, 21–24. Dublin. Royal Irish Academy.

John, D.M. and Williamson, D.B. 2009 *A Practical Guide to the Desmids of the West of Ireland*. Martin Ryan Institute, National University of Ireland, Galway.

Johnson, W.F. 1912 *Chilopoda and Diplopoda*. Clare Island Survey. Part 33. *Proceedings of the Royal Irish Academy* **31**.

Johnson, W.F. and Halbert, J.N. 1912 *Terrestrial Coleoptera*. Clare Island Survey. Part 28. *Proceedings of the Royal Irish Academy* **31**.

Jones, G. 2009 Darwin, Praeger and the Belfast natural history tradition. In Roisín Jones and Martin Steer (eds), *Darwin, Praeger and the Clare Island Surveys*. Dublin. Royal Irish Academy, 12–15.

Kane, W.F. de V. (1912). *Lepidoptera*. Clare Island Survey. Part 26. *Proceedings of the Royal Irish Academy* **31**.

Kelly, T.C. (ed.) (forthcoming) *New Survey of Clare Island. Volume 10: Birds* (in preparation). Dublin. Royal Irish Academy.

Kelly, T.C. (forthcoming) The birds of Clare Island. In T.K. McCarthy, E. Lenihan and J. Breen (eds), *New Survey of Clare Island. Volume 9: Zoology* (in preparation). Dublin. Royal Irish Academy.

Kelly, T.C. and Smiddy, P. (forthcoming) Size matters: island area and breeding rooks *Corvus frugilegus*. In T.C. Kelly (ed.), *New Survey of Clare Island. Volume10: Birds* (in preparation). Dublin. Royal Irish Academy.

Kelly, T.C., Smiddy, P. and Keating, U. (forthcoming) Breeding Birds on the offshore islands of Ireland—some patterns and trends. In T.C. Kelly (ed.), *New Survey of Clare Island. Volume10: Birds* (in preparation). Dublin. Royal Irish Academy.

Kelly, T.C. and Keating, U. (forthcoming) Aspects of the breeding land bird community on Clare Island. In T.C. Kelly (ed.), *New Survey of Clare Island. Volume 10: Birds* (in preparation). Dublin. Royal Irish Academy.

Kelly, T.C., Keating, U., Coombes, R.H. and Smiddy, P. (forthcoming) Waterfowl—ducks, rails and waders on Clare Island. In T.C. Kelly (ed.), *New Survey of Clare Island. Volume 10: Birds* (in preparation). Dublin. Royal Irish Academy.

Kelly, T.C., Smiddy, P., McGreal, E., Winters, P., Keating, U. and Cussen, R. (forthcoming) The equilibrium number of land bird species on Clare Island. *New Survey of Clare Island. Volume 10: Birds* (in preparation). Dublin. Royal Irish Academy.

Kelly-Quinn, M. and Bracken, J.J. 2000 *The Distribution of the Ephemeroptera in Ireland*. Occasional Publication of the Irish Biogeographical Society Number 5.

Kelly-Quinn, M. and Regan, E.C. 2012 *Ireland Red List No. 7: Mayflies (Ephemeroptera)*. National Parks and Wildlife Service, Department of Arts, Heritage and the Gaeltacht, Dublin, Ireland.

Kennedy, B. 2019 EPA unpublished data prepared for the New Survey of Clare Island. Environmental Protection Agency.

Kinahan, G.H., Symes, R.G., Wilkinson, S.B., Nolan, J., Leonard, H. and Baily, W.H. 1876 *Memoir to accompany sheets 73 and 74 (in part) 83 and 84 of the Maps of the Geological Survey of Ireland, including the country around Westport, Eriff Valley, Killary Harbour, and western shores of Lough Mask*. HMSO.

Kinahan, G.H. 1869 The Kelp Industry. Hints for Kelp-burners with regard to Seaweed. *Quarterly Journal of Science* **6** (23), 331–41.

King, A. 2008. Clare Island. *Science Spin* Issue **30**, 13–16.

King, P. 2007 The archaeological excavation of a pre-bog wall (Cat. No. 183) at Bunnamohaun. In C. Manning, P. Gosling and J. Waddell (eds), *New Clare Island Survey Volume 5: Archaeology*, 291–96. Dublin. Royal Irish Academy.

Lawton, C. and McAney, K. (forthcoming) The Mammals of Clare Island. In T.K. McCarthy, E. Lenihan and J. Breen (eds), *New Survey of Clare Island. Volume 9: Zoology* (in preparation). Dublin. Royal Irish Academy.

Lett, H.W. 1912 *Musci and Hepaticae*. Clare Island Survey. Parts 11–12. *Proceedings of the Royal Irish Academy* **31**.

Lewis, S. 1837 *A topographical dictionary of Ireland, Volume 1*. London, S. Lewis & Co.

Lindman, C.A.M. 1922 *Bilder Ur Nordens Flora*.

Lister, G. 1912 Mycetozoa. Clare Island Survey. Part 63. *Proceedings of the Royal Irish Academy* **31**.

Lloyd, C. 1982 The birds of Clare Island, Co. Mayo in June 1982. Unpublished report for the Royal Irish Academy.

Lloyd, C.S. 1982 The Birds of Clare Island, Co. Mayo in June 1982. *Irish Naturalists' Journal* **21**, 212–16.

Long, C.B., MacDermott, C.V., Morris, J.H., Sleeman, A.G. and Tietzsch-Tyler, D. 1992 *Geology of North Mayo: a Geological Description, with accompanying Bedrock Geology 1:100,000 Scale Map, Sheet 6*. Geological Survey of Ireland.

Lynch, Brian 1993 'Old Furrows II'. *Easter Snow. An Island off Ireland*. Edition die horen and Salmon Publishing.

Lyons, W.J. 1914 *Climatology*. Clare Island Survey. Part 6. *Proceedings of the Royal Irish Academy* **31**.

Lysaght, S. 1991 *The Clare Island Survey*. Gallery Books.

Lysaght, S. 1998 *Robert Lloyd Praeger. The Life of a Naturalist*. Dublin. Four Courts Press.

MacCárthaigh, C. 1999 Clare Island Folklife. In C. Mac Cárthaigh and K. Whelan (eds), *New Survey of Clare Island, Volume 1: History and Cultural Landscape*, 41–72. Dublin. Royal Irish Academy.

MacCárthaigh, C. and Whelan, A. (eds) 1999 *New Survey of Clare Island. Volume 1: History and Cultural Landscape*. Dublin. Royal Irish Academy.

MacNeill, J. and Westropp, T.J. 1913 *Place-Names and Family Names*. Clare Island Survey. Part 3. *Proceedings of the Royal Irish Academy* **31**.

Maggs, C.A., Guiry, M.D. and Rueness, J. 1991 *Aglaothamnion priceanum* sp. nov. (Ceramiaceae, Rhodophyta) from the North-eastern Atlantic: Morphology and Life History of Parasporangial Plants. *British Phycological Journal* **26**, 343–52.

Manning, C. 2005 History, survey and analysis of the building. In C. Manning, P. Gosling and J. Waddell (eds), *New Survey of Clare Island, Volume 4*: The Abbey, 7–27. Dublin. Royal Irish Academy.

Manning, C. 2007 Clare Island: an historical outline to 1700. In C. Manning, P. Gosling and J. Waddell (eds), *New Survey of Clare Island. Volume 5: Archaeology*, 9–28. Dublin. Royal Irish Academy.

Manning, C. and Morton, K. 2009 *Clare Island, Co. Mayo—the Abbey and its paintings*. Archaeology Ireland Heritage Guide No. 46.

Manning, C., Gosling, P. and Waddell, J. 2005 (eds) *New Survey of Clare Island, Volume 4: The Abbey*. Dublin. Royal Irish Academy.

Margulis, L. and Schwartz, K.V. 1998 *Five kingdoms. An illustrated guide to the Phyla of life on Earth*. New York. Freeman and Company.

McCall, G. (forthcoming) Non-marine ostracods of Clare Island. In T.K. McCarthy, E. Lenihan and J. Breen (eds), *New Survey of Clare Island. Volume 9: Zoology* (in preparation). Dublin. Royal Irish Academy.

McCarthy, T.K., Creed, K. and Lenihan, E.S. (in preparation) Freshwater fish and their parasites in Clare Island. In T.K. McCarthy, E. Lenihan and J. Breen (eds), *New Survey of Clare Island. Volume 9: Zoology* (in preparation). Dublin. Royal Irish Academy.

McCarthy, T.K., Lenihan, E. and Breen, J. (eds) (forthcoming) *New Survey of Clare Island. Volume 9: Zoology* (in preparation). Dublin. Royal Irish Academy.

McCormack, S. and McCarthy, T.K. (forthcoming) The aquatic Coleoptera of Clare Island. In T.K. McCarthy, E. Lenihan and J. Breen (eds), *New Survey of Clare Island. Volume 9: Zoology* (in preparation). Dublin. Royal Irish Academy.

McCormack, S. and McCarthy, T.K. (forthcoming) Intra-habitat spatial variation in aquatic beetle (Insecta: Coleoptera) assemblages, with particular reference to an oligodystrophic lough on Clare Island. In T.K. McCarthy, E. Lenihan and J. Breen (eds), *New Survey of Clare Island. Volume 9: Zoology* (in preparation). Dublin. Royal Irish Academy.

McDonnell, R., Gormally, M. and McCormack, S. (forthcoming) Snail-killing flies (Sciomyzidae) of Clare Island. In T.K. McCarthy, E. Lenihan and J. Breen (eds), *New Survey of Clare Island. Volume 9: Zoology* (in preparation). Dublin. Royal Irish Academy.

McGrath, A. 2015 *Inventing the Universe*. Hodder and Stoughton.

McGrath, D. and Nunn, J. 2002 The marine gastropod *Osilinus lineatus* (da Costa) on Clare Island: evidence for extinction and recolonisation. In A. Myers (ed.), *New Survey of Clare Island. Volume 3: Marine Intertidal Ecology*, 145–49. Dublin. Royal Irish Academy.

McHugh, R., Mitchel, D., Wright, M. and Anderson, R. 2001 The fungi of Irish grasslands and their value for nature conservation. *Biology and Environment: Proceedings of the Royal Irish Academy* **101B**, 225–42.

Micks, W. 1925 *An account of the constitution, administration and dissolution of the Congested Districts Board for Ireland from 1891 to 1923*. Dublin. Eason.

Millais, J.G. 1904–6 *The Mammals of Great Britain and Ireland*.

Mitchel, D. 2013 Grassland fungi. In D. Synnott (ed.), *New Survey of Clare Island Volume 7: Plants and Fungi*, 253–318. Dublin. Royal Irish Academy.

Molloy, K. 2007 Reconstruction of past vegetation history on Clare Island: palaeoenvironmental investigations undertaken in conjunction with the archaeological excavations. In C. Manning, P. Gosling and J. Waddell (eds), *New Clare Island Survey Volume 5: Archaeology*, 297–310. Dublin. Royal Irish Academy.

Morgan T.H. 1890 The relationships of the sea-spiders. *Biological lectures delivered at the Marine Biological Laboratory of Wood's Hole in the Summer Session of 1890*, 142–67. Boston. Ginn and Company.

Morley, C. 1911 *Hymenoptera*. Clare Island Survey. Part 24. *Proceedings of the Royal Irish Academy* **31**.

Morton, K. 2005 Iconography and dating of the wall paintings. In C. Manning, P. Gosling and J. Waddell (eds), *New Survey of Clare Island, Volume 4: The Abbey*, 97–121. Dublin. Royal Irish Academy.

Morton, K. and Oldenbourg, C. 2005 Catalogue of the wall paintings. In C. Manning, P. Gosling and J. Waddell (eds), *New Survey of Clare Island, Volume 4: The Abbey*, 61–95. Dublin. Royal Irish Academy.

Morton, K. and Oldenbourg, C. 2005 Introduction to the wall paintings. In C. Manning, P. Gosling and J. Waddell (eds), *New Survey of Clare Island, Volume 4: The Abbey*, 47–48. Dublin. Royal Irish Academy.

Murray, D. and Murray, W.A. 2003 A reassessment of Chironomidae (Diptera) of Clare Island, Co Mayo, with first records of *Acamptocladius reissi* Cranston & Saether and *Limnophyes angelicae* Sether (Orthocladiinae) for the Irish faunal checklist. *Bulletin of the Irish biogeographical Society* **27**, 255–69.

Murray, D.A. 2004 Chironomidae of the Co. Mayo Mainland near Clare Island. Unpublished report for the Royal Irish Academy.

Murray, D.A. (forthcoming) The Chironomidae (Diptera: Insecta) of Clare Island and adjacent mainland County Mayo. In T.K. McCarthy, E. Lenihan and J. Breen (eds), *New Survey of Clare Island. Volume 9: Zoology* (in preparation). Dublin. Royal Irish Academy.

Murray, J. 1911a *Arctisoida*. Clare Island Survey. Part 37. *Proceedings of the Royal Irish Academy* **31**.

Murray, J. 1911b *Rotifera Bdelloida*. Clare Island Survey. Part 52. *Proceedings of the Royal Irish Academy* **31**.

Myers, A. (ed.) 2002 *New Survey of Clare Island. Volume 3: Marine Intertidal Ecology*. Dublin. Royal Irish Academy.

Myers, A. and Delany, J. 2002 Overview: the extremely exposed shore at Leckacanny and the high shore tide pools at Portnakilly. In A. Myers (ed.), *New Survey of Clare Island. Volume 3: Marine Intertidal Ecology*, 9–10. Dublin. Royal Irish Academy.

Myers, A. and McGrath, D. 2002 General Introduction to Clare Island intertidal marine studies. In A. Myers (ed.), *New Survey of Clare Island. Volume 3: Marine Intertidal Ecology*, 1–5. Dublin. Royal Irish Academy.

Myers, A.A., McGrath, D. and O'Riordan, R.M. 1990 Clare Island feasibility study: marine intertidal survey. Unpublished report for the Royal Irish Academy.

Newton, S.F., Harris, M.P. and Murray, S. 2015 Census of Gannet *Morus bassanus* colonies in Ireland in 2013–2014. *Irish Birds* 10, 215–20.

Ní Cheanáinn, Á. (1988). *The Heritage of Mayo*.

Nichols, A.R. 1912a *Polyzoa* . Clare Island Survey. Part 53. *Proceedings of the Royal Irish Academy* **31**.

Nichols, A.R. 1912b *Echinodermata*. Clare Island Survey. Part 57. *Proceedings of the Royal Irish Academy* **31**.

Nolan, M. (forthcoming) Spiders (Arachnida: Araneae) from Clare Island. In T.K. McCarthy, E. Lenihan and J. Breen (eds), *New Survey of Clare Island. Volume 9: Zoology* (in preparation). Dublin. Royal Irish Academy.

Nunn, J. 2002 The marine mollusca of the shores of Clare Island and Clew Bay. In A. Myers (ed.), *New Survey of Clare Island. Volume 3: Marine Intertidal Ecology*, 159–83. Dublin. Royal Irish Academy.

Ó Cadhla, Ó. *et al.* 2007 An assessment of the breeding population of grey seals in the Republic of Ireland. *Irish Wildlife Manuals*, No. 34. Dublin. National Parks and Wildlife Service.

O'Connor, J. P. 2015 *A catalogue and atlas of the caddisflies (Trichoptera) of Ireland*. Occasional Publication of the Irish Biogeographical Society Number 11. Published by the Irish Biogeographical Society in association with the National Museum of Ireland.

O'Grady, J. 2008 *The Green Road to the Lighthouse. Memoirs of Clare island and the Lighthouse Service*. Faungloss Press.

O'Malley, M. n.d. (Ballytoughey; born 1915). *Notes on Island Living*. Unpublished manuscript (20pp plus list of people living on Clare Island when she was there (1915–1936)).

Ó Muraíle, J. 1999 The Place-names of Clare Island. *New Survey of Clare Island. Volume 1: History and Cultural Landscape*, 99–141. Dublin. Royal Irish Academy.

O'Riordan, R., Myers, A., McGrath, D., Delany, J., Cussen, R. and Cronin, M. 2002 An extremely exposed shore: Leckacanny, Clare Island. *New Survey of Clare Island. Volume 3: Marine Intertidal Ecology*, 11–27. Dublin. Royal Irish Academy.

Oldenbourg, C. 2005 Conservation of the wall paintings. *New Survey of Clare Island, Volume 4: The Abbey*, 49–60. Dublin. Royal Irish Academy.

Oster, M. 2008 Low congruence between the diversity of waxcaps (*Hygrocybe* spp.) fungi and vascular plants in seminatural grasslands. *Basic and Applied Ecology* **9**, 514–22.

Pack Beresford, D.R. 1911a *Araneida*. Clare Island Survey. Parts 35. *Proceedings of the Royal Irish Academy* **31**.

Pack Beresford, D.R. 1911b *Phalangida*. Clare Island Survey. Parts 36. *Proceedings of the Royal Irish Academy* **31**.

Palmer, D., Johnston, J.D., Dooley, T. and Maguire, C.K. 1989 The Silurian of Clew Bay, Ireland: part of the Midland Valley of Scotland? *Journal of the Geological Society of London* **146**, 385–8.

Patton, J.A. 1999 *The Liverwort Flora of the British Isles*. Harley Books.

Phillips, W.E.A. 1966 *The geology of Clare Island, County Mayo, Ireland*. Unpublished Ph.D. thesis, University of Dublin, Trinity College.

Phillips, W.E.A. 1973 The pre-Silurian rocks of Clare Island, Co. Mayo, Ireland, and the age of the metamorphism of the Dalradian *Journal of the Geological Society of London* **129**, 585–606.

Phillips, W.E.A. 1974 The stratigraphy, sedimentary environments and palaeogeography of the Silurian strata of Clare Island, Co. Mayo, Ireland, and the age of the metamorphism of the Dalradian *Journal of the Geological Society of London* **130**, 19–41.

Phillips, W.E.A. 2001 Caledonian Deformation. In C.H. Holland (ed.), *The Geology of Ireland*, 179–99. Edinburgh. Dunedin Academic Press.

Phillips, W.E.A. and Clayton, G. 1979 The Dinantian clastic succession of Clare Island, County Mayo. *Journal of Earth Sciences. R. Dubl. Soc.* **2**, 115–35.

Phillips, W.E.A., Kennedy, M.J. and Dunlop, G.M. 1969 Geological Comparison of western Ireland and Northeastern Newfoundland. In Kay, M. (ed.), North Atlantic – geology and continental drift, a symposium. *Mem. Am. Ass. Petrol. Geol.* **12**, 194–211.

Praeger, R.L. 1903 The Flora of Clare Island. *Irish Naturalist* xii, 277–94.

Praeger, R.L. 1911 Phanerogamia and Pteridophyta. Clare Island Survey. Part 10. *Proceedings of the Royal Irish Academy* **31**.

Praeger, R.L. 1913 On the buoyancy of the seeds of some Britannic plants. *Scientific Proceedings of the Royal Dublin Society* **15**, 13–62. 277–94.

Praeger, R.L. 1915 General Introduction and Narrative; General Summary. Clare Island Survey. Parts 1, 68. *Proceedings of the Royal Irish Academy* **31**.

Praeger, R.L. 1930 *Beyond Soundings*. Dublin and Cork. The Talbot Press.

Praeger, R.L. 1939 *The Way that I Went. An Irishman in Ireland*. Dublin, Hodges Figgis; London. Methuen.

Praeger, R.L. 1949 *Some Irish Naturalists: A Biographical Notebook*. Dundalk. Dundalgan Press.

Praeger, R.L. 1950 *Natural History of Ireland. A Sketch of its Flora and Fauna*. London. Collins.

Rea, Carleton and Hawley, H.D. 1912 Fungi. Clare Island Survey. Part 13. *Proceedings of the Royal Irish Academy* **31**.

Rindi, F. and Guiry, M.D. 2004 Composition and spatio temporal variability of the epiphytic macroalgal assemblage of *Fucus vesiculosus* Linnaeus on Clare island, Mayo, western Ireland. *Journal of Experimental Marine Biology and Ecology* **311**, 233–52.

Rindi, F. and Guiry, M.D. 2007 The subaerial algae. In M.D. Guiry, D.M. John, F. Rindi and T.K. McCarthy (eds), *New Survey of Clare Island Volume 6: The Freshwater and Terrestrial Algae*. Dublin. Royal Irish Academy, 179–92.

Rindi, F., Guiry, M. and Maggs, C. 2002a Checklist of the benthic marine algae of Clare Island. In A. Myers (ed.), *New Survey of Clare Island. Volume 3: Marine Intertidal Ecology*, 189–205. Dublin. Royal Irish Academy.

Rindi, F., Guiry, M. and Maggs, C. 2002b The benthic marine algae of Clare Island. In A. Myers (ed.), *New Survey of Clare Island. Volume 3: Marine Intertidal Ecology*, 187–88. Dublin. Royal Irish Academy.

Rousellet, Charles F. 1911 *Rotifera (excluding Bdelloida)*. Clare Island Survey. Part 51. *Proceedings of the Royal Irish Academy* **31**.

Ruttledge-Fair, R. 1892 *Confidential Report of Major Ruttledge-Fair to the Congested Districts Board for Ireland: District of Clare Island*. HMSO.

Ryle, T. 2013 Vegetation-environment interactions on Clare Island. In D. Synnott (ed.), *New Survey of Clare Island Volume 7: Plants and Fungi*, 27–177. Dublin. Royal Irish Academy.

Scharff, R.F. 1917 On the Irish pig. *The Irish Naturalist* **26**, 173–85.

Scharff, R.F. 1912 *Reptilia and Amphibia*. Clare Island Survey. Part 18. *Proceedings of the Royal Irish Academy* **31**.

Scourfield, D.J. 1912 *Fresh-Water Entomostraca*. Clare Island Survey. Part 46. *Proceedings of the Royal Irish Academy* **31**.

Seaward, M.R.D. and Richardson, D.H.S. 2013 Lichen flora of Clare Island. In D. Synnott (ed.), *New Survey of Clare Island Volume 7: Plants and Fungi*, 233–51. Dublin. Royal Irish Academy.

Sharp, D. 1895 *Insects, Part 1. The Cambridge Natural History*. London. Macmillan.

Sims, P. 2007 The Diatoms (Bacillariophyta). *New Survey of Clare Island Volume 6: The Freshwater and Terrestrial Algae*, 115–39. Dublin. Royal Irish Academy.

Smith, A.J.E. 2004 *The Moss Flora of Britain and Ireland* (2nd edition). Cambridge University Press.

Smith, A.L. 1911 *Lichenes*. Clare Island Survey. Part 14. *Proceedings of the Royal Irish Academy* **31**.

Southern, R. 1912 *Platyhelmia*. Clare Island Survey. Part 56. *Proceedings of the Royal Irish Academy* **31**.

Southern, R. 1913a *Gephyrea*. Clare Island Survey. Part 49. *Proceedings of the Royal Irish Academy* **31**.

Southern, R. 1913b *Hirudinea*. Clare Island Survey. Part 50. *Proceedings of the Royal Irish Academy* **31**.

Southern, R. 1913c *Oligochaeta*. Clare Island Survey. Part 48. *Proceedings of the Royal Irish Academy* **31**.

Southern, R. 1913d *Nemertinea*. Clare Island Survey. Part 55. *Proceedings of the Royal Irish Academy* **31**.

Southern, R. 1914a *Archiannelida and Polychaeta*. Clare Island Survey. Part 47. *Proceedings of the Royal Irish Academy* **31**.

Southern, R. 1914b *Nemathelmia, Kinorhyncha, and Chaetognatha*. Clare Island Survey. Part 54. *Proceedings of the Royal Irish Academy* **31**.

Southern, R. 1915 *Marine Ecology*. Clare Island Survey. Part 67. *Proceedings of the Royal Irish Academy* **31**.

Stalley, R. 2005 The Abbey in its later Gothic context. In C. Manning, P. Gosling and J. Waddell (eds), *New Survey of Clare Island, Volume 4: The Abbey*, 135–50. Dublin. Royal Irish Academy.

Stalley, R. 1987 The Cistercian Monasteries of Ireland. London and New Haven. Yale University Press.

Stelfox, A.W. 1912 *Land and Fresh-water Mollusca*. Clare Island Survey. Part 23. *Proceedings of the Royal Irish Academy* **31**.

Stephens, J. 1912a *Marine Porifera*. Clare Island Survey. Part 59. *Proceedings of the Royal Irish Academy* **31**.

Stephens, J. 1912b *Coelenterata*. Clare Island Survey. Part 58. *Proceedings of the Royal Irish Academy* **31**.

Stephens, J. 1912c *Fresh-water Porifera*. Clare Island Survey. Part 60. *Proceedings of the Royal Irish Academy* **31**.

Sutherland, H. 1909 *Ireland Yesterday and Today*. Philadelphia. The North American Company.

Sweeney, J. *et al.* (ICARUS: the Irish Climate Analysis and Research Units at NUI Maynooth) 2013 *Co-ordination, Communication and Adaptation for Climate Change in Ireland: an Integrated Approach* (COCOADAPT) (2007-CCRP-2.2.6a). *End of Project Report*. Wexford, Environmental Protection Agency. Chapter 8: Integrating Adaptation: A Case Study of Co. Mayo, 62–91.

Synnott, D. (ed.) 2013 *New Survey of Clare Island. Volume 7: Plants and Fungi*. Dublin. Royal Irish Academy.

Synnott, D. 2013 Bryophytes of Clare Island. In D. Synnott (ed.), *New Survey of Clare Island Volume 7: Plants and Fungi*, 209–32. Dublin. Royal Irish Academy.

Tattersall, W.M. 1912a *Nebaliacea, Cumacea, Schizopoda, and Stomatopoda*. Clare Island Survey. Part 41. *Proceedings of the Royal Irish Academy* **31**.

Tattersall, W.M. 1912b *Marine Isopoda*. Clare Island Survey. Part 43. *Proceedings of the Royal Irish Academy* **31**.

Tattersall, W.M. 1913 *Amphipoda*. Clare Island Survey. Part 42. *Proceedings of the Royal Irish Academy* **31**.

Tudge, C. 2000 *The Variety of Life*. Oxford University Press.

Upton, B.G.J., Phillips, W.E.A. and Stillman, C.J. 2001 Late Palaeozoic Ultramafic Magmatism on Clare Island. In J. Graham (ed.), *New Survey of Clare Island Volume 2, Geology*, 63–74. Dublin. Royal Irish Academy.

Ussher, R.J. 1912 *Aves*. Clare Island Survey. Part 20. *Proceedings of the Royal Irish Academy* **31**.

Vesterholt, J., Boertmann, D. and Tranberg, H. 1999 1998—et usaedvanlig god tar for over-drevssvampe. *Svampe* 40, 36–44.

Viney, M. 1990 Landscape history in a nutshell. *The Irish Times* ('Another Life'), Saturday, 30 June.

Viney, M. 2000 The Irish Times ('Another Life') April.

Viney, M. 2003 *Ireland, a Smithsonian natural History*. Belfast. the Blackstaff Press.

Vullings, W., Collins, J.F. and Smillie, G. (eds) 2013 *New Survey of Clare Island. Volume 8: Soils and Soil Associations*. Dublin. Royal Irish Academy.

Wailes, G.H. and Penard, E. 1911 *Rhizopoda and Heliozoa*. Clare Island Survey. Part 65. *Proceedings of the Royal Irish Academy* **31**.

Wallis Kew, H. 1911 *Pseudoscorpiones*. Clare Island Survey Part 38. *Proceedings of the Royal Irish Academy* **31**.

Ward, H.B. and Whipple, G.C. 1918 *Fresh-Water Biology*. New York. John Wiley.

Wentworth Thompson, D. 1942. *On Growth and Form*. Cambridge University Press.

West, W. 1912 *Fresh-water Algae*. Clare Island Survey. Part 16. *Proceedings of the Royal Irish Academy* **31**.

Westropp, T.J. 1911 *History and Archaeology*. Clare Island Survey. Part 2. *Proceedings of the Royal Irish Academy* **31**.

Whelan, K. 1997 The modern landscape from plantation to present. In F.H.A. Aalen, M. Stout and K. Whelan (eds), *Atlas of the Irish rural landscape*, 67–103. Cork University Press.

Whelan, K. 1999 Landscape and society on Clare Island 1700–1900. In C. Mac Cárthaigh and K. Whelan (eds), *New Survey of Clare Island, Volume 1: History and Cultural Landscape*, 73–98. Dublin. Royal Irish Academy.

White, J. and Doyle, G.J. 1982 The vegetation of Ireland: a catalogue raisonné. *Journal of Life Sciences, Royal Dublin Society* 3, 289–368.

Whitton, B.A. 2007 The blue-green algae (Cyanophyta). In M.D. Guiry, D.M. John, F. Rindi and T.K. McCarthy (eds), *New Survey of Clare Island Volume 6: The Freshwater and Terrestrial Algae*, 141–78. Dublin. Royal Irish Academy.

Whitton BA 2002 Phylum Cyanophyta (Cyanobacteria). *The Freshwater Algal Flora of the British Isles*, 25–122. Cambridge University Press.

Williamson, D.B. 2007 The desmids (Chlorophyta, Zygnematales, in part). In M.D. Guiry, D.M. John, F. Rindi and T.K. McCarthy (eds), *New Survey of Clare Island Volume 6: The Freshwater and Terrestrial Algae*, 27–65. Dublin. Royal Irish Academy.

Wilson, J. 1911 *Agriculture and its History*. Clare Island Survey. Part 5. *Proceedings of the Royal Irish Academy* **31**.

Worthington, R. and Walsh, J.J. 2011 Structure of Lower Carboniferous basins of NW Ireland, and its implications for structural inheritance and Cenozoic faulting. *Journal of Structural Geology* **33**, 1285–99.

Index

Note: page references in bold indicate tables. Those in italics indicate illustrations. Latin family, genus or species names are included in parentheses for reference to the taxonomic index.

Taxonomic Index

Note: page references in bold indicate tables. Those in italics indicate illustrations.